Topics in Organic Electrochemistry

Topics in Organic Electrochemistry

Edited by

Albert J. Fry
Wesleyan University
Middletown, Connecticut

and

Wayne E. Britton
The University of Texas at Dallas
Richardson, Texas

SPRINGER SCIENCE+BUSINESS MEDIA, LLC

Library of Congress Cataloging in Publication Data

Main entry under title:

Topics in organic electrochemistry.

Includes bibliographical references and index.
1. Electrochemistry. 2. Chemistry, Physical organic. I. Fry, Albert J. II. Britton, Wayne E.
QD555.5.T66 1986 547.1′37 85-28178

DOI 10.1007/978-1-4899-2034-8

Originally published by Plenum Press, New York in 1986
MyCopy version of the original edition 1986

Contributors

Wayne E. Britton • Department of Chemistry, University of Texas at Dallas, Richardson, Texas

Marye Anne Fox • Department of Chemistry, University of Texas at Austin, Austin, Texas

Albert J. Fry • Department of Chemistry, Wesleyan University, Middletown, Connecticut

Masamichi Fujihira • Department of Chemical Engineering, Tokyo Institute of Technology, Tokyo, Japan

John C. Kotz • Chemistry Department, State University of New York, Oneonta, New York

Vernon D. Parker • Norwegian Institute of Technology, University of Trondheim, Trondheim, Norway

Preface

Organic electrochemistry is a remarkably diverse science. Many study it for its own sake. Such individuals have traditionally been interested primarily in the mechanisms of reactions of organic substances at electrode surfaces, in developing new synthetic applications, or in electrochemical methods for analyzing mixtures of organic substances. In recent years, however, the field has attracted the attention of individuals with a wider variety of research interests. Physical organic chemists have learned that electrochemistry can afford valuable thermodynamic and structural information on organic systems, and that there exists a wealth of electrochemical methods which can be employed to measure the rates and mechanisms of fast organic reactions occurring at electrodes. Organometallic chemists are beginning to discover—and this is still almost virgin territory—the wide range of reaction mechanisms which can take place upon electrochemical oxidation or reduction of organometallic substrates. Physical chemists are attempting to understand the complex processes which occur at the surfaces of semiconductors, and when light impinges on an electrode surface. Others are studying the ways in which the composition of the electrode surface affects the course of electrode processes, and are devising ways to modify the chemical structure of electrode surfaces to achieve specific purposes. The range of contexts in which electrochemistry is now being used in some way in organic research is truly impressive.

The very pace and breadth of developments in these areas have created problems of their own. A number of important areas in organic electrochemistry, we felt, were at that stage in their development where they were ripe for a critical review. Some of the topics in this volume have not been reviewed in some time, and others not at all. This alone would argue for their presence here. In addition, however, we felt the need for a collection of reviews addressed to the general organic chemical community, who may not be aware of how extensive recent developments in some of these areas have been. The nature of some of these problems is such that frequently they can benefit from insights which organic chemists are uniquely suited to contribute. We hope in this set of authoritative and critical reviews not only to summarize developments in a number of important areas of organic electrochemistry at a level useful to investigators in the respective fields, but also to make this information available in a format accessible to organic chemists not previously familiar with it. We hope thereby to promote some useful interactions between what would otherwise be very disparate constituencies.

A brief survey of the subjects covered in this volume will serve to illustrate the remarkably wide range of subjects studied by modern organic electrochemistry. (We have not included here synthetic electrochemistry, which has been adequately reviewed recently in a number of formats, including the large multi-author work of Baizer and Lund.) Parker discusses the use of electrochemical techniques, including recently developed methods for analysis of the data, for extracting from electrochemical systems information on the rates and mechanisms of organic reactions occurring at electrodes. In chapters also directed toward elucidation of information of interest in physical organic chemistry, Fry examines the electrochemical behavior of nonbenzenoid aromatic hydrocarbons and Britton discusses the effects of conformational change upon electrochemical behavior in a variety of flexible systems, including cyclooctatetraenes and cyclohexanes. Kotz presents a schema for rationalizing the wide variety of reaction mechanisms observed upon electrochemical oxidation or reduction of organometallic substances. Fox reviews the variety of phenomena involved when light, organic substances, and electrode surfaces interact, an area of far-reaching practical implications in an era of diminishing energy resources. Fujihara takes up another subject of great current interest, the effects of experimental conditions, including chemical modifications of many kinds, upon the nature of electrode surfaces and processes occurring there. The ambitious but not unrealistic hope of workers in this field is of course to be able to create new surfaces which are able to effect highly selective transformations. The breadth of subjects covered by the authors is impressive, yet all retain the characteristic orientation of the organic chemist. We hope that we have been successful in making better known the excitement and the many successes and challenges of modern physical organic electrochemistry.

Albert J. Fry
Middletown, Connecticut

Wayne E. Britton
Dallas, Texas

Contents

Chapter 3

The Electrochemistry of Transition Metal Organometallic Compounds

John C. Kotz

Chapter 4

Organic Photoelectrochemistry

Marye Anne Fox

Chapter 5

Structural Effects in Organic Electrochemistry

Wayne E. Britton

Chapter 6

Modified Electrodes

Masamichi Fujihira

1

The Electrochemistry of Nonbenzenoid Hydrocarbons

Albert J. Fry

1. INTRODUCTION

The electrochemical behavior of benzenoid hydrocarbons (naphthalene, anthracene, etc.) has been the subject of a very large number of investigations.[1,2] This is quite understandable and appropriate, inasmuch as studies on such compounds have, in general, provided a variety of insights into the electrochemistry of unsaturated organic compounds, and, in particular, have permitted development of generalizations which are untainted by mechanistic complications arising from the presence of heteroatoms.

The electrochemical behavior of benzenoid hydrocarbons is rather well understood. For example, there is a good correlation between the first oxidation and reduction potentials of benzenoid hydrocarbons and the energy of their highest occupied or lowest unoccupied molecular orbitals (HOMO or LUMO, respectively), even at as crude a theoretical level as simple Hückel molecular orbital (HMO) theory.[3] Furthermore, many of the chemical reactions which often follow the initial electron-transfer are well-characterized mechanistically (see Section 2). Nevertheless, some questions remain. In particular, to what extent does electrochemical behavior exhibited by benzenoid hydrocarbons derive from the stability ("aromaticity") generally associated with such structures, as opposed, for example, to the fact that they are in general highly conjugated structures? How can one separate

---

Albert J. Fry ● Department of Chemistry, Wesleyan University, Middletown, Connecticut 06457.

these two features? Answers exist in principle to these questions, since there are a relatively large number of nonbenzenoid compounds known whose structures formally resemble those of benzenoid aromatics (at least to the extent that they are cyclic and polycyclic conjugated systems) but whose properties may be quite different from their benzenoid relatives. For example, the Kekule structures for naphthalene (**1**) and octalene (**2**) are

1 **2**

quite similar, but the two compounds differ enormously in chemical and physical properties. Naphthalene is very stable to heat, light, and oxygen, reacts with electrophilic reagents with difficulty, and then to afford substitution, not addition products, and sustains a diamagnetic ring current. Octalene, on the other hand, is very reactive and thermally unstable, and, in fact, shows all of the properties normally associated with a polyolefinic substance.[4] Clearly, there is some special element of stability, often referred to as "aromaticity," associated with benzenoid structures, such as those in naphthalene. However, definitions of just precisely what aromaticity is and how to recognize it have proven to be remarkably elusive and difficult problems.[5] For this reason, the practice is followed in the present review of using the terms "benzenoid" for compounds whose unsaturated rings are all six-membered, and "nonbenzenoid" for all other substances. These definitions permit unambiguous classification, without the semantic distinctions implied by the terms "aromatic" and "nonaromatic."

A very substantial body of experimental data on the electrochemical behavior of nonbenzenoid hydrocarbons has accumulated in recent years and is reviewed for the first time herein. The subject matter of the review includes several types of substances. There are, for example, the *annulenes*, cyclic conjugated materials of empirical formula $(CH)_n$. The best known annulene (after benzene, i.e., [6] annulene) is cyclooctatetraene, or [8] annulene (**3**), but a number of annulenes of different ring sizes are known.

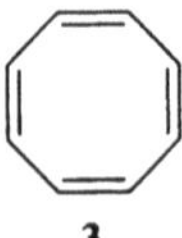

3

We will also consider substances in which two or more rings are fused, for example, azulene (**4**), pentalene (**5**), and heptalene (**6**), substances which differ enormously in chemical and physical properties despite the similarity

4 5 6

of their Kekule structures. Azulene and pentalene also present another problem. These substances, and, indeed, all nonbenzenoid hydrocarbons with at least one ring consisting of an odd number of carbons, are so-called "nonalternant" species, unlike benzenoid compounds, which are "alternant." (For a definition of the terms alternant and nonalternant, see reference 3, p. 45). In general, nonalternant hydrocarbons exhibit nonuniform charge distributions such that, even though the molecule may be electrically neutral overall, some sites carry a net positive charge, while others are negatively charged. Benzenoid (and alternant nonbenzenoid) hydrocarbons, on the other hand, have identical (zero) charge at every position. The nonuniform charge distribution in nonalternant hydrocarbons has a number of consequences, but in particular it should be noted that simple Hückel M.O. theory, which assumes identical Coulomb integrals at every carbon atom, does not handle these substances as well as it does benzenoid hydrocarbons. It has been shown, for example, that the reduction potentials of nonalternant hydrocarbons do not fit correlations between reduction potential and Hückel LUMO derived from studies on benzenoid hydrocarbons.[3,6]

In addition to annulenes and fused-ring substances, we will also examine the electrochemical behavior of so-called "homoconjugated" hydrocarbons, for example substance 7, in which a double bond of the

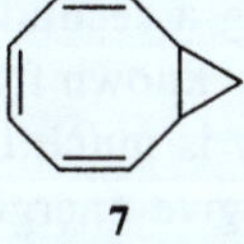

7

parent substance (3) has been replaced by a cyclopropane ring. Since the cyclopropane ring can transmit conjugation,[7] the question arises whether it can also help to delocalize the added electron when the substance is reduced to a radical anion. Electrochemistry can provide answers to such questions, thus helping to define the limits of the phenomenon of homoconjugation.

Finally, it should be noted that the overwhelming preponderance of literature data on the electrochemistry of nonbenzenoid hydrocarbons relates to their reduction. Therefore, in the following discussions, one may assume that reduction processes are being discussed unless it is explicitly stated that oxidation is involved.

2. BENZENOID HYDROCARBONS

It will be useful to review at this point a number of salient features of the electrochemical behavior of benzenoid hydrocarbons, since this will provide a benchmark against which we may compare the behavior of the various nonbenzenoid substances.

In aprotic media, benzenoid hydrocarbons exhibit two one-electron reduction steps at potentials which we may call E_1 and E_2, respectively, and which correspond to stepwise formation of the arene radical anion and dianion, respectively. Cyclic voltammetry shows the radical anion to be relatively long-lived in aprotic media, and, in fact, it is possible to generate these species electrochemically in an electron spin resonance (ESR) cavity and thus directly measure their ESR spectra in such solvents. The second wave usually appears irreversible unless meticulous precautions are taken to remove the last traces of electrophilic impurities, including water, from the solvent, in which case reversible behavior can be observed, at least on the cyclic voltammetric time scale.[8] Heterogenous electron-transfer rates between the electrode and arene are generally quite high, this and the relative stability of the products accounting for the observed electrochemical reversibility.

One striking feature of the electrochemical reduction of benzenoid hydrocarbons in aprotic media is the fact that ΔE, the difference in potential between E_1 and E_2, has about the value (0.55 ± 0.10 V) for all of the very large number of such compounds which have been studied, despite the great differences in their structures. This originally led Hoijtink to suggest that 0.55 V, or *ca.* 13 kcal, represents the energy required to overcome electron-repulsion when adding a second electron to the LUMO to form the dianion.[9] However, it is now known from quantum mechanical calculations that this repulsion energy is much larger. For example, calculations by Dewar's π-M.O. method[10] give energies for the anthracene molecule, radical anion, and dianion of 2856.6, 2651.0, and 2343.4 kcal/mole, respectively.[11] The problem may be more profitably viewed in an alternative manner. If one writes the individual electron-transfer steps as

$$\mathrm{ArH} + e^- \rightleftarrows \mathrm{ArH}^{\overline{\cdot}} \qquad E_1 \tag{1}$$

$$\mathrm{ArH}^{\overline{\cdot}} + e^- \rightleftarrows \mathrm{ArH}^{-2} \qquad E_2 \tag{2}$$

then, by subtraction of Equation 2 from Equation 1, one obtains Equation 3, from which it may be seen that the difference ΔE between the two

$$\mathrm{ArH} + \mathrm{ArH}^{-2} \rightleftarrows 2\mathrm{ArH}^{\overline{\cdot}} \qquad E_1 - E_2 \tag{3}$$

reduction potentials is related to the disproportionation equilibrium of two radical anions into neutral hydrocarbon and dianion. However, computed values of ΔE are much too large. For example, using the values computed by Dewar and quoted above, ΔE for anthracene may be calculated to be 103.1 kcal/mole, or 4.47 V (1 eV = 23.06 kcal/mole), as compared with the actual value of *ca.* 0.5 V.[11] This is general for most benzenoid hydrocarbons. ΔE is computed to be on the order of 5 V or more, while, as we have seen, the experimental values actually cluster about a value of 0.5 V. It is now generally accepted that this discrepancy arises because (a) solvation and ion-pairing effects are not included in the calculation, and (b) such effects should operate to a much greater extent upon the more highly charged dianion, thus driving the equilibrium represented by Equation 3 to the left and decreasing ΔE. Apparently, and unexpectedly, these differential effects operate to about the same extent in a large number of hydrocarbons, such that ΔE is roughly similar in all cases. Other experimental data are consistent with this interpretation; for example, as the cation of the supporting electrolyte is decreased in size, thus becoming better at ion-pairing, ΔE decreases, with E_2 exhibiting more sensitivity to the nature of the electrolyte than does E_1.[1,11,12]

The electrochemical behavior of aromatic hydrocarbons in aprotic solvents usually undergoes striking changes upon addition of proton donors, for example, phenols or carboxylic acids.[1]

As the concentration of the proton donor increases, the height of the first wave increases at the expense of the second until a single two-electron wave is observed at E_1. As Hoijtink first pointed out,[13] this behavior may be understood with the aid of Hückel M.O. theory. Thus, using anthracene (**8**) as an example, one notes that protonation of the radical anion should,

8 —— 1) e^- / 2) H^+ ——> **9** (H H)

and does, take place at the central ring to afford the neutral radical **9**. This (and in general all such species produced by protonation of benzenoid radical anions) is a so-called "odd alternant hydrocarbon." As such, it possesses a nonbonding molecular orbital,[3] and injection of an electron into this orbital should be considerably easier than into the original higher energy antibonding orbital of anthracene itself. In other words, the reduction potential E_3 of 9 will be positive of E_1 for anthracene, so that as soon as the radical anion is formed and protonated at E_1, it is immediately reduced further, resulting in an overall two-electron wave at E_1. Reduction of the neutral radical is generally effected by another molecule of radical anion (Equation 4), rather than by the electrode (Equation 5).[14]

$$\underset{\mathbf{10}}{ArH^{\overline{\cdot}}} + H^+ \rightarrow \underset{\mathbf{11}}{ArH_2\cdot}$$

$$\mathbf{10} + \mathbf{11} \rightarrow ArH + ArH_2^- \quad (4)$$

$$\mathbf{11} + e^- \rightarrow ArH_2^- \quad (5)$$

A number of investigators have pointed out the correlation which exists between the reduction potential and the energy of the lowest unfilled molecular orbital (LUMO), as computed by simple Hückel theory, for a wide variety of benzenoid hydrocarbons.[1,3] The relationship is remarkably good, considering the many approximations in Hückel theory. What turns out to be even more surprising is the relative independence of such correlations of the nature of the solvent in which the reduction potentials are measured. For example, the reduction potentials of benzenoid hydrocarbons have been found to be practically identical in 75 percent and 96 percent dioxane.[3] Reduction potentials measured under such conditions obey the relationship shown in Equation 6.[10] In this equation,

$$E_{1/2} = (2.37 \pm 0.10)m_{m+1} - (0.94 \pm 0.11) \quad (6)$$

$E_{1/2}$ is the polarographic half-wave potential measured (relative to SCE), and m_{m+1}, a negative number, is the energy of the Hückel LUMO, in units of β. Even more striking is the observation that the results in 2-methoxyethanol (Equation 7)[15] and dimethylformamide

$$E_{1/2} = (2.41 \pm 0.09)m_{m+1} - (0.94 \pm 0.07) \quad (7)$$

(Equation 8)[16] follow very similar relationships to those in the aqueous dioxane mixtures, despite the vastly different solvation properties represented by these different media.

$$E_{1/2} = (2.41 + 0.18)m_{m+1} - (0.86 \pm 0.09) \quad (8)$$

A final point of interest may be added with respect to the structure of benzenoid aromatics. Consider the substance pyrene (**12**). It contains a total of sixteen pi-electrons. (a) Should one consider this a Hückel $4n$ system, with all that implies about lack of stability? (b) Is it a set of four independent benzene rings (**12a**)? (c) Is it a substituted biphenyl with two bridging ethylene rings, as implied by Structure **12b**? Or, (d) should one count only the electrons on the perimeter of the molecule, in this case fourteen, making up a Hückel $4n + 2$ system which in turn interacts only weakly with the central double bond (**12c**)? Alternative (a) is clearly untenable, inasmuch

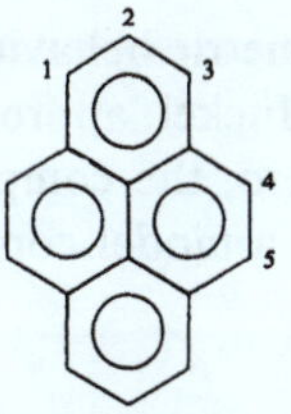

12a 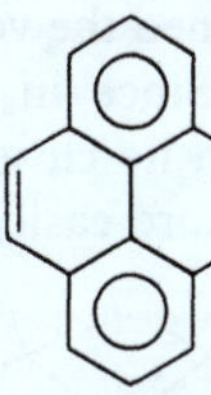 **12b** 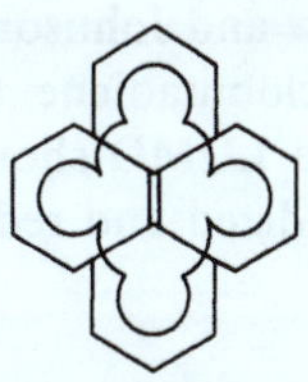 **12c**

as pyrene exhibits all of the chemical and physical properties generally associated with benzenoid species. At the same time, alternative (b) is not satisfactory either, since, for example, pyrene undergoes electrophilic substitution considerably more readily than does benzene, and their spectral properties are quite different.[17] Actually, none of the remaining possibilities is fully satisfactory, either. Clar has argued for the representation **12b**, with high double-bond order between C-4 and C-5.[17] Several properties of pyrene are explainable on this basis: the 4,5-bond is readily catalytically hydrogenated,[18] and in the NMR spectrum of 4-methylpyrene, the methyl group resembles that of propene rather than that of toluene.[19] On the other hand, the perimeter model (**12c**) in which one considers pyrene to be a cyclic fourteen electron array,[20] weakly interacting with the electron-pair of the inner double bond, is useful in understanding other properties of pyrene: the protons, even those at C-4 and C-5, appear at very low field, molecular bromine does not add across the 4,5-bond, and alkali metal reduction of pyrene in either solvents produces a dianion,[21] whose NMR spectrum is consistent with expectations based upon the assumption that the perimeter of this species contains 16 electrons. Similar questions of structure arise when considering fused-ring nonbenzenoid substances and their reduction products, but often, as we shall see, it is more clear how one should regard a given molecule.

3. ANNULENES

The annulenes present a nice series of compounds for electrochemical investigation, inasmuch as their structures differ both in (a) the number of π-electrons, allowing comparisons and correlations based on the Hückel

$4n$ and $4n + 2$ rules,[22] and (b) the molecular geometry of the neutral hydrocarbons and their charged derivatives, permitting one to examine the effects of this parameter upon electrochemical behavior.

3.1. [4] Annulene (Cyclobutadiene)

Breslow and Johnson examined the voltammetric behavior of the substituted cyclobutadiene (**13**).[23] Since in the Hückel approximation, its HOMO and LUMO should both lie close to zero, the compound should be both oxidized and reduced more easily than a model compound, such

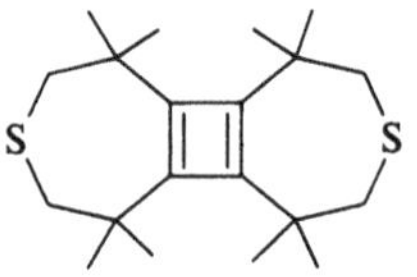

13

as 1,3-butadiene. This turns out to be the case: cyclic voltammetry shows irreversible waves at −2.7 and +0.6 V (SCE) for the reduction and oxidation, respectively, of **13**. Correcting for the inductive effect of the four alkyl substituents, these workers estimated the corresponding potentials of the parent substance as −2.2 and +1.0 V, respectively, as compared to values of −2.8 and +2.3 V for 1,3-butadiene, or +1.6 and −3.0 V for 1,3-cyclohexadiene.

Another way, in principle, to test the prediction from Hückel M.O. theory that cyclobutadiene should undergo facile two-electron reduction or oxidation is to examine the electrochemical behavior of its dibenzo derivative, i.e., biphenylene (**14a**). Reduction by sodium does produce a stable dianion, which Bauld has argued constitutes evidence, *inter alia*, for formation of a six-electron cyclic array of electrons about the central ring,[24] but West and co-workers[25] have disputed this interpretation, argiung in reply that the diproportionation equilibrium of the biphenylene radical anion

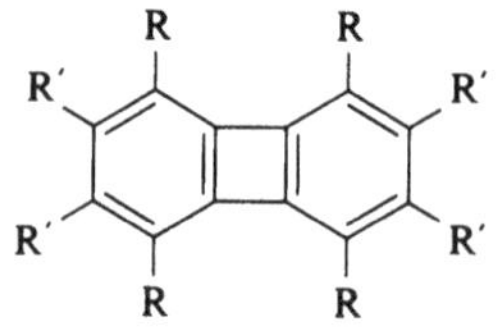

14a, $R = R' = H$
b, $R = H; R' = OCH_3$
c, $R = H; R' = CH_3$
d, $R = CH; R = H$
e, $R = R' = CH_3$

(into neutral hydrocarbon and dianion) is, first of all, quite sensitive to experimental conditions and, secondly, is not unlike the corresponding equilibrium involving anthracene radical anion, indicating that the biphenylene dianion does not exhibit any particularly great stability. Strain alone might contribute to the rather large value of the disproportionation equilibrium constant of biphenylene radical anion; in fact, the 1,2-diphenylcyclobutene radical anion disproportionation constant is similar to that of biphenylene.[26]

There is a literature report[27] on the electrochemical reduction of biphenylene which indicates its reduction potential to occur at a remarkably positive potential (+0.08 V/SCE), but this appears to be an error in interpretation; from the published voltammogram, it appears likely that the peak at +0.08 V is due to a decomposition product of the biphenylene radical cation, not to reduction of biphenylene itself.

Unlike the dianion, there is reason to believe that two-electron oxidation of biphenylene affords a dication exhibiting special stability. Thus, Ronlan and Parker found that biphenylene and its tetramethoxy derivative (**14b**) are both easier to oxidize than the corresponding biphenyls,[28] and Olah and co-workers found (by chemical oxidation of biphenylene) that the dication exhibits an NMR spectrum consistent with its being a delocalized species.[29]

More recently, Hart *et al.* examined the anodic behavior of biphenylene and some alkyl derivatives (**14c–e**).[27] As one would expect, alkyl substitution stabilizes the radical cations and moves the redox potentials to less positive potentials. Unfortunately, these investigators did not extend their studies to potentials positive enough to observe formation of the respective dications.

It has often been suggested that not only are Hückel 4*n* species not stabilized by resonance ("aromatic"), they are on the contrary less stable than one would expect based upon comparisons to model compounds, i.e., that they are "antiaromatic."[30] It has proved a difficult problem to establish the validity of the concept of antiaromaticity, in large part because of disagreements over the choice of proper model componds.[31,32] In the case of cyclobutadiene, several attempts have been made to test the concept of antiaromaticity by electrochemical methods. The idea is to examine the redox behavior of a substance which should afford a cyclobutadiene upon oxidation or reduction, in order to learn whether the redox process is harder to carry out than reductions or oxidations of model compounds. It was found that **15** is 0.3 V harder to oxidize to the corresponding quinone than are the model substances **16** and **17**, presumably because the quinone from **15** would be a cyclobutadiene derivative.[33] Similarly, 1,2-dibromobenzocyclobutene (**18**) is reduced to benzcyclobutadiene (**19**) (presumably; the

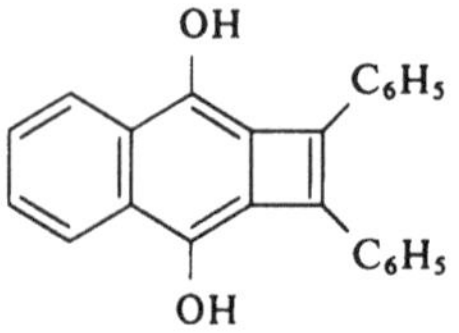

15 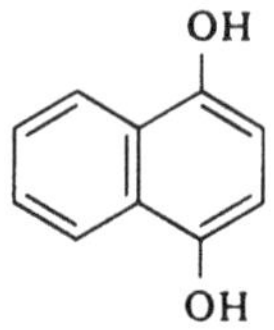 **16** 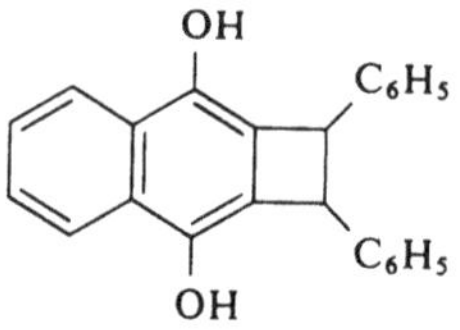 **17**

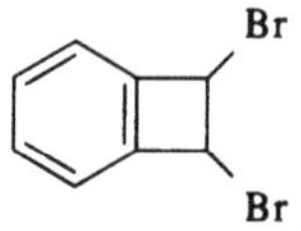

18 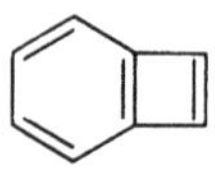 **19**

known dimer of **19** is the actual product isolated) at a potential 0.7 V negative of that required to convert dibromoacenapththene (**20**) to the corresponding alkene.[34] Species **21** is also oxidized to a cyclobutadiene derivative with great difficulty, but at least part of the difficulty must arise from the fact that the final product is a tetracation.[35]

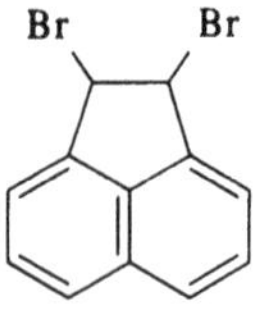

20 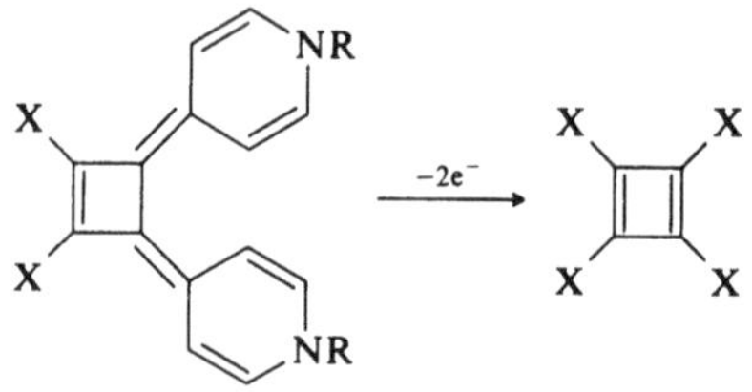 **21**

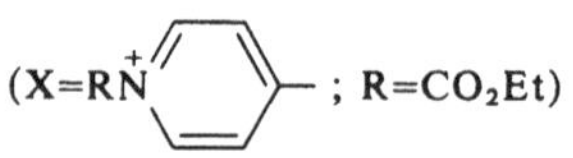

($X=R\overset{+}{N}C_5H_4$— ; $R=CO_2Et$)

Along the same lines, but on the cathodic side, species **22–24** all exhibit a large spacing (0.70–0.90 V) between the first and second reduction steps, which may be interpreted (see p. 20) as indicative of a high degree of

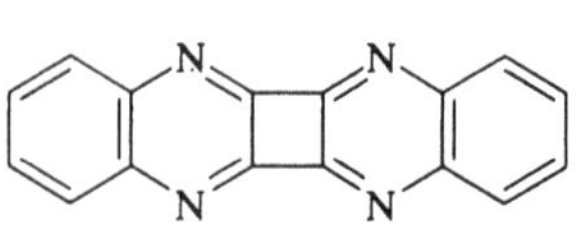

22 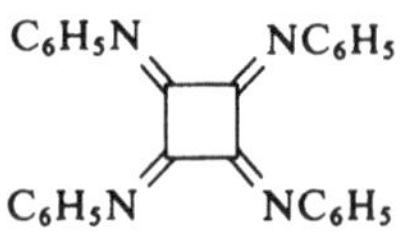 **23** 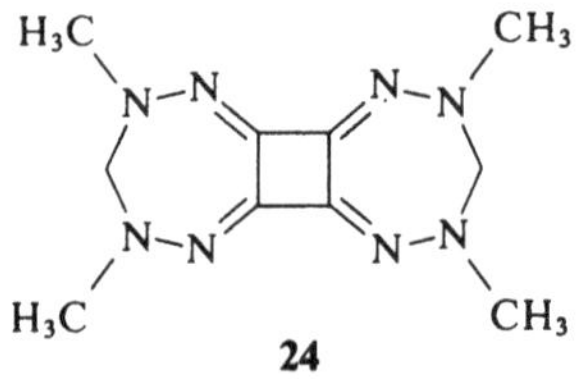 **24**

instability in the corresponding cyclobutadienoid dianions,[36] and dication (**25**) exhibits the remarkably large spacing of 0.93 V upon reduction, illustrating the difficulty of forming the neutral hydrocarbon.[37] Some interesting

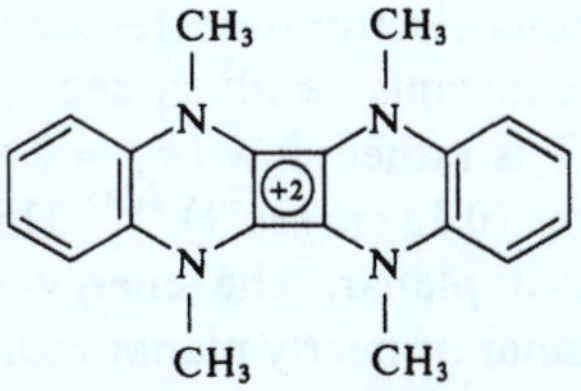

25

examples of this phenomenon were observed by Breslow and Rieke and their co-workers.[38,39] Cyclobutadienoquinones (**26**)–(**28**) are reduced with difficulty, unlike 1,2-cyclobutanedione (**29**). Much easier to reduce is 9,10-phenanthrenequinone; the product (**30**) is benzenoid unlike the cyclobutadienoid products from reduction of **26–28**. The fused-ring quinone **31**

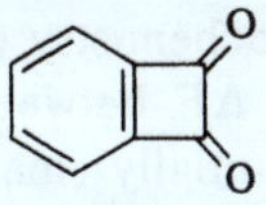

26 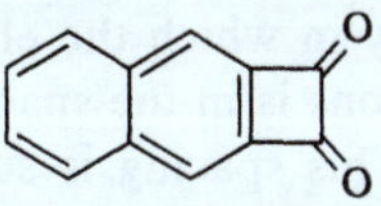 **27** 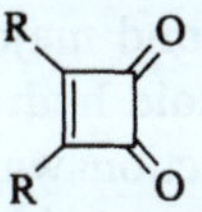 **28** (R = OCH_3 or C_6H_5)

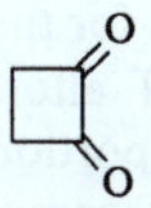

29 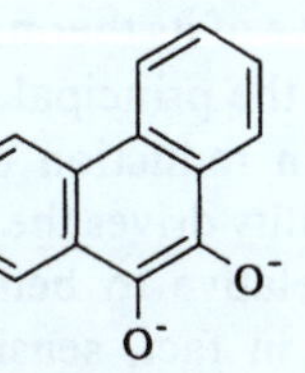 **30** 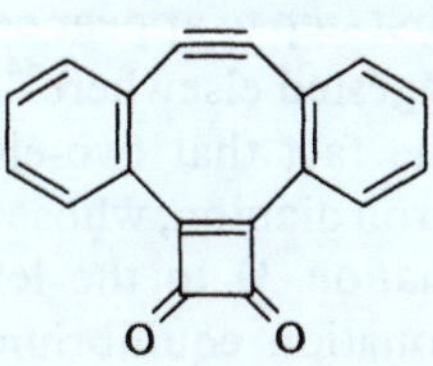 **31**

undergoes relatively facile two-electron reduction, however, despite the nominally cyclobutadienoid character of the product dianion. In this case, however, the latter should be more properly considered (apparently) as a derivative of the cyclooctatetraenedianion, which is known to be stable. The inescapable conclusion from all of these studies is that antiaromaticity is a real concept: there is indeed an energy cost to formation of Hückel $4n$-electron structures.

3.2. [8] Annulene (Cyclooctatetraene)

The most intensively studied annulene is [8] annulene, or cyclooctatetraene (COT) (**3**), together with its many derivatives.[40] The electrochemical behavior of COT differs substantially from that exhibited by benzenoid hydrocarbons. It will be recalled from Section 2 that the two-electron transfer steps to benzenoid hydrocarbons are reversible, i.e., that the two heterogeneous electron-transfer steps are fast (heterogeneous electron-transfer rate constants $\geq 10^3$ cm sec^{-1}). In contrast, the first electron-transfer to COT is rather slow ($k_{,h} = 2 \times 10^{-3}$ cm sec^{-1} in DMF), while the second is larger (0.15 cm sec^{-1}).[41,42] This is generally ascribed to the fact that COT is not planar. The energy necessary to convert the tub-shaped COT to a planar or nearly planar radical anion is thus reflected in the slow first-electron transfer. By the same token, the second electron-transfer requires little or no geometric change, hence, the larger electron-transfer rate constant. (Benzenoid hydrocarbons are planar and require no geometric change upon reduction, hence, both reduction steps are rapid.) Supporting this interpretation is the fact that homogeneous electron exchange between COT and its radical anion is slow because of their differing geometries, but fast between the radical anion and dianion, since both are planar.[43]

The second major way in which the electrochemistry of COT differs from benzenoid hydrocarbons is in the spacing, ΔE, between the first and second reduction waves. This spacing is substantially smaller (<0.3 V)[11] than with benzenoid species, for which it will be recalled that ΔE is *ca.* 0.55 V. Part of this difference probably arises because slow electron-transfer causes the first step to appear negative of its thermodynamic value. However, we have suggested elsewhere[44] that the principal reason for the small value of ΔE is the fact that two-electron reduction of COT affords a Hückel $4n + 2$ electron dianion, whose stability drives the disproportionation equilibrium (Equation 3) to the left (relative to benzenoid compounds). The disproportionation equilibrium is, in fact, sensitive to the nature of the solvent and cation of the supporting electrolyte,[45,46] since ion-pairing and solvation differentially stabilize the COT dianion relative to the radical anion. Indeed, under conditions favorable to ion-pairing, (liquid ammonia[47] or tetrahydrofuran containing sodium ion[48]) Equation 3 is endothermic (E_2 is positive of E_1), and the two waves coalesce into a single two-electron step.

The foregoing concepts, developed from studies upon the parent COT molecule, have been fully confirmed and extended by studies upon the electrochemical reduction of a large number of cyclooctatetraene derivatives (Table 1). The simplest such substance, methylcyclooctatetraene, is reduced with distinctly more difficulty (0.15 V) than COT itself.[49] This is a substantially greater effect than observed upon methyl substitution of benzenoid hydrocarbons. It presumably arises because the electron-releasing methyl group not only inductively destabilizes the methyl-COT radical anion, but also creates a steric barrier to reduction, because of repulsion generated between the methyl group and the two hydrogens vicinal to it, as the ring flattens to form the radical anion. This steric effect is greater in *t*-butyl-COT, which is therefore relatively harder to reduce than methyl-COT (see Table 1).[50] Steric effects become more pronounced with increasing alkyl substitution. Thus, all of the dimethyl-COT isomers are harder to reduce than methyl-COT,[49] and in the extreme, reached with the two isolable, noninterconverting conformers of 1,2,3,4-tetramethyl-COT (**32** and **33**) (which are

32 **33**

noninterconverting precisely because of the difficulty of achieving the planar conformation which separates them), the first reduction step is barely reached before solvent breakdown.[51] Another type of steric impediment to reduction arises when the COT ring is bridged by a chain, whose length is

TABLE 1
Electrochemistry of Cyclooctatetraenes

Compound	n_1 [a]	$-E_1$ [b]	$-E_2$ [b]	Solvent[c]	Electrolyte[d]	Ref.
COT	1	1.64	1.88	DMF	TBAP	41
COT	1	1.62	1.86	DMF	TPAP	42
COT	1	1.64	1.80	DMF	TEAP	54
COT	1	1.74	1.97	DMF	TBAI	61
COT	1	1.76	1.89	AN	TEAP	54
COT	1	1.57	1.76	DMSO	TEAP	11
COT	1	1.606	1.921	HMPA	TBAP	57
COT	1	1.83[e]	1.99[e]	THF	TBAP	49
COT	2	1.48	—	THF	$KALEt_4$	48
COT	2	1.55		NH_3	KI	47
Methyl-COT	1-2	1.98[e]	—	THF	TBAP	49
t-Butyl-COT	1	1.882	1.996	HMPA	TBAP	50
Methoxy-COT	1	1.87	2.05	DMF	TBAI	61
Phenyl-COT	1	1.76	1.85	AN	TEAP	54
Trimethyl-silyl-COT	1	1.640	1.804	HMPA	TBAP	50
Trimethyl-germyl-COT	1	1.651	1.795	HMPA	TBAP	50
Trimethyl-stannyl-COT	1	1.691	1.77	HMPA	TBAP	50
1,2-Dimethyl-COT	1-2	2.18[e]	—	THF	TBAP	49
1,3-Dimethyl-COT	1-2	2.10[e]	—	THF	TBAP	49
1,4-Dimethyl-COT	1-2	2.07[e]	—	THF	TBAP	49
1,5-Dimethyl-COT	1-2	2.00[e]	—	THF	TBAP	49
35		>3.7[f]		HMPA	TBAP?	40
34	1	2.36	2.69	AN	TBAHFP	52

(continued)

TABLE 1—continued

Compound	n_1 [a]	$-E_1$ [b]	$-E_2$ [b]	Solvent[c]	Electrolyte[d]	Ref.
Benzo-COT	1	1.72	1.95	HMPA	TBAP	57
1,2,3,4-Tetramethyl-COT (**33**)	1?	3.6	—	HMPA	TBAP?	51
1,2,3,4-Tetramethyl-COT (**32**)	1?	3.7	—	HMPA	TBAP?	51
1,3,5,7-Tetraphenyl-COT	2	1.67	—	AN	TEAP	54
1,2,4,7-Tetraphenyl-COT	2	1.87	—	AN	TEAP	54
Dibenzo-COT (**39**)	1	1.896	2.016	HMPA	TBAP	57
Dibenzo-COT (**39**)	2	1.95	—	DMF	TBAP	55
40	1	1.53	(204)	DMF	TBAP	55
41	1	1.76	2.16	DMF	TBAP	55
36		>3.0[f]		AN	TBAHFP	52
1,1′-Bycyclo-octatetraenyl	2	1.657	2.324, 2.566	HMPA	TBAP	60
45	2	1.471	—	HMPA	TBAP	57
42	1	−0.79[g]	−0.14[g]	AN	TBAHFP	56

[a] n_1: number of electrons in first (or only) voltammetric wave.
[b] E_1 and E_2: potentials, in volts relative to S.C.E., of first and second voltammetric waves.
[c] DMF: Dimethylformamide; AN: acetonitrile; DMSO: dimethylsulfoxide; HMPA: hexamethylphosphoric triamide; THF: tetrahydrofuran.
[d] TBAP: tetrabutylammonium perchlorate; TPAP: tetrapropylammonium perchlorate; TEAP: tetraethylammonium perchlorate; TBAI: tetrabutylammonium iodide; TBAHFP: tetrabutylammonium hexafluorophosphate.
[e] These values have been corrected from the original literature report, which used 0.5 V as the difference between S.C.E. and the Ag/Ag^+ reference electrode, whereas the correct value is 0.35 V.
[f] Not reducible before solvent discharge.
[g] These potentials are *positive* relative to S.C.E.

too short to permit the ring to achieve planarity. This situation occurs with **34**,[52] whose reduction waves occur at −2.36 and −2.69 V (COT is reduced at −1.61 V), and is most marked in **35** and **36**, neither of which is reducible before solvent breakdown.[40,52]

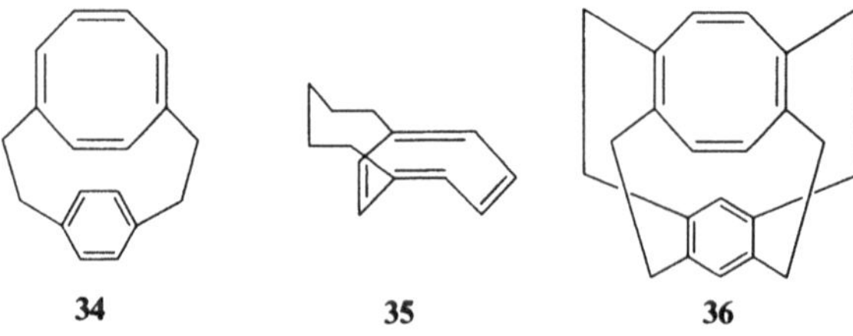

It is interesting to note the variation in reduction potentials exhibited by the dimethyl-COT isomers.[49] There is an inverse linear correlation between the polarographic reduction potentials of these substances and the number of carbons separating the methyl groups. This unusual correlation is not readily interpretable, but is presumably of steric origin.

Phenyl-substituted cyclooctatetraenes also exhibit some interesting features. The steric interaction between the phenyl group and its vicinal neighbors in the planar radical anion of phenyl-COT (**37**) is opposed by its electronic effect: there is substantial resonance interaction between the two rings in the radical anion,[53] and consequently **37** is reduced at exactly the same potential as COT itself.[54] In 1,3,5,7-tetraphenyl-COT (**38**), both effects are presumably amplified, but again they come very close to canceling; **38** is slightly easier to reduce than COT.[54] However, the steric compression between two vicinal phenyl groups in a planar radical anion should be substantial, causing at least one of the phenyl groups to twist out of planarity, and thus reducing its resonance-stabilizing effect. Indeed, 1,2,4,7-tetraphenyl-COT is harder to reduce than **38** or COT.[54] Furthermore, notice that in these systems all of the energy costs of steric overlap of the phenyl groups with their neighbors on the ring are paid in the first step, where ring planarity is achieved. The only effect of the phenyl group upon the second reduction step is its resonance-stabilizing effect upon the dianion. Thus, the second wave of **37** is positive of that of COT, and in the two tetraphenylcyclooctatetraenes, this effect is sufficiently large that E_2 is positive of E_1, i.e., a single two-electron wave is observed.[54]

Throughout the foregoing discussion, a major theme has been the fact that the first electron-transfer to COT and its derivatives is generally slow because of the change in geometry as the tub-shaped neutral molecule is converted to a planar, or nearly planar, radical anion. Would the initial electron-transfer rate be larger if COT were planar? The answer is yes. Bard *et al.* examined the electrochemical behavior of dibenzocyclooctatetraene (DBCOT, **39**) dibenzocyclooctadienediyne (DBCOD, **40**), and dibenzocyclooctatrienyne (DBCOM, **41**).[55] While DBCOT is tub-shaped, DBCOD is

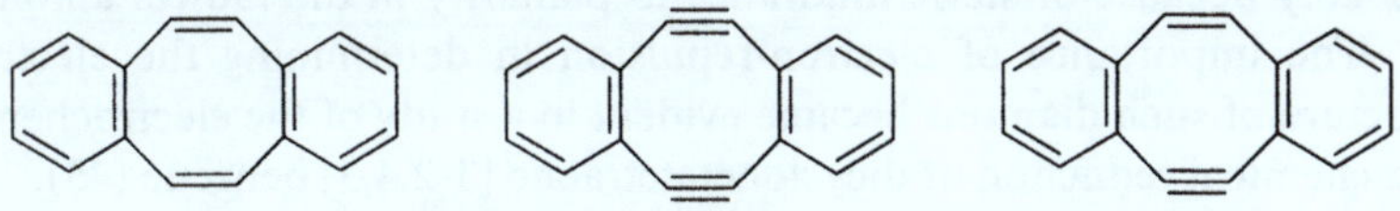

39 **40** **41**

definitely planar, and DBCOM probably so. The rate constants for the first heterogeneous electron-transfer to DBCOD and DBCOM are an order of magnitude larger than to DBCOT, and the rate constant for the second electron-transfer to DBCOT is 50 times larger than the first. Similarly,

Britton found that the heterogeneous electron-transfer rate for the first reduction step of the planar COT derivative (**42**) is about 30 times larger than for COT (56). (The first electron-transfer to **42**, incidentally, occurs at the remarkably positive potential of +0.79 V vs. SCE, making it possibly the most powerful organic oxidant yet discovered.)

F_2 F_2 F_2 F_2 F_2 F_2 F_2 F_2

42 **43**

An intriguing problem arises with the reduction of benzoannulenes, for example, benzo-COT (**43**) or *sym*-dibenzo-COT (**39**), to their respective dianions. Consider the dianion derived from **43**. Is it more properly viewed as a substituted COT dianion (**44a**) in which the individual rings retain their Huckeloid character, or as a bridged [12] annulene dianion (**44b**)?

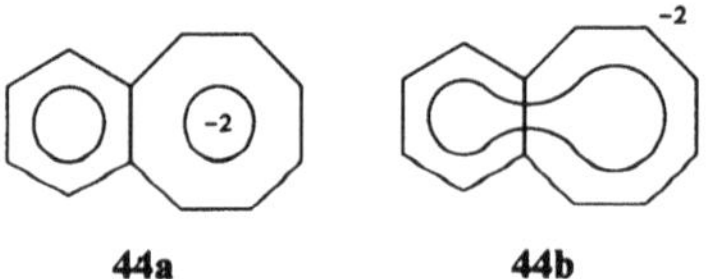

44a **44b**

The evidence, from NMR spectroscopy, seems clear on this point: the benzene ring carries substantial negative charge, and the dianion is best described as structure **44b**.[57] This is presumably at least partly due to the fact that electron-repulsion in the dianion is minimized by spreading the charge over the whole system, rather than concentrating it in one ring. The electrochemical behavior of **39** and **43** is conventional in other respects: the reduction potentials become more negative in the order COT, **43**, **39**, probably because of steric hindrance to planarity in the radical anions.

The importance of electron-repulsion in determining the electronic structure of such dianions became evident in a study of the electrochemical and chemical reduction of dicyclooctatetraeno [1,2,4,5] benzene (**45**).[57] The

45

46a 46b

47

substance undergoes reversible two-electron reduction at a platinum electrode in hexamethylphosphoramide, implying facile conversion to a rather stable dianion. Is the latter an equilibrating pair of benzo-COT dianions (**46a–46b**), or is it a fully delocalized planar species (**47**)? The latter structure has the dual disadvantages of having a Hückel antiaromatic 20-electron periphery and requiring additional energy to flatten the second COT ring. Nevertheless, the proton NMR spectrum of the dianion is more consistent with structure **47** than with **46a–46b** pair; apparently electron-repulsion is minimized in the dianion by spreading the charge over all three rings, thus overcoming the general tendency to form Hückel $4n + 2$ structures. Compound **45** is, incidentlly, reduced at a remarkably positive potential.[57] Paquette ascribed this to the fact that the two electrons are added to a degenerate pair of nonbonding molecular orbitals, in the Hückel approximation. A somewhat more sophisticated modified Hückel computation, in which Coulomb and exchange integrals are adjusted iteratively to self-consistency,[22] shows that these two orbitals are both weakly antibonding and nondegenerate (-0.08 and -0.25β),[58] but does not change the essential argument that reduction of **45** places two electrons into a relatively low-energy orbital. It is not at all obvious, however, why electron-transfer to **45** should be reversible, i.e., that the heterogeneous electron-transfer rate from the electrode to **45** should be larger than to COT, since formation of dianion **47** involves flattening of *two* COT rings.

It was also found that **45** can be reduced (chemically) to a tetranion. Unfortunately, this interesting species proved to be too insoluble for NMR spectral examination; with a 22-electron ($4n + 2$) periphery, it ought to exhibit a dia-magnetic ring current, like other $4n + 2$ tetraanions prepared more recently.[59] Bicyclooctatetraenyl (**48**) can also be reduced to a

48

tetraanion,[60] but this species is more properly considered as two coupled COT dianions, inasmuch as the two rings are twisted from planarity. The two rings of **48** accept their first electron independently of each other.

A number of other substituted COTs have been examined. Paquette and co-workers examined the polarographic behavior of a series of Group IV-trimethylsubstituted COTs (**49**) (Table 1).[50] All of the compounds are harder to reduce than COT itself, presumably because of the electron-donating properties of the substituents; in fact, there is a rough linear correlation between the Hammett σ_p constants of the various substituents and the deviation of their reduction potentials from that of COT.

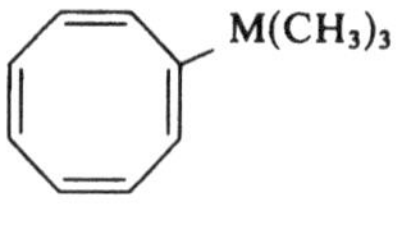

$M(CH_3)_3$

49a, M = C
b, M = Si
c, M = Sn
d, M = Ge

While heterocyclic compounds lie outside the scope of this review, it is appropriate to mention here the electrochemical behavior of 2-methoxyazocine (**50**) and its derivatives.[61] Unlike COT, these substances

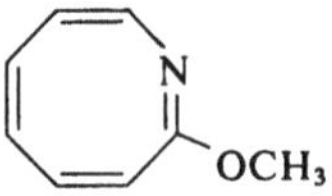

N
OCH_3

50

exhibit a single two-electron wave. As we have seen, this can occur in COT derivatives if either the first wave moves negative of that of COT because of steric and/or electronic resistance to ring-flattening, or the second wave moves positive because of increased stability in the dianion relative to COT. While the latter effect may be operating with **50** because of the presence of the electronegative nitrogen atom in the ring, there does seem to be a substantial barrier to ring-flattening. This shows up in cyclic voltammetry, where it is seen that the reduction is reversible chemically (i_{pa}/i_{pc} = unity), but with the cathodic peak a full volt negative of the anodic peak.[61] This suggests an equal degree of resistance to geometrical change upon reoxidation of the dianion back to the neutral starting material. Similar effects to these and to steric effects observed with **43** were also noted in a series of benzo derivatives of **50**.[62]

Little work has been carried out upon the anodic oxidation of cyclooctatetraene, which should undergo facile two-electron oxidation to a Hückel $4n + 2$ dication (**51**). COT is known to undergo anodic oxidation in acetic acid-sodium acetate to afford *cis*-fused bycyclo [4.2.0] derivates (**52** and **53**) and the tropilidene (**54**),[63] all of which could hypothetically arise from **51**, but other mechanisms could also account for these products.

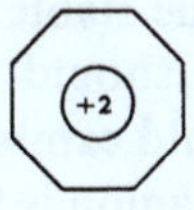

51 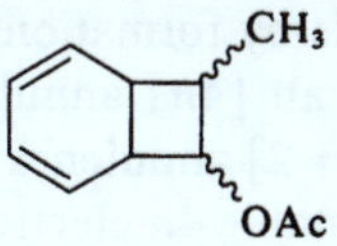 **52** 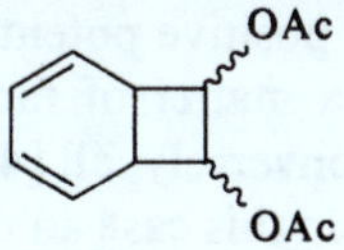 **53** 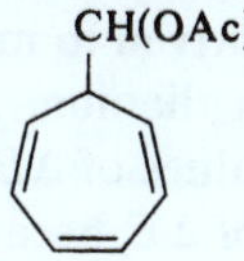 **54**

3.3. [10] Annulene

Although the radical anion of a methano-bridged [10] annulene has been prepared electrolytically in an ESR cavity,[64] the electrochemical reduction of [10] annulene or any of its derivatives does not appear to have been otherwise reported. The stable [10] annulene (**55**) and its tetramethoxy derivative (**56**) do, however, undergo anodic oxidation to products (**57** from **55**; **58** from **56**) which are quite analogous to those characteristic of electrochemical oxidation of benzenoid substances.[65]

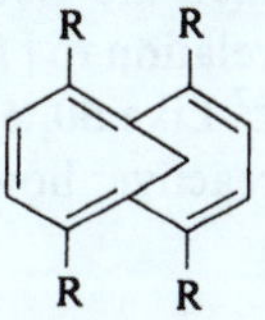

55, R = H
56, R = OCH_3
57 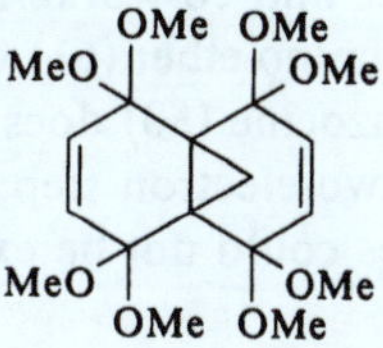 **58** 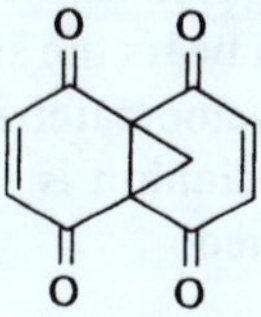

3.4. [12] Annulene

The polarographic behavior of the bridged [12] annulenes **59** and **60** has been examined.[66] Each exhibits two one-electron waves in aprotic media, and both steps can be shown to be reversible by cyclic voltammetry. The *methano* bridge holds each system in a roughly planar geometry, hence,

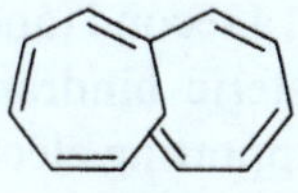

59 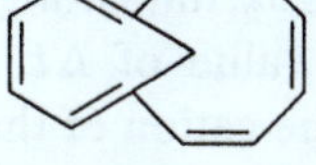 **60**

there is no reorganizational barrier to electron-transfer, unlike the first step in COT reduction. Consequently, the initial electron-transfer to these compounds is fast. Like COT, however, the reduction of these substances affords $4n + 2$ electron dianions, the NMR spectrum of which exhibit the expected diamagnetic ring current. Again like COT, the spacing ΔE (in volts) between the first and second reduction steps is smaller (0.2–0.3 V) than with benzenoid species. As we have suggested,[44] this is because the second reduction step is driven to more positive potentials by formation of the stable $4n + 2$ electron dianion. As a matter of fact, all [$4n$] annulenes should exhibit small values of ΔE; conversely, all [$4n + 2$] annulenes should exhibit large values of ΔE, because in this case an unstable $4n$ electron dianion is formed in the second step. As we shall see, this generalization holds up remarkably well; the only apparent exception in literature is one of the geometrical isomers of the parent [12] annulene.[66] However, both this substance and its dianion are nonplanar, making comparison with other substances difficult.

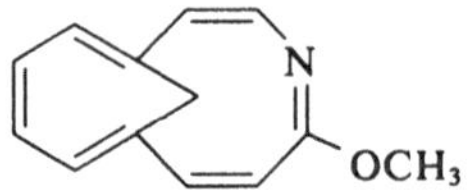

N

OCH_3

61

Paquette and co-workers prepared and examined the electrochemical behavior of imino ether (**61**) which bears the same relation to [12] annulene as methoxyazocine (**50**) does to cyclooctatetraene.[67] Like **50**, **61** is reduced in a single two-electron step; the dianion is very reactive, however, hence its properties could not be examined.

3.5. [14] Annulene

A surprisingly large number of [14] annulene derivatives have been synthesized, and the electrochemical behavior of most of them has been examined. Following up an earlier study[12] on the parent substance **62a**, Fry *et al.* examined the polarographic reduction of a series of 15,16-disubstituted-15,16-dihydropyrenes (**62**).[44] Each compound exhibits two one-electron waves in aprotic media, and as expected (*vide supra*), the spacing ΔE between the two waves is substantially larger (>0.7 V) than with benzenoid hydrocarbons. The value of ΔE increases as the sizes of R_1 and R_2 increase in compounds **62**, indicating that a second (and probably secondary) reason for the large value of ΔE is steric hindrance by R_1 and R_2 to ion-pairing between the cation of the supporting electrolyte and the radical

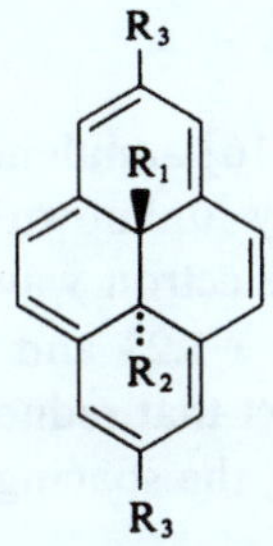

62a 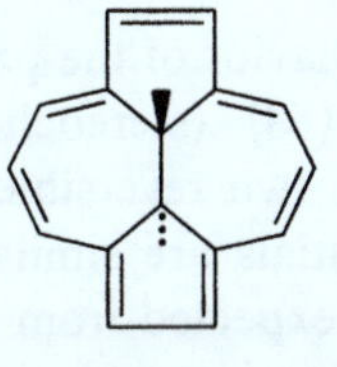 63 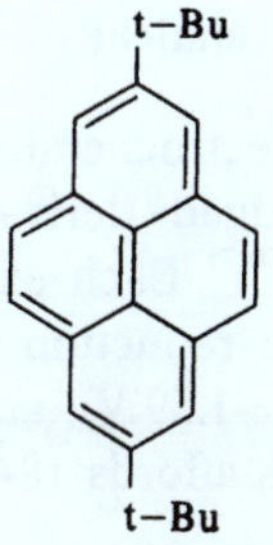 64

	R_1	R_2	R_3
a	Mr	Me	H
b	Me	C_6H_5	H
c	Me	Me	t-Bu
d	Me	Et	t-Bu
e	Me	n-Pr	t-Bu
f	Et	Et	t-Bu
g	Et	n-Pr	t-Bu
h	n-Pr	n-Pr	t-Bu
i	n-Bu	n-Bu	t-Bu

anions and dianions of **62** and **63**. The groups R_1 and R_2 are positioned above the center of the pi-system and are, therefore, ideally positioned to impede ion-pairing and solvation, unlike, for example, the two *t*-butyl groups in 2,7-di-*t*-butylpyrene (**64**), which, though large, are located on the periphery of the ring system and, therefore, exert relatively little effect upon the electrochemistry of the pyrene system.

15,16-Dimethyl-15,16-dihydroazupyrene (**63**) is structurally closely related to the dihydropyrene (**62a**). It is, therefore, not surprising that they are reduced at almost identical potentials (**62a**: −2.07 and −2.86 V; **63**: −2.07 and 2.81 V; all potentials relative to S.C.E.).[44,68]

Gerson and co-workers examined the electrochemical behavior of a series of bridged [14] annulenes (**65** and **66**).[69] These substances exhibit electrochemical behavior similar to that of the dihydropyrenes studied by Fry *et al.* including the expectedly large values of ΔE.

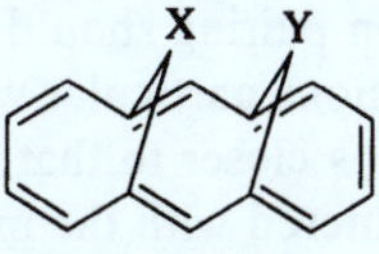

65a, X = Y = O
b, X = O: Y = CH_2

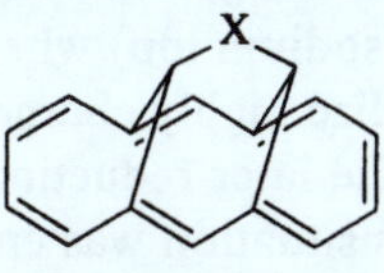

66a, X = CH_2
b, X = CO
c, X = CH_2CH_2

3.6. [16] Annulene

The electrochemical behavior of the parent [16] annulene (**67**) and its doubly-bridged derivative (**68**) (stereochemistry unknown) have been examined.[70,71] Each exhibits two reversible one-electron waves in aprotic media. The reduction potentials are similar (**67**, −1.23 and −1.52 V; **68**, −1.31 and −1.59 V), and as expected from the fact that reduction of these compounds affords 18-electron ($4n + 2$) dianions, the spacing ΔE is small (<0.3 V).

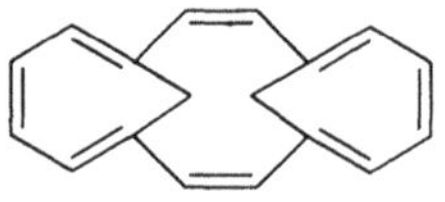

68

3.7. Larger Annulenes

Mullen, Nakagawa *et al.*[59] examined the cyclic voltammetric behavior of a number of large-ring [$4n + 2$] annulenes, for example, the 22-electron species (**69**). The substances generally exhibit broad 4-electron waves, indicating that four electrons can all be added at about the same potential.

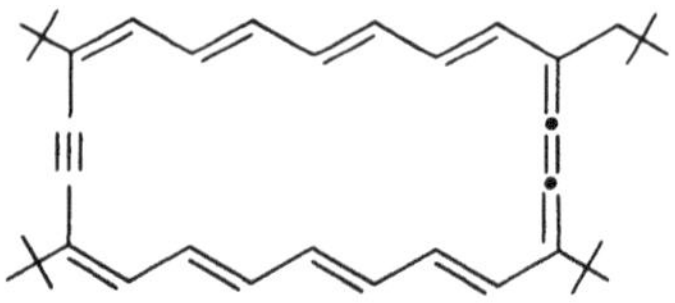

69

There must, therefore, be some special stability associated with the corresponding tetraanions, presumably because the latter are also $4n + 2$ species. This, however, raises the question why the [14] annulenes already discussed do not form tetraanions. This is probably because electron-repulsion in the tetraanion becomes less important as ring size increases, but, more importantly, because the larger annulenes were examined under conditions (tetrahydrofuran, sodium ion) where ion-pairing should be important. The latter should stabilize highly charged anions particular well, and thus drive the potentials of the later reduction steps closer to that of the first.

An analogous situation was encountered with the unusual cyclic compound **70**.[72] Since this substance can be viewed as a [24] annulene, one would anticipate (*vide supra*) that the spacing between the first and second reduction steps should be small. It is; in fact, **70** exhibits a single reversible two-electron wave in acetonitrile containing tetrabutylammonium perchlor-

70

ate.[72] No other reduction wave is observable before solvent breakdown, yet **70** can be reduced by lithium metal in tetrahydrofuran not only to the dianion, but further to the tetraanion.[73] As with the large annulenes discussed previously, ion-paring must drive the formation of the tetraanion to a potential sufficiently positive that it can be effected by lithium metal.

4. HOMOAROMATICITY

There has been great interest among organic chemists in the phenomenon of "homoconjugation," in which a saturated carbon is introduced into a conjugated chain, and one then asks whether the conjugative properties of the chain are retained in the new substance. The term homoconjugation also includes situations in which a carbon-carbon double bond is replaced by a cyclopropane ring. When applied to cyclic resonance-stabilized species, the concept is known as "homoaromaticity." There have been a number of attempts to obtain experimental electrochemical support for these concepts. Paquette and co-workers[74] examined the polarographic behavior of compounds **71–75**. Reduction of **71** takes place in two steps

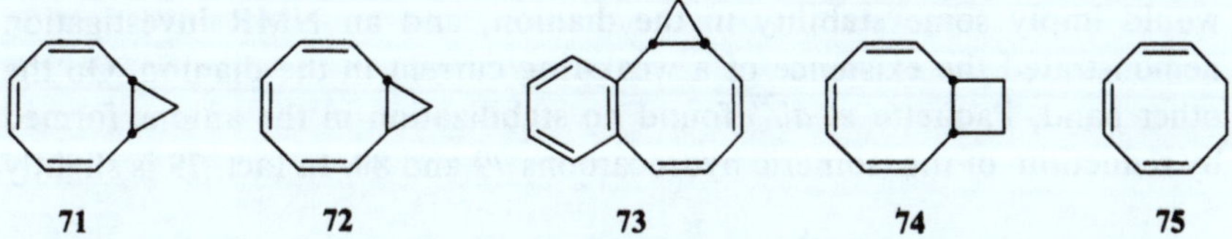

which occur at potentials (−2.55, −2.79 V) distinctly negative of those of cyclooactatetraene (−1.98, −2.3 V). On the other hand, **71** is easier to reduce than either **74** (−2.83 V) or **75** (−2.77 V), suggesting a degree of homoconjugative stabilization in the radical anion derived from **71**. Similarly, reduction of **73** is more difficult than that of **43**, but easier than **75**. The results on the trans isomer **72** were less conclusive because its radical anion is very short-lived; however, its reduction potential is similar to that of **71**. In the same vein, Miller found that dibenzonorcaradiene (**76**) is harder to reduce than phenanthrene (**77**), but easier to reduce than biphenyl, implying homoconjugative stabilization in the radical anion.[75]

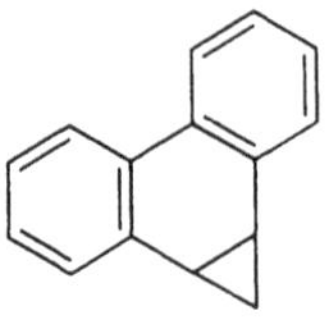

76 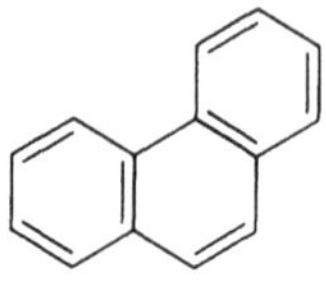 **77**

These results are not surprising; since the cyclopropane ring can transmit conjugation, but not as well as a double bond, homoaromatic stabilization will in general be real, but weaker than aromatic stabilization. However, what about a situation in which the homoconjugation involves insertion of a single saturated carbon atom, not a cyclopropane ring, into the chain? We know from studies on species, such as homotropylium ion, that this situation can also result in homoconjugative stabilization. Electrochemical investigations in this area appear to be restricted to substances in which more than one saturated carbon has been introduced into the ring. The experimental results appear to suggest that homoconjugation energies are marginal at best in such situations. Huber, Mullen *et al.* examined the electrochemical behavior of the tetraene **78**.[76] The cyclic

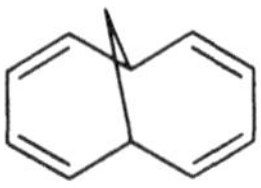

78

voltammogram appears to correspond to a two-electron reduction, which would imply some stability in the dianion, and an NMR investigation demonstrated the existence of a weak-ring current in the dianion. On the other hand, Paquette *et al.*[77] found no stabilization in the anions formed by reduction of the isomeric hydrocarbons **79** and **80**. In fact, **79** is slightly

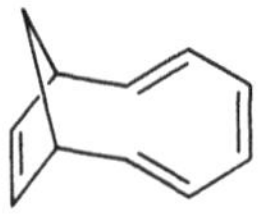

79 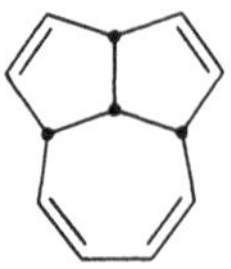 **80**

harder to reduce than a simple conjugated triene such as 1,3,5-cyclooctatriene. Apparently, one does not derive homoconjugative stabilization when even minor deviations from the "optimum" geometry are involved, which is probably reasonable in view of the small energy gain.

5. FUSED-RING HYDROCARBONS

It has already been pointed out (Section 2) that fused-ring species raise conceptual questions of structure. Are the properties of substance **81** those

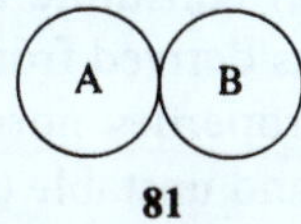

81

of the individual constituents, or are they submerged in a larger structure whose properties are different from those of A or B? Specifically, do the properties of **81** derive from the total number of electrons about the perimeter of the molecule? This is a difficult question, inasmuch as a number of properties, such as strain, electron-repulsion, and degree of nonplanarity, vary with ring size—yet are not considered if one simply counts the total number of peripheral pi-electrons. There is some evidence that fused-ring species such as **81** choose the best alternative, whether this is a fully delocalized $4n + 2$-electron perimeter structure or a localized structure in which rings A and B tend to retain their individuality.[78] Thus, naphthalene (**1**) has little need for peripheral delocalization because the individual rings are already Hückel $4n + 2$ species. However, the real situation is rarely this clear cut. For example, butalene (**82**) appears to be more stable than its respective cyclobutadiene ring constituents,[79, 80] but nevertheless is far less

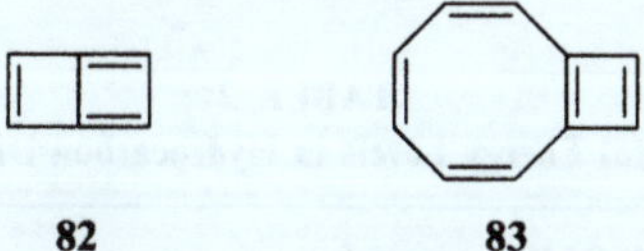
82 **83**

stable (because of ring strain) than benzene, even though both have six peripheral electrons. Octalene (**2**) and the bicyclic pentaene (**83**) each have two $4n$-electron rings, with $4n + 2$ electrons about the perimeter of the molecule. There would appear to be a considerably energetic advantage to delocalization of the electrons about the perimeter. However, to gain this stabilization, both substances would have to be planar. The chemical properties of these compounds give no evidence for stability of the sort one would normally expect from a Hückel $4n + 2$ species.[4,81] It appears that the energetic cost of ring-flattening is greater than any possible gain from resonance. We have encountered a similar situation earlier with the dicyclooctatetraenobenzene (**45**), in which the two COT rings are nonplanar, even though a planar structure would have an 18-electron periphery. While a clear-cut, generally applicable answer cannot be given to the questions raised at the outset of this section, it will be useful to keep this discussion in mind as we take up the electrochemical behavior of a series of fused-ring hydrocarbons.

The symmetrical fused-ring substances pentalene (**5**), naphthalene (**1**), heptalene (**6**), and octalene (**2**) constitute a very interesting homologous series. As we shall see, dianions derived from this series do not exhibit the characteristic alternation in properties normally observed in monocyclic species, where stable $(4n + 2)$ and unstable $(4n)$ structures alternate as ring sizes are incremented successively by two carbons. How can one understand this failure to exhibit an alternation of properties? We could compute the energy of the highest filled orbital (HOMO) of the dianion by the conventional Hückel method (Table 2). However, as was pointed out earlier (Section 1), Hückel calculations do not handle charged species very well because of the assumption that all Coulomb integrals are identical. Modified Hückel calculations, in which the Coulomb integrals are adjusted in an iterative manner to self-consistency with charge densities (Section 3.2)[22] tell an interesting story (Table 2).[81] Both Hückel methods predict that the dianion HOMO energies decrease when proceeding from naphthalene to octalene; pentalene falls out of sequence. The HOMOs of all three nonbenzenoid compounds are substantially lower energy than that of naphthalene, indicating that all three dianions should form with relative facility. (As we shall see, this is indeed the case of heptalene and octalene.) Note also that the LUMO of octalene dianion is of lower energy than that of heptalene

TABLE 2
Orbital Energy Levels in Hydrocarbon Dianions

	HOMO[a] energy[b]		LUMO[a] energy[b]	
Dianion of	Hückel[c]	Modified Hückel[d]	Hückel[c]	Modified Hückel[d]
Pentalene	0.0	−0.41	−1.41	−1.79
Naphthalene	−0.62	−0.92	−1.00	−1.21
Heptalene	−0.31	−0.49	−0.70	−1.01
Octalene	−0.18	−0.29	−0.45	−0.70

[a] HOMO: highest occupied molecular orbital; LUMO: lowest unoccupied molecular orbital.
[b] Energy given in units of β.
[c] Standard Hückel computation.
[d] Iterative adjustment of Coulomb and exchange integrals to self-consistence. See refs. 22 and 78.

dianion. It should, therefore, be easier to convert octalene into a tetraanion than it would heptalene (or pentalene and naphthalene, for that matter). This turns out also to be the case; treatment of octalene with excess lithium affords the dianion at first, but upon prolonged exposure to lithium, the dianion is reduced further to a tetraanion.[82]

5.1. Pentalene

Pentalene (**5**) presents a problem in structure, as do many of these compounds. Is it a perturbed (and planar) cyclooctatetraene? If so, how does the transannular bond affect its structure and behavior? Johnson considered **5** as a planar COT derivative.[83] He postulated that, like COT, **5** should undergo facile two-electron oxidation and reduction to stable $4n+2$ species (dication and dianion, respectively). Since the parent substance is unknown, the actual measurements were made upon the tri-*t*-butyl derivative (**84**). After corrections based upon estimates of the probable

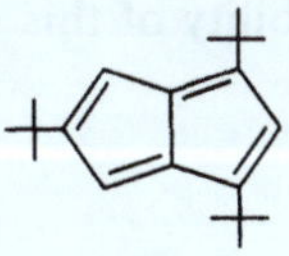

84

effect of the *t*-butyl groups upon the oxidation and reduction potentials of pentalene, and of the energy necessary to flatten the COT ring, it was concluded that pentalene *is* easier to oxidize than COT, and also that the spacing between the first oxidation and reduction potentials is smaller than for COT, suggesting good stability for the anion, as well. It proved impossible to convert **84** to a dication or dianion, however. It was suggested that this is because it would be difficult to accommodate a high degree of charge in a small system; however, COT, with the same number of carbons, forms a dianion readily.[40] Modified Hückel calculations (Table 2) predict that the dianion should form even more readily than that of heptalene; since it does not, one might be tempted to conclude that a double negative charge on an eight-carbon system is intrinsically unfavorable. This is not so, however—we know that COT forms a dianion readily. The difficulty of forming the pentalene dianion remains obscure.

No difficulty was encountered in converting the dibenzopentalenes (**85**) into stable dications and dianions.[84] The spectroscopic properties of these species indicated them to be delocalized $4n+2$ electron species, whose stability was manifested by a relatively small spacing between the first oxidation and reduction waves, as with COT.

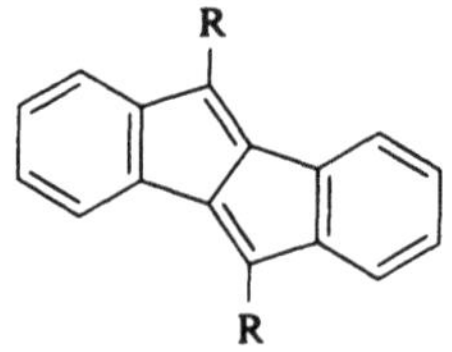

85a, R = H
b, R = CH_3

5.2. Heptalene

Heptalene (**6**), an unstable polyolefinic substance,[85] can be reduced by lithium at −80° to a delocalized dianion exhibiting a diamagnetic ring current.[66] It exhibits two reversible polarographic waves at −1.45 and −2.15 (SCE) in aprotic media,[66] but the spacing between the two waves is rather large for a species forming a stable dianion. As was noted at the introduction to this section, modified molecular orbital calculations (Table 2) nicely explain the formation and stability of this dianion.

5.3. Octalene

Octalene (**2**) has 14 electrons in its periphery, but like a number of related compounds (**45** and **83**, for example), peripheral delocalization involving this $4n + 2$ pi-electron system would require that the system adopt a planar geometry. The potential energy gain from delocalization is less than the torsional and van der Waals' repulsions, which would be present in a planar structure, and therefore, it turns out that the molecule adopts a nonplanar geometry.[4] The substance has polyolefinic properties, like its relative cyclooctatetraene. Like the latter, the spacing between the first and second reduction potentials is small. (E_1 and E_2 are −1.67 and −1.70 V [SCE] in dimethylformamide containing tetrabutylammonium perchlorate.)[82] However, the second reduction step is a *three*-electron process. Thus, octalene may be reduced to a *tetraanion* electrochemically. In fact, the voltammetry suggests that the dianion and trianion are *easier* to reduce than the monoanion, implying a very special stability to the tetraanion. Chemical experiments corroborate these conclusions: upon reaction of octalene with lithium, a dianion is first formed, but is slowly converted to the tetraanion reaction with excess lithium.[82] In the latter experiment, it is not possible to stop reduction at the radical trianion stage,[86] again demonstrating the fact that the octalene tetraanion is a very stale structure. The ability of this small (fourteen carbons, two rings) system to accommodate a charge of −4 is very striking and casts doubt upon electron-repulsion alone as the reason for the inability of substances such as pentalene to form dianions.

5.4. Azulene

The thermal stability of azulene (**4**) and its chemical properties (substitution rather than elimination upon reaction with electrophiles, etc.)[87] indicates that it possesses considerable resonance stabilization relative to the hypothetical bicyclic polyene of the same formula. We may inquire, however, how closely the electrochemical behavior of azulene parallels that of naphthalene, its benzenoid isomer. As it happens, the two exhibit considerably different behavior. As we have seen (Section 2), benzenoid hydrocarbons exhibit two one-electron waves in aprotic media, but a single two-electron wave number protic conditions. Azulene and a number of its derivatives, however, exhibit a one-electron polarographic wave in mixed aqueous-organic colvents.[88] At least one additional wave is observed, whose height corresponds to uptake of three electrons, but it is difficult to interpret this fact, since no product studies were carried out. The first wave of azulene and its derivatives in protic media is quite interesting. First of all, it is a one-electron wave; this contrasts sharply with the electrochemical behavior of benzenoid hydrocarbons, which exhibit a two-electron wave under the same conditions. Furthermore, addition of a twenty-fold excess of phenol to a dimethylformamide solution of azulene has no effect upon the one-electron reversible cyclic voltammetric wave exhibited by azulene under these conditions.[89] It appears, therefore, that the azulene radical anion (**86**)

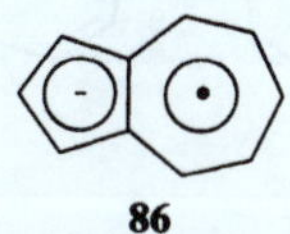

86

is much more stable against protonation than the corresponding benzenoid (viz., naphthalene) species, presumably because of localization of the charge in the five-membered ring. (This hypothesis is supported by modified Hückel calculations, which place ca. 70 percent of the negative charge on **86** in the five-membered ring.)[81] Interestingly, substitution of benzoic acid for phenol in the preceding experiment results in complete disappearance of the anodic cyclic voltammetric peak, presumably because benzoic acid is a strong enough acid to protonate **86** rapidly on the cyclic voltametric time scale.[89]

Azulene is a relatively highly basic hydrocarbon.[90] This manifests itself electrochemically in the form of a new wave at ca. −0.4 V (SCE) in strong acid,[91] i.e., much more positive potential than the usual azulene wave at

−1.6 V; presumably the new wave involves reduction of the azulene conjugate acid (**87**).

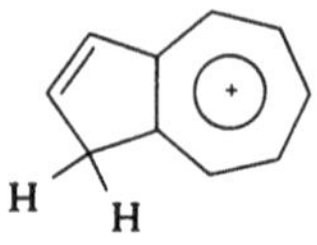

87

The biazulenyls (**88–91**) exhibit very interesting and instructive behavior upon electrochemical and chemical reduction. It has been found that the voltammetric behavior of 5,5′-biazulenyl (**88**) in aprotic solvent differs considerably from that of the 6,6′-isomer (**89**).[92] Although each compound exhibits two reversible waves in cyclic voltammetry, the two waves for **88**,

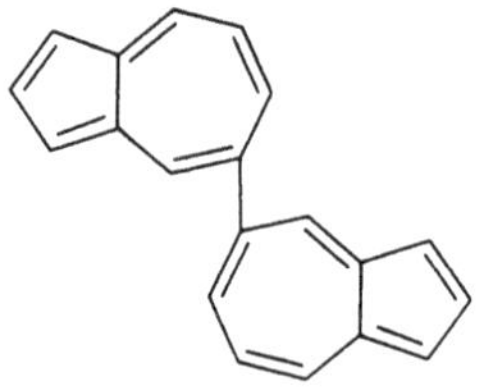

88 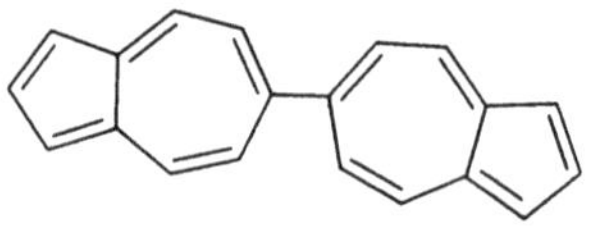 **89**

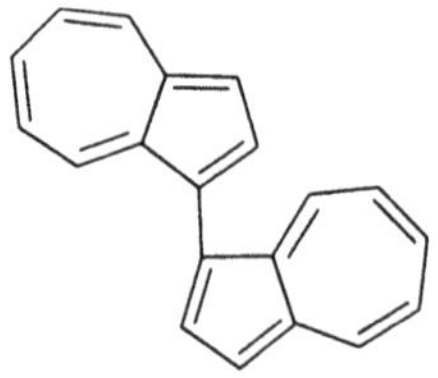

90 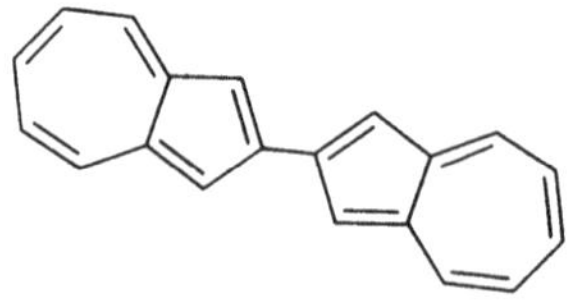 **91**

at −1.50 and −1.77 V (SCE), bracket that of azulene itself (−1.63), while those of **89** (−1.19 and −1.39 V) are both well positive of azulene. The behavior of **88** is characteristic of systems in which two identical electroactive moieties are reduced relatively independently of each other to a product in which the two reduced units interact only weakly. Such behavior has been observed, for example, with *meta*-dinitrobenzene, since the unpaired electron in nitroarene radical anions is largely localized on the nitrogen atom, and the two reduced groups cannot interact by resonance. The behavior of **89**, on the other hand, is reminiscent of systems in which two reduced moieties interact strongly with each other to produce a resonance-stabilized species. The behavior of *para*-dinitrobenzene exemplifies such behavior nicely: the two nitro groups interact strongly with each other in the radical anion and dianion (**92**) to afford stabilization of quinoid type.[93] [Interestingly, similar behavior is observed with the dinitro-compound (**93b**) (polarographic half-wave potentials −1.23 and −1.43 V, respectively, compared with the mononitro compound (**93a**) at −1.43 V);[94] this appears to be one of a very few electrochemical studies made upon a substituted nonbenzenoid hydrocarbon where the focus was upon the behavior of the substituent and not the ring system.]

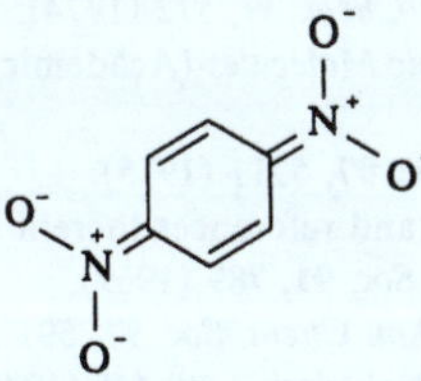

92

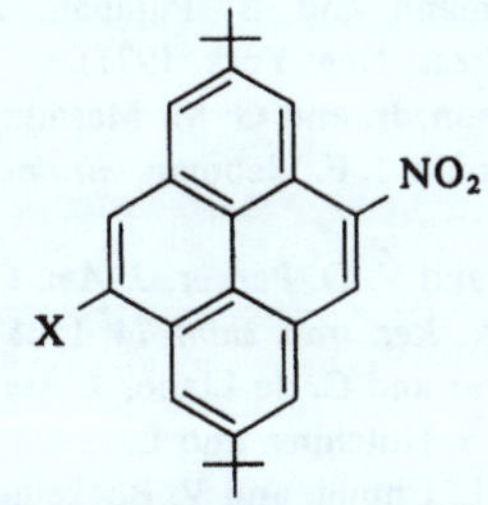

93a, X = H
b, X = NO_2

The sharply contrasting electrochemical behavior of (**88**) and (**89**) carries over into the ESR spectra of their corresponding radical anions. The radical anion of (**88**) is very unstable, and its spectrum resembles that of the parent azulene radical anion, while the radical anion of (**89**) is much more stable, and exhibits hyperfine coupling parameters which are distinctly different from those of the azulene radical anion. Similarly, **90** affords a radical anion which is short-lived, while **91** affords a long-lived radical anion.[95] Why should **89** and **91** behave so differently from azulene and the superficially similar species **88** and **90**? It turns out that these differences can be explained quite nicely by the Hückel M.O. theory. The LUMO of azulene has a large coefficient at atoms 2 and 6. The two azulene units can interact strongly, therefore, in the radical anions and dianions derived from **89** and **91**, which are connected at these positions; otherwise (**88** and **90**), the behavior is that of two isolated azulene radical anion units.

ACKNOWLEDGMENTS. The hospitality of the Chemistry Department of the University of Arizona, especially Professor George S. Wilson, during a sabbatical leave during which this manuscript was completed, is gratefully acknowledged. Financial support was provided by Wesleyan University. Mr. Peter Fox generously carried out a number of the computations referred to herein.

REFERENCES

1. M. E. Peover in *Electroanalytical Chemistry*, edited by A. J. Bard (Marcel Dekker, New York, 1967), Vol. 2, Chap. 1.
2. A. J. Fry, *Synthetic Organic Electrochemistry* (Harper and Row, New York, 1972), Chap. 7.
3. A. Streitwieser, Jr., *Molecular Orbital Theory For Organic Chemists* (J. Wiley and Sons, New York, 1961), Chap. 7.
4. E. Vogel, H.-V. Runzheimer, F. Hogrefe, B. Baasner, and J. Lex, *Angew. Chem., Int. Ed. Engl.* **16**, 871 (1977).
5. E. D. Bergmann and B. Pullman, *Aromaticity, Pseudo-aromaticity, Antiaromaticity* (Academic Press, New York, 1971).
6. A. G. Anderson, Jr. and G. M. Masada, *J. Org. Chem.* **39**, 572 (1974).
7. A. Greenberg and J. F. Liebman, *Strained Organic Molecules* (Academic Press, New York, 1978).
8. B. S. Jensen and V. D. Parker, *J. Am. Chem. Soc.* **97**, 5211 (1975).
9. G. J. Hoijtink, *Rec. trav. chim.* **74**, 1525 (1955), and references therein.
10. M. J. S. Dewar and C. de Llano, *J. Am. Chem. Soc.* **91**, 789 (1969).
11. A. J. Fry, C. S. Hutchins, and L. L. Chung, *J. Am. Chem. Soc.* **97**, 591 (1975).
12. A. J. Fry, L. L. Chung, and V. Boekelheide, *Tetrahedron Lett.* 445 (1974).
13. G. J. Hoijtink, J. van Schooten, E. deBoer, and W. I. Aalbersberg, *Rec. trav. chim.* **73**, 355 (1954).
14. M. Fujihira, H. Suzuki, and S. Hayano, *J. Electroanal. Chem.* **33**, 393 (1971).
15. L. Bergman, *Trans. Faraday Soc.* **50**, 829 (1954).
16. A. Streitweiser, Jr. and I. Schwager, *J. Phys. Chem.* **66**, 2316 (1962).
17. E. Clar, *Polycyclic Hydrocarbons* (Academic Press, New York, 1964), Vols. 1 and 2.
18. P. P. Fu, H. M. Lee, and R. G. Harvey, *J. Org. Chem.* **45**, 2797 (1980).
19. E. Clar, B. A. McAndrew, and U. Sanigok, *Tetrahedron* **26**, 2099 (1970).
20. J. R. Platt, *J. Chem. Phys.* **22**, 1448 (1954).
21. K. Mullen, *Helv. Chim. Acta* **59**, 1357 (1976).
22. K. Mullen, *Chem. Rev.* **84**, 603 (1984).
23. R. Breslow, R. W. Johnson, and A. Krebs, *Tetrahedron Lett.* 3443 (1975).
24. N. L. Bauld and D. Banks, *J. Am. Chem. Soc.* **87**, 128 (1965).
25. R. Waack, M. A. Doran, and P. West, *J. Am. Chem. Soc.* **87**, 5508 (1965).
26. F. Jachimowicz, G. Levin, and M. Szwarc, *J. Am. Chem. Soc.* **99**, 5977 (1977).
27. H. Hart, A. Teuerstein, and M. A. Babin, *J. Am. Chem., Soc.* **103**, 903 (1981).
28. A. Ronlan and V. D. Parker, *Chem. Commun.*, 33 (1974).
29. G. A. Olah and G. Liang, *J. Am. Chem. Soc.* **99**, 6045 (1977).
30. R. Breslow, *Acc. Chem. Res.* **6**, 393 (1973).
31. N. L. Bauld, T. L. Welsher, J. Cessac, and R. L. Holloway, *J. Am. Chem. Soc.* **100**, 6920 (1978).
32. M. R. Wasielewski and R. Breslow, *J. Am. Chem. Soc.* **98**, 4222 (1976).
33. R. Breslow, O. R. Murayama, S.-I. Murahashi, and R. Grubbs, *J. Am. Chem. Soc.* **95**, 6688 (1973).
34. R. D. Rieke and P. M. Hudnall, *J. Am. Chem. Soc.* **95**, 2646 (1973).
35. M. Horner and S. Hunig, *Angew. Chem. Intern. Ed. Engl.* **16**, 410 (1977).
36. K. Hesse, S. Hunig, H. J. Bestmann, G. Schmid, E. Wilhelm, G. Seitz, R. Matusch, and K. Mann, Chem. Ber. **115**, 795 (1982).
37. S. Hunig and H. Putter, *Angew. Chem. Intern. Ed. Engl.* **12**, 149 (1973).

38. R. D. Rieke, C. K. White, L. D. Rhyne, M. S. Gordon, J. F. W. McOmie, and N. P. Hacker, *J. Am. Chem. Soc.* **99**, 5387 (1977).
39. H. N. C. Wong, F. Sondheimer, R. Goodin, and R. Breslow, *Tetrahedron Lett.* 2715 (1976).
40. L. A. Paquette, *Tetrahedron* **31**, 2855 (1975).
41. B. J. Huebert and D. E. Smith, *J. Electroanal. Chem.* **31**, 333 (1971).
42. R. D. Allendoerfer and P. H. Rieger, *J. Am. Chem. Soc.* **87**, 2336 (1965).
43. T. J. Katz, *J. Am. Chem. Soc.* **82**, 3785 (1960).
44. A. J. Fry, J. Simon, M. Tashiro, T. Yamato, R. H. Mitchell, T. W. Dingle, R. V. Williams, and R. Mahedevan, *Acta Chem. Scand.* **37B**, 445 (1983).
45. M. A. Fox and K.-ud-Din, *J. Phys. Chem.* **83**, 1800 (1979).
46. F. J. Smentowski and G. R. Stevenson, *J. Phys. Chem.* **73**, 340 (1969).
47. W. H. Smith and A. J. Bard, *J. Electroanal. Chem.* **76**, 19 (1977).
48. H. Lehmkuhl, S. Kintopf, and E. Janssen, *J. Organomet. Chem.* **56**, 41 (1973).
49. L. A. Paquette, S. V. Ley, R. H. Meisinger, R. K. Russell, and M. Oku, *J. Am. Chem. Soc.* **96**, 5806 (1974).
50. L. A. Paquette, C. D. Wright III, S. G. Traynor, D. L. Taggart, and G. D. Ewing, *Tetrahedron* **32**, 1885 (1976).
51. L. A. Paquette, J. M. Photis, and G. D. Ewing, *J. Am. Chem. Soc.* **97**, 3538 (1975).
52. J. E. Garbe and V. Boekelheide, *J. Am. Chem. Soc.* **105**, 7384 (1983).
53. G. R. Stevenson, J. G. Concepcion, and L. Echegoyen, *J. Am. Chem. Soc.* **96**, 5452 (1974).
54. R. D. Rieke and R. A. Copenhafer, *J. Electroanal. Chem.* **56**, 409 (1974).
55. H. Kojima, A. J. Bard, H. N. C. Wong, and F. Sondheimer, *J. Am. Chem. Soc.* **98**, 55670 (1976).
56. W. E. Britton, J. P. Ferraris, and R. L. Soulen, *J. Am. Chem. Soc.* **104**, 5322 (1982).
57. L. A. Paquette, G. D. Ewing, S. Traynor, and J. M. Gardlik, *J. Am. Chem. Soc.* **99**, 6115 (1977).
58. A. J. Fry, unpublished computation.
59. K. Mullen, W. Huber, T. Meul, M. Nakagawa, and M. Iyoda, *J. Am. Chem. Soc.* **104**, 5403 (1982).
60. L. A. Paquette, G. D. Ewing, and S. G. Traynor, *J. Amer. Chem. Soc.* **98**, 279 (1976).
61. L. B. Anderson, J. F. Hansen, T. Kakihana, and L. A. Paquette, *J. Am. Chem. Soc.* **93**, 161 (1971).
62. L. A. Paquette, L. B. Anderson, J. F. Hansen, S. A. Lang, Jr., and H. Berk, *J. Am. Chem. Soc.* **94**, 4907 (1972).
63. L. Eberson, K. Nyberg, M. Finkelstein, R. C. Peterson, S. D. Ross, and J. J. Uebel, *J. Org. Chem.* **32**, 16 (1967).
64. F. Gerson, K. Mullen, and E. Vogel, *Helv. Chim. Acta* **54**, 2731 (1971).
65. W. Bornatsch and E. Vogel, *Angew. Chem. Intern. Ed. Engl.* **14**, 420 (1975).
66. F. M. Oth, K. Mullen, H. Konigshofen, J. Wassen, and E. Vogel, *Helv. Chim. Acta* **57**, 2387 (1974).
67. L. A. Paquette, H. C. Berk, and S. V. Levy, *J. Org. Chem.* **40**, 902 (1975).
68. A. J. Fry, K. Mullen, and J. A. Simon, unpublished research.
69. F. Gerson, K. Mullen, and E. Vogel, *J. Am. Chem. Soc.* **94**, 2924 (1972).
70. J. F. M. Oth, H. Baumann, J.-M. Gilles, and G. Schroder, *J. Am. Chem. Soc.* **94**, 3498 (1972).
71. D. Tanner, O. Wennerstrom, and E. Vogel, *Tetrahedron Lett.* 1221 (1982).
72. K. Ankner, B. Lamm, B. Thulin, and O. Wennerstrom, *Acta Chem. Scand.* **32B**, 155 (1978).
73. W. Huber, K. Mullen, and O. Wennerstrom, *Angew. Chem. Intern. Ed. Engl.* **19**, 624 (1980).
74. L. B. Anderson, M. J. Broadhurst, and L. A. Paquette, *J. Am. Chem. Soc.* **95**, 2198 (1973).
75. R. D. Allendoerfer, L. L. Miller, M. E. Larschied, and R. Change, *J. Org. Chem.* **40**, 97 (1975).

76. W. Huber, K. Mullen, R. Busch, W. Grimme, and J. Heinze, *Angew. Chem. Intern. Ed. Engl. Supp.*, 566 (1982).
77. L. A. Paquette, M. J. Kukla, S. V. Ley, and S. G. Traynor, *J. Am. Chem. Soc.* **99**, 4756 (1977).
78. A. Minsky, A. Y. Meyer, K. Hafner, and M. Rabinovitz, *J. Am. Chem. Soc.* **105**, 3975 (1983).
79. R. Breslow, J. Napierski, and T. C. Clarke, *J. Am. Chem. Soc.* **97**, 6275 (1975).
80. M. J. S. Dewar and W.-K. Li, *J. Am. Chem. Soc.* **96**, 5570 (1974).
81. A. J. Fry and P. C. Fox, unpublished computations.
82. K. Mullen, J. F. M. Oth, H.-W. Engles, and E. Vogel, *Angew. Chem. Intern. Ed. Engl.* **18**, 229 (1979).
83. R. W. Johnson, *J. Am. Chem. Soc.* **99**, 1461 (1977).
84. I. Willner, J. Y. Becker, and M. Rabinovitz, *J. Am. Chem. Soc.* **101**, 395 (1979).
85. H. J. Dauben, Jr. and D. J. Bertelli, *J. Am. Chem. Soc.*, **83**, 4659 (1961).
86. W. Huber, *Tetrahedron Lett.* 3595 (1983).
87. K. Hafner, *Angew. Chem.* **70**, 419 (1958).
88. L. H. Chopard-dit-Jean and E. Heilbronner, *Helv. Chim. Acta.* **36**, 144 (1953).
89. A. J. Fry, unpublished research.
90. A. E. Sherndal, *J. Am. Chem. Soc.* **37**, 167, 1537 (1915).
91. P. Zuman, *Z. Physikal. Chem.* (Leipzig), *Sonderheft*, 243 (1958).
92. F. Gerson, J. Lopez, and A. Metzger, *Helv. Chim. Acta.* **63**, 2135 (1980).
93. J. Q. Chamber, III and R. N. Adams, *J. Electroanal. Chem.* **9**, 400 (1965).
94. A. J. Fry and J. Simon, unpublsished research.
95. Y. Ikegami and S. Seto, *Bull. Chem. Soc. Japan* **43**, 2409 (1970).

2

Electrochemical Applications in Organic Chemistry

Vernon D. Parker

1. INTRODUCTION

The initial product of charge transfer to an organic species at an electrode is either an ion radical or a neutral radical depending upon whether the substrate is a molecule or an ion. In either case, the reaction seldom ends at this stage. Although there are exceptions, the ion radicals or radicals are usually highly reactive intermediates, and the charge transfer is the first in a sequence of coupled reactions finally leading to the product which can be isolated after macroscale electrolysis.

The mechanisms of the reactions of electrode-generated intermediates may be of interest for a number of different reasons. From the point of view of the chemist interested in developing an electrochemical synthesis, some knowledge of the mechanism of the reaction may be essential in choosing the experimental conditions in order to optimize the process. At the other end of the spectrum is the physical or physical organic chemist whose primary interests lie in the chemistry of the reactive intermediates and regard the electrode reaction as a convenience in their preparation.

The study of the mechanisms of electrode reactions has been actively pursued during the past two decades. Theoretical calculations have been carried out for a large number of mechanism types, and criteria for mechanism analysis have emerged. At the same time, digital electronic equipment

Vernon D. Parker ● Norwegian Institute of Technology, University of Trondheim, Trondheim, Norway.

suitable for the acquisition of precise electrochemical data have become readily available. Thus, at the present time, the art of electrode mechanism analysis is highly developed and can readily be implemented into non-specialist laboratories.

In this chapter, electrode mechanism will refer to the mechanisms of the reactions of intermediate B generated from substrate A in charge transfer reaction (1). Furthermore, the

$$A \pm e^- \rightleftarrows B \tag{1}$$

rate of the heterogeneous reaction (1), in relation to the time scale of the measurements, will be considered to be large. This type of charge transfer reaction is commonly referred to as Nernstian. Also, the electrode is regarded as an electron source or sink, and it is assumed that there are no specific interactions between the electrode and substrates, intermediates, or products. Thus, the reactions of interest take place in homogeneous solution. In fact, little is known about the mechanisms of surface reactions in which either reactants, intermediates, or products are chemisorbed to the electrode. At the present time, little can be done with such systems and adsorption can be an insurmountable problem in mechanism studies.

The primary objective of this chapter is to present the electrochemical methods and their use in mechanism analysis in such a way that the nonspecialist reader obtains a clear picture of what is done in such studies. In order to achieve this objective several examples from the recent literature are discussed in some detail. All of the examples have at least one common aspect. In each case, mechanisms have been postulated in the initial studies of the reaction and later, on the basis of further work, these mechanisms have been challenged and new mechanisms set forth. In the opinion of the author this reflects a healthy situation and serves as a reminder that mechanisms are never proven but rather can be shown to be consistent with the knowledge at hand. This close scrutiny of published mechanisms by other workers should also have the beneficial effect of inducing a considerable degree of caution in mechanism assignment. The overall result should be that fewer mistakes will be made, and those that are will be corrected.

Only the results of theoretical calculations will be discussed. For a general treatment of the theoretical approach the reader is referred to the monograph by MacDonald.[1] More specific information on the numerical solutions of integral equations common to electrochemical methods can be found in the chapter by Nicholson.[2] The most commonly used method for the calculation of the theoretical electrochemical response is digital simulation, which has been well reviewed by Feldberg,[3,4] Prater,[5] Maloy,[6] and Britz.[7]

A number of abbreviations for techniques and reaction schemes are used in this chapter. Definitions are given where the terms are first used and, for quick reference, are included in a Glossary at the end of the chapter.

2. CHEMICAL REACTIONS COUPLED TO CHARGE TRANSFER

The electrochemical generation of reactive species has a unique feature. The electrochemical response accompanying the charge transfer serves as a highly sensitive probe of the homogeneous reactions of the intermediate. The electrode potential is a direct measure of the free energy of reversible reactions and can also be used as a kinetic probe. The current passed during the process or upon reversal of the polarity of the electrode can provide precise kinetic data. Some of the types of equilibria and kinetic processes amenable to study by electrochemical methods will be outlined in this section.

2.1. Equilibrium Processes

A number of different equilibria of the primary intermediate (B) have been studied. The most common measurement involves the determination of the reversible potential (E_{rev}) for the electrode process, and how this is affected by the coupled equilibrium.

If B is an ion radical, the difference in E_{rev} for the first and second charge transfer to or from the substrate (A) is directly related to the disproportionation equilibrium constant (K_{disp}). The relationship between ΔE_{rev} and K_{disp} for reaction (2) is given by Equation 3. To use Equation 3 E_{rev} must be

$$2B \xrightleftharpoons{K_{disp}} C + A \tag{2}$$

$$K_{disp} = \exp(\Delta E_{rev}/(RT/F)) \tag{3}$$

expressed to reflect a reduction, and ΔE is equal to $E^2_{rev} - E^1_{rev}$ where the superscripts refer to the second and first charge transfers, respectively. Extensive studies of equilibria (2) have been reported, where B are anion radicals[8,9] and cation radicals.[10,11]

The acquisition of electrode potential data for thermodynamic purposes requires measurement techniques capable of giving very precise results. Although reasonably precise data can be obtained using polarography or cyclic voltammetry, the method of choice is phase selective second harmonic ac voltammetry (SHAC). Precise values of E_{rev} can be obtained during SHAC measurements even when the product of charge transfer is undergoing rapid followup reactions.[12,13] SHAC gives a very well defined measure of E_{rev} where the rapidly rising or falling in phase or quadrature component of the ac current passes through zero. The measurement of E_{rev} during the oxidation of 9,10-diphenylanthracene in acetonitrile both in the absence (a) and presence (b) of pyridine is illustrated in Fig. 1.[14] The level of precision that can be achieved using SHAC is illustrated by measurements on a second type of equilibrium described in the following paragraph.

When the generation of B is accompanied by a reversible reaction with an additional reactant (X), E_{rev} is shifted either in a positive (reduction) or negative (oxidation) direction to a degree depending upon the magnitude of K_4.

$$B + X \overset{K_4}{\rightleftharpoons} B/X \tag{4}$$

$$\Delta E_{rev} = RT/F \ln(1 + K_4[X]) \tag{5}$$

Equation (5) was used by Peover and Davies[15] to determine ion-pair association constants of anion radicals with counter ions from E_{rev} shifts. The additional reactant, X, need not be an ion. The application of SHAC to the determination of the equilibrium constant for the association of acetophenone anion radical with water (K_6) is illustrated by the data in

$$PhCOCH_3^{\overline{\cdot}} + H_2O/CH_3CN \overset{K_6}{\rightleftharpoons} PhCOCH_3^{\overline{\cdot}}/H_2O + CH_3CN \tag{6}$$

Table 1.[16] Water and other hydroxylic compounds are strongly bound to

TABLE 1
Reversible Potential Shifts and Equilibrium Constants for the Association of Acetophenone Anion Radical with Water in Acetonitrile[a]

$[H_2O]/M$	ΔE_{rev}[b]/mV	K/M^{-1}
0.0694	22.1	20.2
0.139	35.0	21.6
0.278	48.4	20.9
0.556	64.7	21.5

[a] In solvent containing Bu_4NBF_4 (0.1 M) at 293.3 K. Data from reference 16.
[b] Measurements by phase selective second harmonic ac voltammetry at 300 Hz with a dc sweep rate of 40.0 mV/s.

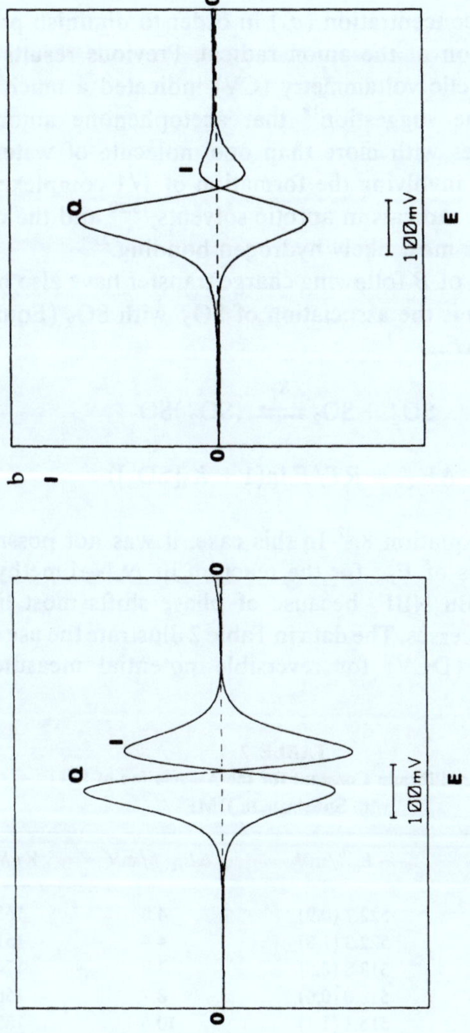

Figure 1. Phase selective second harmonic AC voltammograms for the oxidation of 9,10-diphenylanthracene in acetonitrile (a) and in the presence of pyridine (b).

the common aprotic solvents, such as acetonitrile, by hydrogen bonding,[17] so that, in this case, practically no free water exists in solution. The reversible potentials were measured with a precision of ±0.1 mV.[16] Application of Equation 5 on the data obtained at $[H_2O]$ ranging from 0 to 0.556 M resulted in $K_6 = 21.1 \pm 0.6\ M^{-1}$ at 293.3 K. These measurements were made at low (0.1 mM) substrate concentration (c_A) in order to diminish problems with the rapid dimerization of the anion radical. Previous results obtained at higher c_A[18] using cyclic voltammetry (CV) indicated a much greater E_{rev} shift and led to the suggestion[19] that acetophenone anion radical in acetonitrile associates with more than one molecule of water. Equilibria such as Equation 6, involving the formation of 1/1 complexes, appear to be general for anion radicals in aprotic solvents,[18,20] and the driving force for the association is most likely hydrogen bonding.[21]

Other equilibria of B following charge transfer have also been studied. A pertinent example is the association of $SO_2^{\overline{\cdot}}$ with SO_2 (Equation 7) and the dependence of ΔE_{rev}

$$SO_2^{\overline{\cdot}} + SO_2 \underset{}{\overset{K_7}{\rightleftharpoons}} (SO_2^{\overline{\cdot}})SO_2 \tag{7}$$

$$\Delta E_{rev} = RT/F \ln(1 + K_7[SO_2]) \tag{8}$$

on K_7 is given by Equation 8.[22] In this case, it was not possible to make SHAC measurements of E_{rev} for the reaction in *N,N*-dimethylformamide (DMF) containing Bu_4NBF_4 because of phase shifts most likely due to some adsorption processes. The data in Table 2 illustrate the use of derivative cyclic voltammetry (DCV) for reversible potential measurements.[23] A

TABLE 2
The Equilibrium Constant for the Association of $SO_2^{\overline{\cdot}}$ with Substrate in DMF[a]

$10^4[SO_2]/M$	$-E_p$ [b]/mV	ΔE_{rev} [c]/mV	K/M^{-1}
2.11	522.7 (0.9)	4.0	1631
3.16	522.3 (1.6)	4.4	1513
4.21	518.8 (0.5)	7.9	1749
5.26	518.0 (0.6)	8.7	1568
6.32	516.1 (1.1)	10.6	1655
7.37	514.2 (0.8)	12.5	1743
8.42	513.8 (1.6)	12.9	1588

[a] Measurements by DCV at 293 K in solvent containing Bu_4NBF_4 (0.1 M). Data from reference 23.
[b] Referred to a bias potential of −600 mV vs. Ag/Ag^+ measured at 100 V/s. The numbers in parentheses are the standard deviations in 5 replicate measurements.
[c] The values giving the best fit to Equation (8).

voltage sweep rate (ν) of 100 V/s was used and the precision was about ±1 mV, which resulted in $K_7 = 1635\ M^{-1}$ at 293 K with error limits of ±5.4 percent. The precision observed in the DCV measurements[23] is not that intrinsic in the method since electrode potentials can be measured to about ±0.1 mV by DCV as well.[24]

Numerous other examples of equilibrium measurements could be cited. Those that have been discussed give a good indication of the state of the art in E_{rev} measurements at the present time.

2.2. Kinetic Processes

Since the reactions of electrode generated intermediates frequently involve sequential steps, either heterogeneous or homogeneous, it is convenient to use a short-hand notation to describe the processes. The notation commonly used is that due to Testa and Reinmuth[25] in which *E* indicates heterogeneous charge transfer and *C* a homogeneous chemical reaction.

The first reactions treated theoretically involved simple mechanisms, the kinetics of which are controlled by a single rate-determining step. Thus, in the early work the *CE*,[26,27] the *EC*,[28] the *ECE*,[28,29] and the *EC*(dim)[30] mechanisms were examined and the theoretical responses were deduced. Hawley and Feldberg[31] pointed out an important variation of the *ECE* mechanism in which the second electron transfer does not take place at the electrode but rather in a homogeneous reaction. We have labeled this mechanism ECE_h to emphasize the homogeneous nature of the second *E*. These mechanisms are illustrated in Table 3.

TABLE 3
Simple Electrode Mechanisms and the EC Terminology

Symbols	Mechanism	Steps
CE	$Z \xrightarrow{k} A$	C
	$A \pm e^- \rightleftarrows B$	E
EC	$A \pm e^- \rightleftarrows B$	E
	$B \xrightarrow{k} C$	C
ECE	$A \pm e^- \rightleftarrows B$	E
	$B \xrightarrow{k} C$	C
	$C \pm e^- \rightleftarrows D$	E
ECE_h	$A \pm e^- \rightleftarrows B$	E
	$B \xrightarrow{k} C$	C
	$C + B \rightleftarrows D + A$	E_h

The kinetics of the simple mechanisms, as well as those of more complex reactions can be studied by electrochemical methods. The choice of the kinetic method is usually dictated somewhat by the rate of the homogeneous chemical reactions. First order reactions with rate coefficients as high as about $10^5\ s^{-1}$ can be studied by direct methods. A direct electrochemical method is one in which the response of the intermediate B is observed either by reversing the polarity of the electrode or by employing spectroscopic monitoring of the electrode-solution interface. More rapid reactions require the use of indirect electrochemical methods. The indirect methods are those in which the kinetics of the homogeneous reactions are deduced from their effect on the primary charge transfer process. The basis for the indirect methods is that the homogeneous reactions of B disturb the relative proportions of A and B at the electrode surface (which can be calculated from the Nernst equation) and require that current flow in order to maintain the equilibrium condition dictated by the electrode potential.

The remainder of this chapter is devoted to the study of the kinetics and mechanisms of electrode processes.

3. DIRECT ELECTROCHEMICAL METHODS

Quite a number of different direct electrochemical methods are applicable to the study of the kinetics and mechanisms of the reactions of electrode-generated intermediates. Since the time-gates of the methods do not differ appreciably, the choice of method can usually be made with convenience and availability of instrumentation given strong consideration. The two most convenient methods are cyclic voltammetry and double potential step chronoamperometry (DPSC). These two methods will be discussed in some detail and much less will be said about other methods.

3.1. Cyclic and Derivative Cyclic Voltammetry

It is safe to say that Nicholson and Shain's treatment of CV[28] has had a greater impact upon electrode mechanism studies than any other one publication. Most of the discussion in this section follows from their work.

A cyclic voltammogram for the reversible reduction of a substance is illustrated in Fig. 2. The characteristic features of the CV for a Nernstian process are as follows: (i) the peak potential separation ($E_p^{\text{red}} - E_p^{ox}$) is close to $59/n$ mV at 298 K, where n is the number of electrons transferred. The exact value depends upon the potential at which the sweep is reversed, called the switching potential, E_λ; (ii) the peak current ratio ($I_p^{ox}/I_p^{\text{red}}$) is equal to 1.0. As will be discussed later, this criterion is somewhat subjective

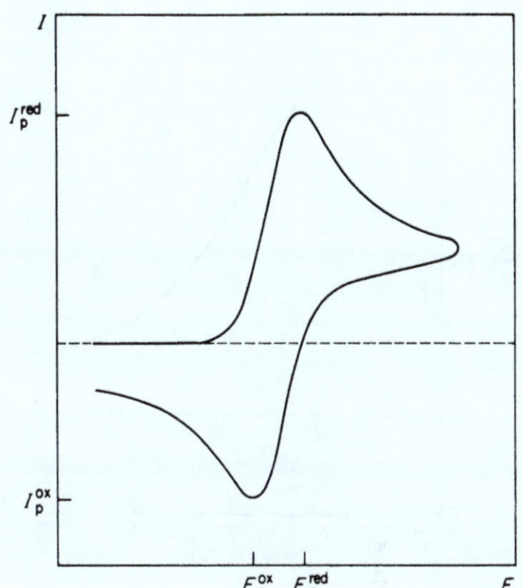

Figure 2. Illustration of a cyclic voltammogram for a reversible redox system.

due to the difficulty in establishing the base line from which to measure the current on the return scan; (iii) the peak current measured on the forward scan increases linearly with $\nu^{1/2}$; and (iv) the peak width, $(E_{p/2} - E_p)$, where $E_{p/2}$ is the potential at which $I = \frac{1}{2}I_p$, is equal to $2.199RT/nF$.

Nicholson and Shain[28] presented their theoretical results for the *CE*, *EC*, and *ECE* and catalytic mechanisms in the form of theoretical working curves which can serve as criteria for mechanism analysis. Their analysis included the effect of ν on (i) the current function which is proportional to $I_p/\nu^{1/2}$, (ii) on I_p^{ox}/I_p^{red} for reduction processes, and on (iii) the half-peak potential. The behavior of a Nernstian system in the absence of kinetic complications is in each case a horizontal straight line. When a homogeneous reaction of *B* takes place, the response will deviate from that for the no-reaction case and will follow the curve for the pertinent mechanism. Thus, using the working curves, it is possible to assign a mechanism and to evaluate the rate constant for the process providing that the rate of the process is such that the kinetic behavior can be observed in the realm of accessible ν, which is of the order of 0.01 to 1000 V/s. In order to make working curves of more general applicability, it is common practice to construct them in dimensionless form.[32] Rather than use ν as the abscissa as described above, it is more useful to have the normalized rate constant as the abscissa, which consists of the rate constant divided by ν and multiplied by the appropriate function of c_A. For a first order reaction of *B* this is k/ν, while that for a second order reaction of *B* is kc_A/ν. A typical

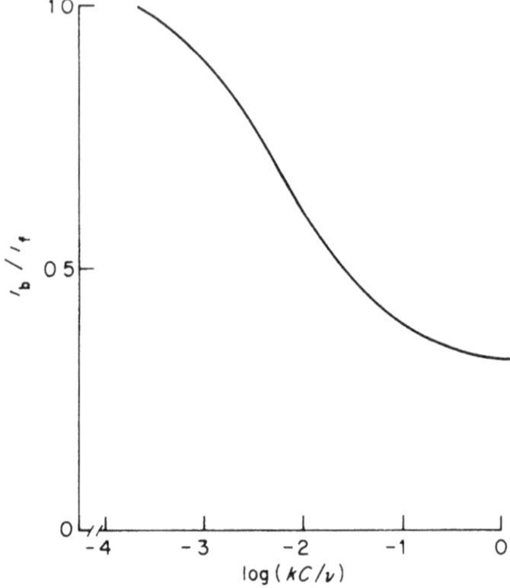

Figure 3. Theoretical working curve for the CV analysis of the EC(dim) mechanism.

working curve for CV analysis of the EC(dim) mechanism is illustrated in Fig. 3.

When CV is used as a direct electrochemical measurement technique, the observable is the peak current ratio, and the variable is ν. There is no difficulty in evaluating the current on the forward scan but how to establish the baseline for the return scan is troublesome. Nicholson[33] has suggested a graphical method for this evaluation. However, the uncertainty in the peak current ratio is generally such that CV can only give qualitative kinetic results.

The baseline problem in CV can be solved by employing the first derivative of the response. The theoretical relationships for DCV were studied by Perone and co-workers[34-36] a number of years ago, but little application was made until the method was developed by Ahlberg and Parker.[37,38] The first derivative of a cyclic voltammogram for a reversible electron transfer is illustrated in Fig. 4. The voltammogram is presented in time ($\sim E$) as the X axis for clarity. The same baseline is taken for measurement of I'_b as for I'_f and the observable of interest in kinetic studies is the ratio of the two derivative peaks, R'_I equal to I'_b/I'_f.

The DCV method was developed using the protonation of anthracene anion radical by phenol as a model reaction. The reaction was treated as an ECE_h mechanism, and several thousand experiments were conducted in which variations were made in the substrate concentration, the phenol concentration, ν and E_λ. It was concluded that rate constants could be evaluated to ± 5 percent even when all possible extremes of the variables were employed.[38]

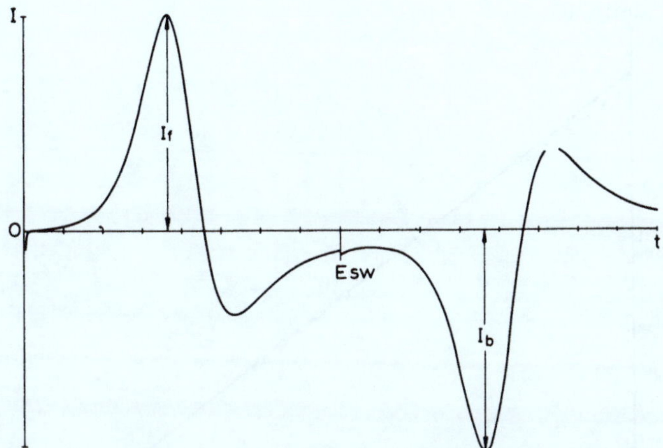

Figure 4. Derivative cyclic voltammogram for a reversible charge transfer.

The DCV working curve for the ECE_h mechanism is shown in Fig. 5, and in Fig. 6 a ln-ln plot of the same data is shown. The significance of Fig. 6 is that between $R'_I = 0.25$ and 0.70, the working curve is linear.[38] Thus, a fit of experimental data to theoretical data can be obtained simply by the measurement of a slope. Theoretical data for a number of other electrode mechanisms were examined, and linear relationships were observed in these cases, as well.[39] The DCV slope, $d \ln R'_I / d \ln(1/\nu)$, for

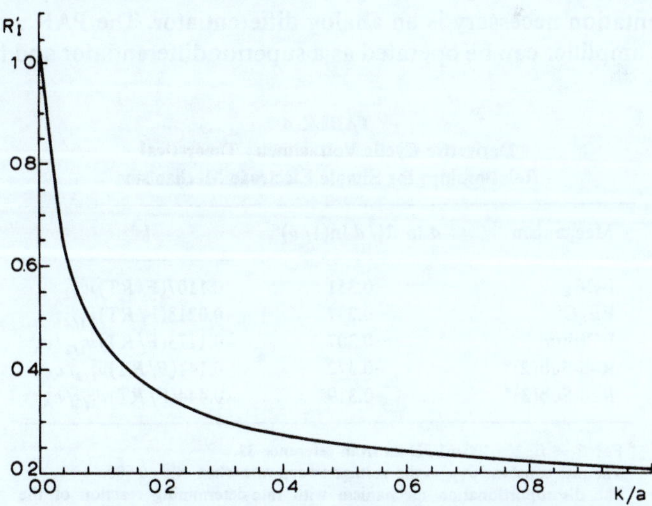

Figure 5. DCV working curve for the ECE_h mechanism (reprinted from E. Ahlberg and V. D. Parker, *J. Electroanal. Chem.* **121**, 73 (1981) with permission).

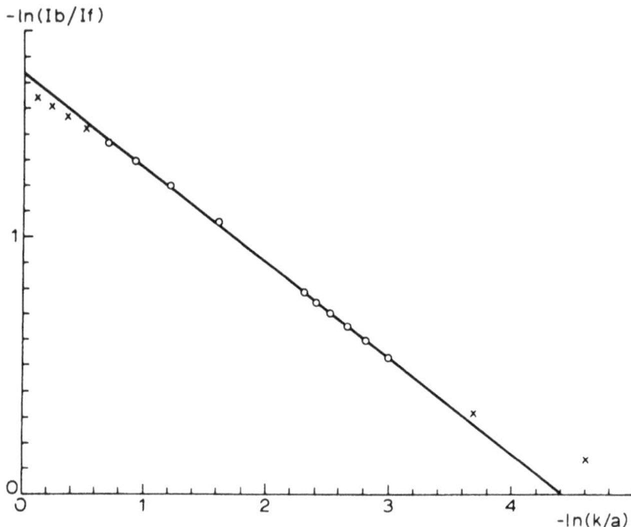

Figure 6. The ln-ln plot of the DCV working curve for the ECE_h mechanism (reprinted from E. Ahlberg and V. D. Parke, *J. Electroanal. Chem.* **121**, 73 (1981) with permission).

the mechanisms considered along with linear equations to evaluate rate constants are given in Table 4. The relationships for rate constants involve $\nu_{1/2}$, defined as ν when R_I' equals 0.500.

Once the capability of carrying out DCV measurements is available, kinetic measurements by ordinary CV are obsolete. The only additional instrumentation necessary is an analog differentiator. The PAR model 189 selective amplifier can be operated as a superior differentiator and has been

TABLE 4
Derivative Cyclic Voltammetry Theoretical Relationships for Simple Electrode Mechanisms

Mechanism	$d \ln R_I'/d \ln(1/\nu)$[a]	k[b]
ECE_h	−0.351	$0.1107(F/RT)\nu_{1/2}$
EE_hC[c]	−0.237	$0.0218(F/RT)\nu_{1/2}$
EC(dim)	−0.307	$0.1173(F/RT)\nu_{1/2}/c_A$
Rad-Sub(2)[d]	−0.372	$0.141(F/RT)\nu_{1/2}/c_A$
Rad-Sub(3)[e]	−0.319	$0.444(F/RT)\nu_{1/2}/c_A^2$

[a] For $E_\lambda - E_{rev} = 300$ mV. Data from reference 39.
[b] The rate constant. $\nu_{1/2}$ is the voltage sweep rate when $R_I' = 0.500$.
[c] The disproportionation mechanism with rate-determining reaction of the doubly charged ion.
[d] Second order radical-substrate coupling.
[e] Third order radical-substrate coupling.

in use in the author's laboratory for the past eight years.[40] Experimentally, the procedure for a DCV kinetic analysis involves first finding the conditions where R'_I falls in the range from 0.25 to 0.70. Then the value of $d \ln R'_I / d \ln(1/\nu)$ can be compared with theoretical slopes for mechanism analysis. Measurements should then be carried out under a wide range of the experimental variables, especially the concentration of substrate and additional reactants, as is feasible. After the investigator is satisfied that the fit of experimental to the theoretical data for a particular mechansim is sufficiently credible, the rate constants can be evaluated from relationships, such as those in Table 4.

3.2. Double-Potential Step Chronoamperometry

DPSC involves stepping the electrode potential from a value where no electrode processes take place to one where the charge transfer of interest takes place at a diffusion-controlled rate, and then back to a potential where only the reverse reaction takes place. The potential limits can be set conveniently by adjusting the starting potential so that the CV for the process is centered in the potential interval. Potential steps of 400 to 1000 mV are commonly used. The electrode potential perturbation along with the current response for a Nernstian process in the absence of kinetic complications is illustrated in Fig. 7. The main features of interest are the value of the current at τ (I_f) and 2τ (I_b), where τ is the potential step width. For the reversible Nernstian system I_b/I_f is equal to $(1 - 2^{-1/2})$ or 0.2929, which reflects the fact that about 70 percent of B is lost by diffusion out into the bulk of the solution. The method was introduced for electrode kinetic studies by Schwarz and Shain[41] who presented theoretical results for the EC mechanism.

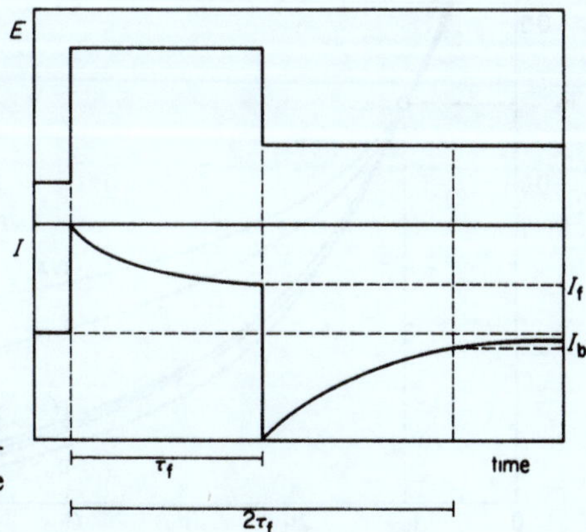

Figure 7. Double-potential step chronoamperometry potential profile and response for a reversible electron transfer.

The most extensive theoretical results have been published by Bard and co-workers.[42] The procedure recommended was to first determine $t_{1/2}$, defined as the pulse width, where R_I is equal to 0.50 by variation of τ in the appropriate region. The normalized current ratio R_I, equal to $I_b/0.2929 I_f$, is used so that values ranging from 1, for the no reaction case, to 0, corresponding to complete reaction, can be observed. After $t_{1/2}$ has been evaluated, data are gathered at multiples of $t_{1/2}$ and compared with theoretical working curves for mechanism assignment. When the mechanism assignment is deemed credible, the pertinent kinetic constants can be evaluated from $t_{1/2}$ values. Working curves for several mechanisms are demonstrated in Fig. 8, and pertinent data for the evaluation of rate constants are gathered in Table 5.

The practical considerations for conducting a DPSC kinetic study differ very little from those discussed for DCV. The first step in the investigation must include some electrode potential measurements, and these are most conveniently carried out by CV. Then the potential step width is selected, keeping in mind that it is only the A/B electrode reaction that is being analyzed. If electroactive species are formed during the potential step, the step width must be adjusted to exclude electrode processes of these species. As indicated earlier, the potential limits may be selected so that the CV for

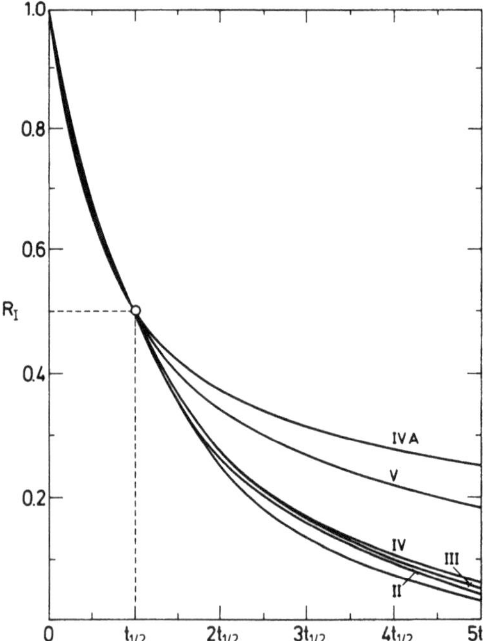

Figure 8. DPSC working curves for a variety of electrode mechanisms. I = EC, II = ECE_h, III = 2nd order EC, IV = Radical-Substrate Dimerization and V = EC(dim). (Adapted from W. V. Childs, J. T. Maloy, C. P. Keszthelyi and A. J. Bard, *J. Electrochem. Soc.* **118**, 874 (1971).)

TABLE 5
Double-Potential Step Chronoamperometry Theoretical Relationships for the Rate Constants of Simple Mechanisms

Mechanisms[a]	k^b
EC	$0.405/\tau_{1/2}$
ECE	$0.272/\tau_{1/2}$
EC(dim)	$0.830/\tau_{1/2}c_A$
Rad-Sub(2)	$0.678/\tau_{1/2}c_A$

[a] See Table 4 for mechanism designation.
[b] Data from reference 42, $\tau_{1/2}$ is the pulse width when $R_I = 0.500$.

the *A*/*B* couple falls in the center of the potential interval. Potential step widths ranging from about 1 ms to 100 ms are practical without any special considerations. Data can be gathered using analog oscilloscopic recording. However, a digital data retrieval system is highly recommended both from the standpoint of precision in the measurements, as well as for convenience.

3.3. Other Techniques

It is apparent from the discussions of DCV and DPSC that there is little difference in the nature of the data which can be obtained by the two techniques. In both cases, a current ratio is obtained which can be related to rate constants of the reactions of intermediate *B* by comparison with theoretical data. The mechanism assignment is made in exactly the same way from working curves for the two methods. The DCV working curve is more conveniently used, since it reduces to linear form when expressed in a ln-ln plot. There are a number of other techniques, such as rotating ring disc electrode voltammetry[32] and double-potential step spectro-electrochemical analysis,[43] which give data in the same format as DCV and DPSC. These techniques differ from the ones already discussed only in the experimental methods used in the measurements. There is little difference in the time resolution for all of these methods, and the mechanism analysis in all cases involves the comparison of experimental data with theoretical working curves.

One aspect of spectroelectrochemistry[43] differs substantially from all the others which involve the measurement of current or potential. The method can be used as a relaxation technique.[44] In this type of experiment, a potential step is used to generate *B*, and then the electrode reaction is discontinued by breaking the electrical circuit. The homogeneous reactions of B are then monitored by the decrease in intensity of spectral bands.

4. INDIRECT ELECTROCHEMICAL METHODS

The indirect electrochemical methods may involve measurement of either the electrode potential or the current. Only those which involve electrode potential measurements will be discussed here. Those involving the measurement of current, such as rotating disc electrode voltammetry and chronoamperometry, are applicable under the same conditions as the direct methods and have been more or less replaced by the latter.

Before discussing the indirect methods it is instructive to consider the conditions under which the methods are applicable in a little more detail. In order to do this, we can refer to the zone diagram for the *EC* mechanism[45] shown in Fig. 9. The zones correspond to regions on a log-log graph, where the ordinate corresponds to the dimensionless heterogeneous rate constant Λ and the abscissa to the dimensionless homogeneous rate constant λ. The rate constants are defined in Equations 9 and 10, where k^0 is the heterogeneous

$$\Lambda = k^0(RT/nF)^{1/2}(\mathrm{D}\nu)^{-1/2} \tag{9}$$

$$\lambda = k(RT/nF)/\nu \tag{10}$$

rate constant, k is the rate constant for the EC mechanism, and D is the diffusion coefficient of A. The zones of interest to Nernstian systems are DP (pure diffusion), the KO, and the KP (pure kinetic). In the DP zone, the chemical reaction is insignificant, and the CV of the process will be that described earlier for the Nernstian system without kinetic complications.

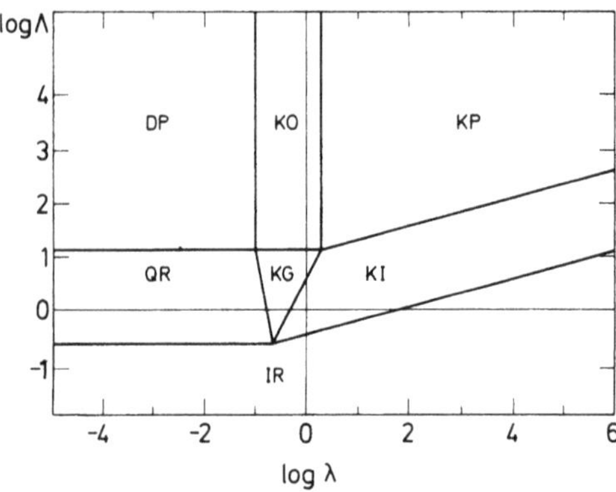

Figure 9. Zone diagram for the EC mechanism.

In the KO zone, the chemical reaction begins to exert its effect on the electrode response. In this zone, the CV peak current ratio is less than 1, the peak potential is not constant, and the shape of the wave is changing. The KO zone is the one in which the direct electrochemical methods are applicable. In the KP zone, a peak cannot be observed on the reverse scan during CV; the peak potential varies linearly with log ν, and the shape of the wave is invariant throughout the zone and characteristic of the mechanism of the reaction. The indirect electrochemical methods are applicable in the KP zone.

The IR zone corresponds to an irreversible charge transfer, and the response does not reflect the kinetics of the homogeneous reactions. The kinetics of the homogeneous reactions are reflected in the response in zones KG and KI, but the analysis is more difficult due to partial control by the charge transfer kinetics. The QR (quasi-reversible) zone is the one in which heterogeneous kinetics can be studied by CV.

In the following paragraphs, the methods described all involve linear sweep voltammetry (LSV) as the primary measurement. In this discussion, the LSV waves, which are current-potential curves, will be of the purely kinetic type (KP zone). Under these conditions, the shape of the wave gives an indication of the mechanism of the reaction. The effect of variables, such as substrate concentration and ν, on the position of the wave along the potential axis can also be used in mechanism analysis and reflects the rate of the reactions following charge transfer.

4.1. Linear Sweep Voltammetry (LSV)

The theoretical relationships involving the shape and the position of the LSV wave under purely kinetic conditions are due primarily to the research groups of R. S. Nicholson, J. M. Savéant, and I. Shain. A large number of calculations were carried out on a variety of electrode mechanisms involving a single rate-determining step. The first publications dealt with the *CE*,[27] the *EC*,[28] the *ECE*,[28] and the *EC*(dim)[30] mechanisms.

In order to demonstrate how the theoretical LSV data were presented and can be used we can consider Equations 11 and 12 for the *EC* and *EC*(dim) mechanisms, where k refers

$$E_p = E_{rev} - 0.780(RT/nF) + (RT/2nF)\ln((RT/nF)k/\nu)$$
$$-dE_p/d\log\nu = (\ln 10)RT/nF; \qquad dE_p/d\log c_A = 0 \tag{11}$$

$$E_p = E_{rev} - 0.902(RT/nF) + (RT/3nF)\ln((2RT/3nF)kc_A/\nu)$$
$$-dE_p/d\log\nu = (\ln 10)RT/3nF; \qquad dE_p/d\log c_A = (\ln 10)RT/3nF \tag{12}$$

to the corresponding first- or second- order rate constant, respectively. Thus, the LSV analysis involves the measurement of the peak potential as a function of ν and c_A. For a one-electron transfer at 298 K, $(\ln 10)RT/nF$ is equal to 59.2 mV. The mechanistic criteria are the slopes $dE_p/d \log \nu$ and $dE_p/d \log c_A$. From these equations, it is obvious that it is only necessary to know the value of E_{rev} in order to evaluate the rate constants. This is a very important advantage, since in a mechanism analysis, it is the mechanism that is of primary interest, and the rate constants only have meaning after the mechanism has been established with some certainty. Therefore, LSV analysis can be used even for very rapid processes for which E_{rev} cannot be measured. Another characteristic of the LSV wave that is generally reported with theoretical data is the peak width, $E_p - E_{p/2}$. Peak widths are a measure of the shape of the LSV wave, but this is given in much more detail by the other methods discussed later in this section.

In connection with theoretical studies on electrodimerization, Nadjo and Savéant[46] took into account the possibility of acid-base reactions, which can affect the rate of the overall reaction. This then gives a third mechanism criterion the effect of Z^0, the proton donor concentration on E_p, $dE_p/d \log Z^0$. A comprehensive table of the various possible electrodimerization mechanisms and the corresponding theoretical results was presented.[46] A similar theoretical study of the possible electrocyclization mechanisms was reported by Andrieux and Savéant.[47]

In order to conduct an LSV mechanism analysis, there are a number of considerations that must be kept in mind. First and foremost, it must be remembered that the theoretical relationships are for purely kinetic waves, i.e., the KP zone (Fig. 9). This means that $\log \Lambda$ must be of the order of 1 or greater, depending upon the mechanism.[45] This is a severe restriction on the range of ν that can be used. Heterogeneous charge transfer rate constants involving organic compounds are usually not greater than about 1 cm/sec.[48] If $\log \Lambda$ is 1, this corresponds to a maximum value of close to 1 V/s for ν, in order that the process give rise to a purely kinetic wave. Secondly, the precision of the measurements are of great importance, since the range of ν is subject to the restrictions already mentioned.

It has been recommended that ν be restricted to 1 V/s or less during LSV analysis.[49] The measurement of an LSV peak potential is associated with considerable error, $\pm(3 - 10)$ mV. Since the LSV variables, ν and c_A, usually are restricted to a ten-fold or smaller range, a much higher degree of precision is desirable. The method of choice is to differentiate the current-potential response. Using analog differentiation the precision of E_p can be improved to the order of ± 0.1 mV.[49] If instrumentation is not available for obtaining the first derivative of the response, precise data can be obtained by measuring $E_{p/2}$, the potential where I is half of I_p.[50] Using digital data processing precision of ± 0.1, mV can be realized, and $E_{p/2}$ can be measured to ± 1 mV using XY recording with a 20 mV/cm expansion on the potential axis.[50] It has also been shown[49] that ν as low as 10 mV/s can be employed without having a detrimental effect on the precision of the movements.

Since precision is such an important factor in the LSV analysis, great care must be taken with the experimental aspects. A procedure which can be strongly recommended is to first test the analysis on a Nernstian system without any kinetic complications. Under these conditions E_p is independent of the variables. This test analysis should be carried out using the same ν as will later be used in a kinetic analysis. The cause for any deviation in E_p can then be evaluated, and correction factors can be determined, if necessary.[49]

4.2. Convolution Potential Sweep Voltammetry (CPSV)

One disadvantage of E_p and $E_{p/2}$ measurements is that a single data point is taken for an entire LSV wave. It was pointed out by Imbeaux and Savéant[51] that CPSV offers the possibility to use more of the information present in an LSV wave to improve the accuracy and to simplify mechanism analysis. CPSV involves the transformation of the LSV wave, by evaluating convolution integrals, to the familiar S-shaped voltammogram observed during polarography or rotating disc electrode voltammetry. After computing the convolution integrals, the data are in a form suitable for logarithmic analysis.* The LSV wave (ψ) upon transformation gives rise to the S-shaped curve ($I\psi$), and the logarithmic analysis of $I\psi$ leads to straight-line $\ln(1 - I\psi)/\psi^{2/3}$ which has a slope equal to RT/nF. The slope is independent of both ν and c_A.

The procedure followed in CPSV mechanism analysis is to test for particular mechanisms by conducting the appropriate logarithmic analysis on the convoluted wave. For example, 16 different mechanisms of electrocyclization were considered and found to give rise to 3 different logarithmic analyses.[52] Thus, one must know the theoretical equation for the LSV wave for the particular mechanism, and, from this, the appropriate logarithmic analysis can be deduced. The slopes of the straight lines from the analysis are integral fractions of RT/nF.

Details of the method and its application can be found in the original literature.[51-54]

* The original literature (reference 51) should be consulted for details of the mathematical analysis.

4.3. Normalized Potential Sweep Voltammetry (NPSV)

NPSV developed by Aalstad and Parker[55] is a somewhat simpler method than CPSV, which gives the same type of information. It involves a three-dimensional analysis of the LSV wave, where the normalized current (I/I_p) is taken as the Z axis, theoretical electrode potential data as the X axis, and experimental electrode potential data as the Y axis. The potential axes are defined relative to $E_{p/2}$. The method is illustrated by Fig. 10. The projection of the wave onto the X-Y plane results in a straight line of unit slope and zero intercept, if the theoretical and experimental data describe

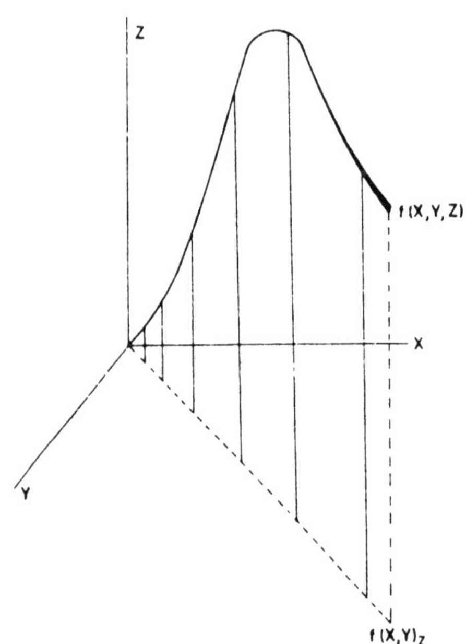

Figure 10. An illustration of a NPSV voltammogram. The Z axis is the normalized current, the X axis is theoretical electrode potential data, and the Y axis is experimental electrode potential data (reprinted from B. Aalstad and V. D. Parker, *J. Electroanal. Chem.* **122**, 183 (1981) with permission).

the same process. In practice, NPSV simply involves the linear correlation of experimental vs. theoretical electrode potentials at particular values of I_N, the normalized current.

A significant feature of NPSV analysis is that linear relationships were observed when theoretical electrode potential data for Nernstian charge transfer were taken as the X axis, and theoretical electrode potential data for various electrode mechanisms were taken as the Y axis. The slopes of the straight lines are indications of the mechanisms of the electrode processes that the data describe.

4.4. Linear Current-Potential Analysis (LCP)

In order to carry out a proper CPSV or NPSV analysis, digital data processing and a digital computer are necessary. LCP, described by Aalstad and Parker,[50] is capable of giving the same type of mechanistic information and has the advantage that the analysis can be carried out using X-Y recording of current-potential curves. An analysis of theoretical data for Nernstian charge transfer and for a number of purely kinetic processes revealed that a very nearly linear region of the LSV wave exists, where I_N ranges from 0.50 to 0.75. The slope of the current-potential curve in this region is a direct reflection of the mechanism of the process. The data in Table 6 give a comparison of the numerical values of the slopes for LCP, LSV, and CPSV analyses.[50] The significant feature of the data is that the slopes vary by a factor of 2 for all three of the methods. The only difference in precision that can be expected for LCP as compared to CPSV is that LCP involves a smaller potential interval. The manner in which the LCP slope, dE/dI_N, can be determined from an X-Y recording of an LSV wave is illustrated in Fig. 11. The scale expansions on the recorder should be

TABLE 6
A Comparison of LCP, LSV, and CPSV Slopes

Mechanism	LCP^a/mV	LSV^b/mV decade^{-1}	$CPSV^c$/mV
EC	68.9	29.6	59.2
ECE_h	68.9	29.6	59.2
EC(dim)	52.8	19.7	39.4
EEC	35.1	14.8	29.6

[a] From reference 50, all data refer to 298 K. dE/dI_N.
[b] $-dE^p/d \log v$.
[c] The appropriate logarithmic analysis.

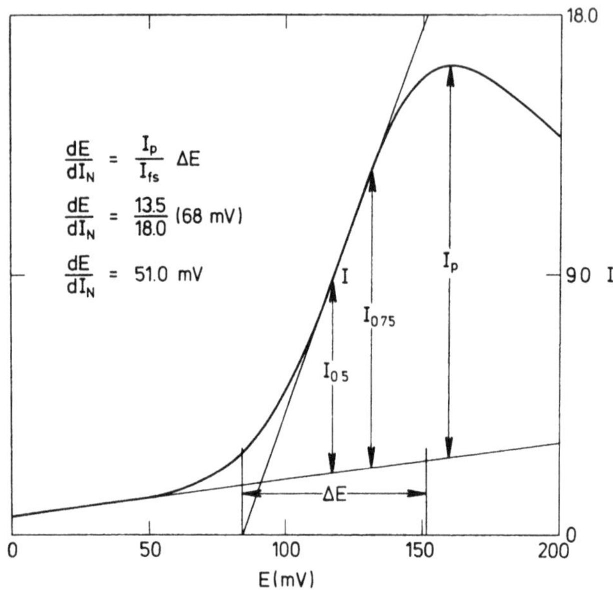

Figure 11. Illustration of the linear current-potential analysis from an X-Y recording of an LSV wave.

adjusted so that I_p is as large as feasible fraction of full scale (fs). The baseline is first extended beyond the peak, and I_p/I_{fs} is determined. The difference in potential (ΔE) between the intersections of the extension of the straight line between I_N equal to 0.50 and 0.75 with the zero and the fs lines is then determined, and dE/dI_N is equal to $(I_p/I_{fs})\Delta E$. The voltammogram shown in Fig. 11 is for an EC(dim) mechanism, $(dE/dI_N)_{\text{theory}} =$ 52.8 mV.

5. THE "REACTION ORDER APPROACH"

The analysis of electrode mechanisms by comparing experimental data to theoretical working curves for the various possible mechanisms is in principle a very sound and reliable procedure. However, there are severe disadvantages which limit the utility of that approach. One disadvantage is that the theoretical working curves for a number of mechanisms may be very similar. This is evident from the DPSC working curves shown in Fig. 8. In fact, in order to differentiate between mechanisms using the theoretical working curves, it may be necessary to analyze data taken at times corresponding to several half-lives of the reactive intermediate. This can be a severe disadvantage, since problems due to adsorption are often minimal at short reaction times but become severe as the time-gate is increased. A further disadvantage inherent in the approach is that either theoretical data must be available for the particular mechanism of interest or the calculations must be feasible. The feasibility of carrying out calculations greatly decreases with the number of adjustable parameters necessary to describe the mechanism. For example, the mechanism described by Equations 13 and 14 requires a theoretical working curve

$$B \underset{k_{-13}}{\overset{k_{13}}{\rightleftharpoons}} C_{13} \tag{13}$$

$$C_{13} \xrightarrow{k_{14}} P_{14} \tag{14}$$

for every value of $K_{13}c_A$.[56] Thus, even for this relatively simple mechanism the theoretical working curve can be changed almost at will by the variation of parameters. Needless to say, when two different mechanisms with different sets of working curves are under consideration, the curves from one set can be made to cross over the other. At this time, theoretical working curves for more complex mechanisms* have not been published, and it is obvious that with even more adjustable parameters the curves would be essentially meaningless for mechanism analysis.

In this section, an alternative to the theoretical working curve will be discussed. The "Reaction Order Approach" involves the determination of the apparent rate law for an electrode mechanism directly from experimental data.[57] The method has been applied to both direct[57,58] and indirect[59,60] measurement techniques for both simple[57,59] and complex[58,60] mechanisms.

5.1. Applied to Direct Methods

The primary experimental quantity in the "Reaction Order Approach" is V_c defined as the value of the experimental variable (V) necessary to hold the experimental observable (O) at a constant value (c). During DCV analysis ν is the variable and R'_I is the observable, while the corresponding quantities in DPSC are τ^{-1} and R_I, respectively.

For homogeneous reactions of B generated in electrode reaction (1), the quantity of mechanistic significance is the reaction order $R_{A/B}$ defined as the sum of the reaction orders in $A(R_A)$ and in $B(R_B)$. When using the direct measurement techniques such as DCV and DPSC, R_A and R_B are not directly separable, and this can be a disadvantage in mechanism analysis. However, it is usually possible to distinguish between the various possibilities on the basis of other considerations.

* Involving more than three adjustable rate parameters.

The reaction orders which can be determined directly from experimental data are $R_{A/B}$ and R_X, where X is an additional reactant usually present in excess. The pertinent relationships between V_c and the reaction orders are Equation 15 for $R_{A/B}$ and Equation 16 for R_X. For the

$$R_{A/B} = 1 + d \log V_c / d \log c_A \quad (15)$$

$$R_X = d \log V_c / d \log c_X \quad (16)$$

simple electrode mechanisms $d \log V_c / d \log c_A$ is a constant and Equation 15 holds at all values of c_A. However, in the case of competing or complex mechanisms $d \log V_c / d \log c_A$ is not necessarily constant, and when this is the case $R_{A/B}$ can only be determined for a specific narrow range of c_A. The fact that $d \log V_c / d \log c_A$ is not constant is adequate evidence that the mechanism in question does not fall into the *simple* classification.[57]

The description of the "Reaction Order Approach" outlined in the previous paragraphs is in essence that which was put forth in the original paper on the subject.[57] However, there has been some misunderstanding of the method. Savéant and co-workers[61,62] claim that it only applies to reactions with a single rate-determining step, i.e., simple mechanisms, and fails when applied to competing or complex mechanisms. Because of these misconceptions, it was necessary to develop the "Reaction Order Approach" in detail for competing and complex electrode mechanisms.[59]

Several complex and competing mechanisms of intermediate B along with the corresponding rate laws are summarized in Table 7. Rate laws for limiting cases are given, as well, and "Reaction Order Approach" quantities which will be discussed in the following paragraphs are listed.

All competing mechanisms of differing $R_{A/B}$ and most complex electrode mechanisms reduce to two limiting cases at the extremes of c_A. The limiting cases are obvious from the rate laws, and these are given in Table 7 for the mechanisms considered. In all cases, $R_{A/B}$ changes by one unit in going from one extreme to the other. Plots of log V_c vs. log c_A curve upward for the competing mechanisms, reflecting that $R_{A/B}$ increases with increasing c_A. Conversely, in the case of the complex mechanisms treated in Table 7, plots of log V_c vs. log c_A curve downward in the transition region as $R_{A/B}$ is decreasing with increasing c_A. The apparent $R_{A/B}$ and R_X in any realm of concentration along with the curvature in the corresponding log V_c vs. log $c_{A(\text{or } X)}$ plots are sufficient to establish the form of the rate law for an unknown electrode mechanism and to make definitive mechanistic conclusions. Thus, the "Reaction Order Approach" provides a powerful mechanistic probe, which relies directly on experimental data without resort to comparison with theoretical data or to numerical calculations.

TABLE 7
The "Reaction Order Approach" Treatment of Competing and Complex Electrode Mechanisms

Number	Mechanism[a]	Rate equation	Limiting rates: Low c_A	Limiting rates: High c_A	$\log(V_c)_{rel}$	m
1	$B \xrightarrow{k_{17}} P_{17}$ $2B \xrightarrow{k_{18}} P_{18}$	$(k_{17} + k_{18}c_B)c_B$	$k_{17}c_B$	$k_{18}c_B^2$	$\log(n+1)/2$	3
2	$2B \xrightarrow{k_{18}} P_{18}$ $B + A \underset{k_{-19}}{\overset{k_{19}}{\rightleftharpoons}} C_{19}$ $C_{19} \xrightarrow{k_{20}} P_{20}$	$(k_{18} + k_{20}K_{19}c_A)c_B^2$	$k_{18}c_B^2$	$k_{20}K_{19}c_Ac_B^2$	$\log n(n+1)/2$	3
3	$B \underset{k_{-21}}{\overset{k_{21}}{\rightleftharpoons}} C_{21}$ $C_{21} + B \xrightarrow{k_{22}} P_{22}$	$(k_{21}k_{22}/(k_{-21} + k_{22}c_B))c_B^2$	$k_{22}K_{21}c_B^2$	$k_{21}c_B$	$\log 2n/(n+1)$	3
4	$B + A \underset{k_{-23}}{\overset{k_{23}}{\rightleftharpoons}} C_{23}$ $C_{23} + B \xrightarrow{k_{24}} P_{24}$	$(k_{23}k_{24}/(k_{-23} + k_{24}c_B))c_B^2c_A$	$k_{24}K_{23}c_B^2c_A$	$k_{23}c_Bc_A$	$\log 2n^2/(n+1)$	6

[a] Rate constants are numbered consecutively and do not refer to equations in the text.

It has been shown that the log V_c-log c_A curves in the transition regions can be constructed using simple arithmetical procedures and theoretical data for the limiting cases.[59] As an example, the competing mechanism 1 (Table 7) can be analyzed. There will be some value of c_A, c_A^0, where either experimental or theoretical data indicate that *on the average* for the time interval of the experiment, $k_{17} = k_{18}c_B^0$, where c_B^0 is the average value of c_B. Expressing c_A in multiples (n) of c_A^0 and taking into account that $k_{17} = k_{18}c_B^0$, the rate law can be expressed as in Equation 17. In rate law (Equation 17), the only variable is the

$$\text{rate} = (1 + n)n(k_{18}c_B^0)^2 \tag{17}$$

relative concentration n, and the variations in the rate with c_A are a consequence of the changes in $n(n + 1)$. There are three distinct cases for rate law (Equation 17) depending upon the magnitude of n, and these are listed below:

$$\textit{Case 1} \quad n \ggg 1, \quad \text{rate} = n^2k_{18}(c_B^0)^2, \qquad R_B = 2 \tag{18}$$

$$\textit{Case 2} \quad n \lll 1, \quad \text{rate} = nk_{18}(c_B^0)^2, \qquad R_B = 1 \tag{19}$$

$$\textit{Case 3} \quad n \simeq 1, \quad \text{rate} = n^{1.5}k_{18}(c_B^0)^2, \qquad R_B \sim 1.5 \tag{20}$$

There will be a rather wide concentration range on either side of $n = 1$, where fractional reaction orders will be observed.

The changes in $\log(V_c)_{\text{rel}}$ with $\log n$, where the subscript indicates relative values, were calculated by equating $\log(V_c)_{\text{rel}}$ to the log of the appropriate $f(n)$ given in Table 7 for mechanism 1 and assigning $(V_c)_{\text{rel}} = 1$ when $n = 1$. In general, $f(n)$ consists of n raised to some power either multiplied or divided by $(n + 1)/2$ depending upon whether the slope of the curve increases or decreases with increasing n, respectively. The exponent of n in $f(n)$ corresponds to $d \log V_c / d \log c_A$ for the limiting case corresponding to the low-concentration limit. This method of calculating the curves was tested by digital simulation of the response for mechanism 1, as well as, by comparison with theoretical working curves for two of the other mechanisms, and excellent correspondence was found.[59] It was pointed out that it is not critical that the correspondence be exact, since the margin for error is small, and all methods of calculating the curves involve approximations.[59]

Thus, using the "Reaction Order Approach," it is possible to evaluate rate constants for competing and complex mechanisms without carrying out numerical calculations on the response in the transition region. This is accomplished by finding the value of c_A^0 in experimental units, which gives the best fit of the $\log(V_c)_{rel} - \log n$ data for the mechanism in question. Once the value of c_A^0 is determined, the corresponding kinetic constants are readily evaluated using the linear equations for the two limiting cases, assuming very small and very large values of n. The method is demonstrated on a practical example in a following section. The linear equations for a number of limiting mechanisms for both DCV and DPSC are given in Tables 4 and 5.

5.2. Applied to Linear Sweep Voltammetry

In the first paper dealing with the application of the "Reaction Order Approach" to LSV, it was pointed out that the rate law for an electrode process can be derived directly from the LSV slopes.[58] For the generalized rate law (Equation 21), the superscripts

$$\text{rate} = kc_A^a c_B^b c_X^x c_I^i \qquad (21)$$

refer to the reaction orders, $a = R_A$, $b = R_b$, etc. In Equation 21, X is an additional reactant, and I is a substance formed during the reaction which further participates, usually in an inhibiting manner. When E_p is expressed in dimensionless form accomplished by division of the value measured in volts by $(\ln 10)RT/F$, the relationships between reaction orders and the LSV slopes are given by Equations 22-24. The parentheses

$$R_B = (dE_p/d \log \nu)^{-1} - 1 \qquad (22)$$

$$R_A = 1 - R_B + (R_B + 1)\, dE_p/d \log c_A - (R_I) \qquad (23)$$

$$R_X = (R_B + 1)\, dE_p/d \log c_X \qquad (24)$$

around R_I in Equation 23 are to emphasize that this term is usually zero. These equations were verified by comparison with theoretical results for a number of simple electrode mechanisms.

It was proposed that Equations 22-24 could be used to derive the rate laws for complex and competing mechanisms directly from experimental data without resorting to theoretical calculations.[58] Because of the criticisms of the "Reaction Order Approach," which have appeared,[61,62] it has been necessary to outline the procedure for LSV analysis in somewhat more detail.[60]

The same example can be treated as before, i.e., mechanism 1 in Table 7. The equations for the limiting cases at 298 K are Equations 25 and 26 which correspond to the EC[28] and

$$E_p - E_{\text{rev}} = -1.134 + 1/2 \log k_{17}/\nu \tag{25}$$

$$E_p - E_{\text{rev}} = -0.981 + 1/3 \log k_{18} c_A/\nu \tag{26}$$

$EC(\text{dim})$[30] mechanisms. At any given value of ν, E_p is independent of c_A for reaction (Equation 17) and directly dependent on $\log c_A$ for reaction (Equation 18). Under these conditions, plots of E_p vs. $\log c_A$ will have slopes of 0 and 1/3 as is evident from Equations 25 and 26, respectively, and intersect at $c_A = 3.29\, k_{17}^{3/2}/\nu^{1/2} k_{18}$. In the intermediate range of c_A, a plot of E_p vs. $\log c_A$ will increase in slope from 0 to 1/3. The curvature in the LSV slope for all of the mechanisms is thus of the same nature as was observed for the direct techniques in the transition region and can be used in the same way for mechanistic considerations.

In order to define the $E_p - \log c_A$ curve in the transition region, it is not necessary to include E_{rev} in the analysis. It is convenient for the slopes of the two limiting cases to differ by one unit, which directly reflects that R_B changes by one in going from one limiting case to the other. For mechanism 1, this is the case when $(3\, E_p)_{\text{rel}}$ is plotted vs. the appropriate function of c_A. Our starting point is where the apparent value of R_B is 1.5. We can assign a value of 0 to $(3\, E_p)_{\text{rel}}$ at this point on the curve, and, since c_A^0 is unknown and must be determined experimentally, the concentration scale can be expressed in n, i.e., multiples of c_A^0. The function of n which is to be equated to $(3\, E_p)_{\text{rel}}$ can be formulated so that the coordinates of the starting point are (0, 0) and is consistent with the known concentration dependence of the limiting cases. In this manner, it is readily shown that $(3\, E_p)_{\text{rel}} = \log(V_c)_{\text{rel}} = \log(n+1)/2$ in the transition region for mechanism 1.

In general, for competing and complex mechanisms the "Reaction Order Approach" applied to LSV involves Equation 27

$$(mE_p)_{\text{rel}} = \log f(n) \tag{27}$$

where m is a common multiple of the equations describing the limiting cases so that the slopes differ by 1 unit. Under these conditions $f(n)$ is the same as for the corresponding curve for the direct techniques. The values of m for each of the mechanisms in Table 7 are listed.

5.3. Determination of Activation Energies

Since V_c is directly related to the apparent rate constant for an electrode mechanism, the apparent activation energy for the process can be determined without knowing the mechanism and hence the rate constants using Arrhenius-like Equation 28.[63]

$$\ln V_c = -E_a/RT + c \quad (28)$$

In Equation 28, c is a constant without physical significance which reflects the fact that the values of the rate constants must be known in order to determine the Arrhenius A factor.*

5.4. Determination of Kinetic Isotope Effects

Isotopic substitution in A or X can give rise to kinetic isotope effects. Considering the case of deuterium kinetic isotope effects arising from isotopic substitution in X, the relationships for the direct techniques and for LSV are Equations 29 and 30, respectively.[59]

$$(k_{app})_H/(k_{app})_D = (V_c)_H/(V_c)_D \quad (29)$$

$$\log((k_{app})_H/(k_{app})_D) = ((E_p)_H - (E_p)_D)(dE_p/d \log \nu)^{-1} \quad (30)$$

6. PUTTING IT ALL TOGETHER, PRACTICAL EXAMPLES

The examples chosen for discussion in connection with the application of the electrochemical methods illustrate most of the different aspects of electrode mechanism analysis discussed in the previous sections. All of the processes have been studied in more than one laboratory, and the data available for analysis are extensive. The work described is recent, and it would be somewhat premature to predict that the final words have been spoken with regard to any of the mechanism assignments. At least one of the studies of each process has been carried out in the author's laboratory. In some cases we have challenged the mechanisms proposed by other workers, while in others our mechanisms have been criticized. The overall result of the challenges and countercharges is that in all cases the present picture is closer to the truth than that presented in the first publications.

6.1. Reactions of 9-Diazofluorene Anion Radical

The initial investigation by McDonald, Hawley, and co-workers[64] provided convincing evidence that 9-diazofluorene ($Fl = N_2$) undergoes 1 e^- reduction in aprotic solvents accompanied by the formation of the dianion

* In the case of DCV, ν_c/T must be used for V_c in (28).

of fluorenone azine (Fl = N-N = Fl). The point of discussion which stirred up some controversy was the proposal that Fl = $N_2^{\overline{\cdot}}$ undergoes rapid loss of N_2 to give the carbene anion radical according to Equation 31, and that $Fl^{\overline{\cdot}}$ is long-lived under the conditions of low-sweep-rate cyclic voltammetry.

$$Fl = N_2^{\overline{\cdot}} \rightarrow N_2 + Fl^{\overline{\cdot}} \xrightarrow{Fl=N_2} (Fl = N\text{-}N = Fl)^{\overline{\cdot}} \tag{31}$$

The evidence presented for the latter was the appearance of a reversible CV wave when measurements were carried out in DMF at ambient temperature and 75 mV/s.[64] This proposal was challenged by Bethell and co-workers,[65] who were able to show by low-temperature CV that the reversible couple under question was actually associated with a second reversible electron transfer, which was masked by the primary reduction wave. This requires that if the assignment of the first wave was correct that the other one must correspond to the reversible reduction of $Fl^{\overline{\cdot}}$ to Fl^{2-}, which was deemed highly improbable. Furthermore, it was shown that at −50°C the intermediate in question was not the first one formed but was due to the decomposition of a shorter lived species.

The first kinetic study of the reaction was reported by Parker and Bethell[66] and consisted of an LSV analysis. The LSV data for the reaction in acetonitrile are given in Table 8. It is worthwhile to consider this data in some detail, since it is a good indication of the current state of LSV measurements. The E_p were measured by analog differentiation of the current-potential curves with a precision of ± 0.2 mV. Measurements were made at four values of ν ranging from 0.10 to 1.00 V/s, and these were known to about ± 1 percent. The data resulted in four values of $dE_p/d \log \nu$ (20.7 ± 1.7 mV/decade) and four values of $dE_p/d \log c_A$ (19.4 ± 1.4 mV/decade). Application of Equations 22 and 23 on this data with $R_I = 0$ results

TABLE 8
LSV Analysis of the Dimerization of Fl = $N_2^{\overline{\cdot}}$ in Acetonitrile[a]

	$-E^P$/mV				
(Fl = N_2)/mM	0.10 Vs^{-1}	0.20 Vs^{-1}	0.40 Vs^{-1}	1.00 Vs^{-1}	$dE/d \log \nu$
0.20	218.0	221.0	227.3	238.2	20.6
0.40	211.4	216.6	222.3	232.4	21.0
0.80	205.1	210.9	218.7	(230.2)[b]	22.6
1.00	203.3	208.3	213.1	222.0	18.5
$dE/d \log c_A$	−21.0	−17.9	−18.3	−19.6	

[a] Data from reference 66. Measurements at 296 K.
[b] Not used in correlations.

TABLE 9
DPSC Kinetic Study of the Dimerization of $Fl = N_2^{\overline{\cdot}}$ in Acetonitrile[a]

$[Fl = N_2]$/mM	$(\tau_{1/2}/ms)$	$(1/\tau_{1/2}c_A/M^{-1}\,s^{-1})$	$(10^{-5})k/M^{-1}\,s^{-1}$
0.25	13.93	287	2.47
0.50	7.51	266	2.29
1.00	3.61	277	2.38
2.00	2.07	242	2.08
			2.31 (0.17)

[a] Data from reference 67. Measurements at 296 K.

in $R_A = 0$ and $R_B = 2$. On the basis of this analysis, the rate law was suggested to be Equation 32 corresponding to the simple dimerization (Equation 33).

$$\text{rate} = k_{33}[Fl = N_2^{\overline{\cdot}}]^2 \tag{32}$$

$$2\,Fl = N_2^{\overline{\cdot}} \xrightarrow{k_{33}} (Fl\text{-}N = N\text{-}Fl)^{2-} + N_2 \tag{33}$$

A more detailed analysis of the LSV wave for the process supports this mechanism assignment. The LCP analysis described in Fig. 12 was conducted on the reduction of $Fl = N_2$ in acetonitrile. The observed value of dE/dI_N was 51.0 mV, which is in good agreement with the theoretical value for reaction (Equation 33), i.e., 52.8 mV.

In the presence of Me_4NBF_4 as supporting electrolyte, it was possible to carry out direct kinetic measurements on the decomposition of $Fl = N_2^{\overline{\cdot}}$ (previous studies had involved Bu_4NBF_4). The reaction was studied by DPSC, and the data in Table 9 verify rate law (Equation 32) and result in $k_{33} = 2.31(\pm 0.17) \times 10^5\ M^{-1}\,s^{-1}$ in acetonitrile at 23°C.[67] The kinetics were studied over a 40 K temperature range which resulted in an Arrhenius plot of poor linearity, correlation coefficient = 0.97, and an apparent activation energy of 2.2 kcal/mol.

The low apparent activation energy along with the observation of at least two intermediates during low-temperature CV[65] led to the proposal that reaction (Equation 33) does not consist of a simple one-step dimerization with the expulsion of N_2, but involves several sequential reactions. The intermediate which was originally believed to be $Fl^{\overline{\cdot}}$ was suggested to be the *cis* isomer of $(Fl = N\text{-}N = Fl)^{\overline{\cdot}}$, which slowly isomerizes to the more stable *trans* isomer.

6.2. Reactions of 4-Methoxybiphenyl Cation Radical

The anodic oxidation of 4-methoxybiphenyl results in the formation of the corresponding dimer, 4,4′-dimethoxyquaterphenyl.[68] The kinetics of the coupling reactions of 4-methoxybiphenyl cation radical were studied in acetonitrile by DCV.[69] The data for measurements over a 64-fold range of c_A are listed in Table 10. The third column labeled $\nu_{1/2}/c_A$ is a "Reaction Order Approach" test for $R_{A/B} = 2$, while that labeled $\nu_{1/2}/c_A^2$ is a test for $R_{A/B} = 3$. The data indicate at the high extreme of c_A, $\nu_{1/2}/c_A$ approaches a constant value, while at the low c_A extreme $\nu_{1/2}/c_A^2$ is close to being constant. Thus, these measurements indicate that the data are in the transition region between two limiting cases. It was suggested that competing mechanisms were involved, Equations 34 and 35 dominating at low c_A and Equation 36 dominating at the higher c_A.

$$R^{\dagger} + R \underset{k_{-34}}{\overset{k_{34}}{\rightleftharpoons}} R^{+} - R^{\bullet} \tag{34}$$

$$R^{+} - R^{\bullet} + R^{\dagger} \xrightarrow{k_{35}} R^{+} - R^{+} + R \tag{35}$$

$$2R^{\dagger} \xrightarrow{k_{36}} R^{+} - R^{+} \tag{36}$$

It should be pointed out that there were problems with adsorption which, besides making the measurements somewhat difficult, caused severe restrictions upon the data analysis. The data do not fit the DCV theoretical working curves[39] because of this problem which was especially severe in the high-concentration range.[70] Therefore, this represents a case where the "Reaction

TABLE 10
DCV Kinetic Data for the Coupling Reactions of 4-Methoxybiphenyl Cation Radical in Acetonitrile[a]

c_A/mM	$\nu_{1/2}$/Vs^{-1}	$\nu_{1/2}/c_A$	$\nu_{1/2}/c_A^2$
8.0	48.3	6.04	0.755
4.0	23.6	5.90	1.48
2.0	9.63	4.82	2.41
1.0	3.40	3.40	3.40
0.5	1.34	2.68	5.36
0.25	0.49	1.96	7.84
0.125	0.11	0.88	7.04

[a] Data from reference 69. All measurements were at 295 K.

Order Approach" was necessary in order to make any mechanistic conclusions.

Amatore and Savéant[62] carried out calculations corresponding to the data in Table 10 and point out that the data are inconsistent with Aalstad, Ronlán, and Parker's mechanism proposal.[69] On the other hand, the data were found to be consistent with mechanism (34) + (35) and rate law (Equation 37) with

$$\text{rate} = (k_{34}k_{35}/(k_{-34} + k_{35}[R^{\ddagger}]))[R^{\ddagger}]^2[R] \quad (37)$$

a change in rate-determining step in going from the high to low concentration of R.

A re-examination of the data of Parker[58] using the more detailed "Reaction Order Approach" described in a previous section confirmed that the original mechanism assignment[69] is not consistent with the data. The data in Table 11 is a comparison of the rate and equilibrium constants assuming mechanism (34) + (35) calculated by the "Reaction Order Approach" and evaluated by the comparison of the data in Table 10 with theoretical working curves.[62] In using the "Reaction Order Approach," it was found[58] that the best fit of the experimental data to the log $V_c/\log(n)$ data was obtained when $n = 1/7$ was assigned to the lowest value of c_A equal to 1.25×10^{-4} M. This resulted in a value of 8.75×10^{-4} M for c_A^0. The values of k_{34} and $k_{35}K_{34}$ were evaluated at very large and very small values of n, respectively, from the corresponding linear equations[39] for the limiting cases. The two sets of these constants obtained by the "Reaction Order Approach" and from theoretical working curves[62] are identical within experimental error. Some error is expected, since both procedures involve curve fitting, and there is obviously some scatter in the data, i.e., the second from last entry in the last column of Table 10 deviates significantly from the curves.

TABLE 11
A Comparison of Rate and Equilibrium Constants for the Dimerization of 4-Methoxybiphenyl Cation Radical in Acetonitrile

Quantity	"Reaction Order Approach"	Theoretical data
k_{34}	$4.14 \times 10^4\ M^{-1}\ s^{-1}$	$4.07 \times 10^4\ M^{-1}\ s^{-1}$
k_{-34}	$(2.98 \times 10^6\ s^{-1})^a$	
k_{35}	$(10^{10}\ M^{-1}\ s^{-1})^a$	
K_{34}	$(1.39 \times 10^{-2}\ M^{-1})^a$	
$k_{35}K_{34}$	$1.39 \times 10^8\ M^{-2}\ s^{-1}$	$1.65 \times 10^8\ M^{-2}\ s^{-1}$

[a] Estimated values based on the assumption of diffusion control for reaction (35).

The additional details given in Table 11, i.e., the values of k_{-34}, k_{35}, and K_{34}, were obtained by assuming that reaction (35) is diffusion controlled with a rate constant of $10^{10}\ M^{-1}\ s^{-1}$ which appears reasonable for the electron exchange reaction. This provides the estimate of K_{34}, $1.39 \times 10^{-2}\ M^{-1}$, which also appears reasonable, since K_{34} must be small in order for rate law (Equation 37) to apply at all. Another useful relationship to check for consistency is Equation 38, which serves as an estimate of c_B^0. The value

$$c_B^0 = k_{-34}/k_{35} = k_{34}/k_{35}K_{34} \tag{38}$$

obtained applying equation 38 is $2.98 \times 10^{-4}\ M$, which once again is reasonable, since c_A^0 was found to be equal to $8.75 \times 10^{-4}\ M$.

6.3. The Anion Radical-Proton Donor Complex Reactions

The inclusion of water in aprotic solvents, such as acetonitrile or DMF, brings about an increase in the rate of dimerization of a number of anion radicals, especially those derived from activated olefins.[42,71,72] It was proposed that the mode of the rate enhancement was specific solvation of the anion radicals by water and that the coupling of anion radicals is made easier when they are surrounded by more water molecules.[71] The hydrodimerization of diethylfumarate anion radical in DMF was studied in some detail in order to gain more information on the effect of water on the kinetics of these reactions.[73] It was observed that the reaction was first order in water, and a small $(k_{app})_H/(k_{app})_D$ was found when the reaction was carried out in the presence of D_2O. The magnitude of the kinetic isotope effect was taken as an indication that proton transfer is not involved in the rate-determining step of the reaction. A two-step mechanism was proposed involving pre-equilibrium (Equation 39) followed by coupling reaction (Equation 40). In a more detailed study, K_{39}

$$\text{DEF}^{\overline{\cdot}} + H_2O \overset{K_{39}}{\rightleftarrows} \text{DEF}^{\overline{\cdot}}/H_2O \tag{39}$$

$$\text{DEF}^{\overline{\cdot}}/H_2O + \text{DEF}^{\overline{\cdot}} \xrightarrow{k_{40}} \text{dimer} \tag{40}$$

was determined to be equal to $1.24\ M^{-1}$ at 273 K, and ΔH_{39}^0 was found to be equal to -5.9 kcal/mol by second harmonic ac measurements. The apparent activation energy for the process was observed to be 1.9 kcal/mol. Thus the activation energy for reaction (Equation 40) was deduced to be equal to 7.8 kcal/mol.

The hydrodimerization of $\text{DEF}^{\overline{\cdot}}$ had previously been studied in DMF in the presence of alkali metal ions, and pre-equilibrium mechanism (41) +

(42) was observed to account

$$\mathrm{DEF}^{\bar{\cdot}} + M^+ \underset{}{\overset{K_{41}}{\rightleftarrows}} \mathrm{DEF}^{\bar{\cdot}}/M^+ \tag{41}$$

$$\mathrm{DEF}^{\bar{\cdot}}/M^+ + \mathrm{DEF}^{\bar{\cdot}} \xrightarrow{k_{42}} \text{dimer} \tag{42}$$

for the kinetic data.[74,75] The apparent activation energy for reactions (41) + (42) where M^+ is Na^+ was observed to be equal to 4.1 kcal/mol.[76] It was not possible to measure K_{41}, since there was no observable change in E_{rev} with $[Na^+]$ ranging from 0 to 13 mM. A maximum value of about 0.5 M^{-1} was estimated for K_{41}.

It was not possible to detect the simple dimerization of $\mathrm{DEF}^{\bar{\cdot}}$, i.e., reaction (Equation 43).[76] It was concluded that if (43)

$$2\,\mathrm{DEF}^{\bar{\cdot}} \xrightarrow{k_{43}} \text{dimer} \tag{43}$$

takes place at all, k_{43} can be no greater than about 10 $M^{-1}\,s^{-1}$. Thus, for the three dimerization mechanisms, the activation energies for the coupling steps were deduced to be in the following order, $(E_a)_{42} > 4.1$ kcal/mol, $(E_a)_{40} = 7.8$ kcal/mol and $(E_a)_{43} > 11$ kcal/mol. These activation energies are of interest to the discussion of the following example.

6.4. Dimerization of 9-Substituted Anthracene Anion Radicals

The reversible dimerization of anthracene anion radicals substituted with electron-withdrawing groups in the 9-position was first reported by Yildiz and Baumgärtel.[77] The mechanism of the reactions was examined in more detail by Hammerich and Parker who concluded that the two coupled equilibria shown in Scheme 1 are involved.[78] The first corresponds to the formation of a complex involving two anion radicals which are not covalently bonded to each other and the second to reversible bond formation. Amatore, Pinson, and Savéant[61] presented a hard critique of this work[78] and concluded that a simple mechanism (44) is involved.

$$2\,\text{AN-}X^{\bar{\cdot}} \underset{k_{-44}}{\overset{k_{44}}{\rightleftarrows}} {}^{-}X\text{-AN-AN-}X^{-} \tag{44}$$

It was further concluded that Hammerich and Parker's mechanism assignment was based on artefacts in the kinetic treatment and on the neglect of results from classical chemical kinetics.[61]

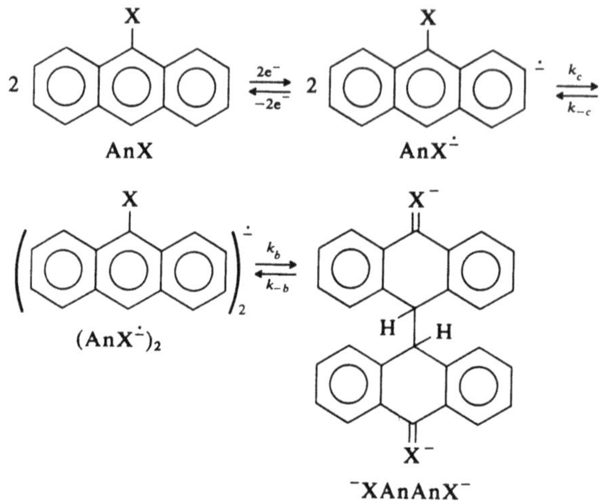

X = NO_2,CN,CHO

SCHEME I

Hammerich and Parker later pointed out that the basis for Amatore, Pinson, and Savéant's mechanism assignment was that the experimental data fit the theoretical working curve for mechanism (44).[79] They point out that Amatore, Pinson, and Savéant neglected to consider that the two mechanisms in question are indistinguishable by theoretical working curves as long as K_c and K_{-b} are small, i.e., $(\text{AN-}X^{\overline{\cdot}})_2$ is present in low concentrations. The two rate laws, (Equation 45) and (Equation 46), differ only by

$$\text{rate} = k_{44}(c_{\text{AN-}X^{\overline{\cdot}}})^2 - k_{-44}c_{(^-X\text{-AN-AN-}X^-)} \tag{45}$$

$$\text{rate} = k_b K_c(c_{\text{AN-}X^{\overline{\cdot}}})^2 - k_{-c}K_{-b}c_{(^-X\text{-AN-AN-}X^-)} \tag{46}$$

the meaning of the constants multiplying the concentration terms. Therefore, calculations cannot help in making a mechanism assignment.

Hammerich and Parker found a way in which to simplify the kinetics.[79] It had previously been observed that acetic acid does not interfere with the kinetics of some other dimerizations of activated olefin anion radicals.[80,81] Thus, it was found that HOAc does not interfere with the kinetics of the formation of the dimers of $\text{AN-}X^{\overline{\cdot}}$ in DMF and in DMSO at concentrations up to about 0.1 *M*. Furthermore, the dissociation of the dimer was shown not to take place in the presence of HOAc which implies the rapid protonation of the dimer dianion.[79]

The primary evidence which made it necessary to propose the two-step mechanism (Scheme 1) consisted of the observations that the apparent activation energies for the processes were lower than expected for reaction (44). At this point it is appropriate to recall that the activation energy for the dimerization of $DEF^{\overline{\cdot}}$, Equation 43, has been estimated to be 11 kcal/mol or greater. No other activation energies are available for anion radical dimerization. However, it is known that the coupling of $DEF^{\overline{\cdot}}/H_2O$ with $DEF^{\overline{\cdot}}$ has an E_a of 7.8 kcal/mol.[76] Also, a related coupling, that of 4-methoxybiphenyl cation radical with substrate,[70] has an activation energy of 11 kcal/mol. Thus, Hammerich and Parker's interpretation[78] that E_a values less than those for translational diffusion are inconsistent with the simple dimerization, i.e., forward reaction (44), seems to be reasonable.

The activation enthalpies for the dimerization reactions of AN-$CN^{\overline{\cdot}}$ and AN-$NO_2^{\overline{\cdot}}$ in DMF in the presence of low concentrations of HOAc were found to be 1.2 and −0.2 kcal/mol, respectively. The data, obtained by the DCV method,[63] for the dimerization of AN-NO_2 are shown in Fig. 12.

Hammerich and Parker state that the important point to be made regarding the activation enthalpies is that they are significantly smaller than those expected for reactions controlled by translational diffusion.[79] Such deviations from "normal behavior" have been reported in a number of other cases involving reactive intermediates, including the dimerization of phenoxy radicals[82-84] and nitroxide radicals,[85] the proton transfer between 2,4-dinitrophenol and aliphatic tertiary amines,[86] several photochemical reactions,[87-89] a Diels-Alder reaction,[90] and the reaction between methyl phenyl ketene with 1-phenyl-ethanol.[91] It was pointed out[79] that the general conclusion arrived at in all these chemically different cases is that the results can only be satisfactorily explained if it is assumed that the reaction in

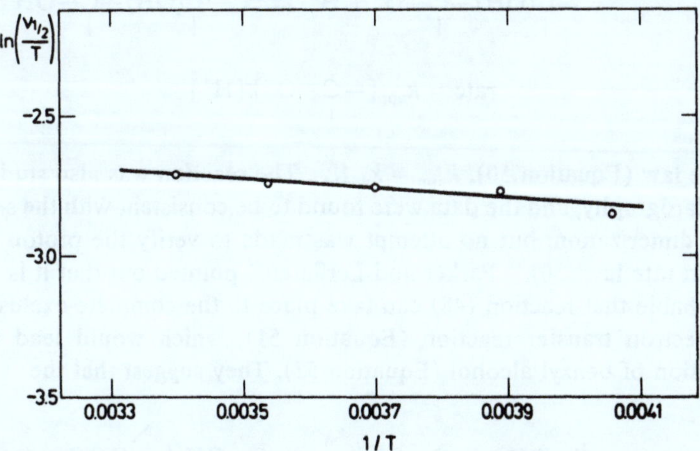

Figure 12. Activation energy determination for the dimerization of AN-$NO_2^{\overline{\cdot}}$.

question proceeds through at least one intermediate, in many cases called a complex, which is present in low concentration and is in equilibrium with the starting material.

Thus, it would appear that these activation energy arguments[78,79] are sound and do not involve a neglect of the classical results of chemical kinetics as proposed by Amatore, Pinson, and Savéant.[61] The two-step mechanism for the dimerization of 9-substituted anthracene anion radicals (Scheme 1) accounts for all of the data available on these systems, and at this time appears to be the most satisfactory mechanism.

6.5. Dimerization of Benzaldehyde Anion Radical

The kinetics and mechanism of the dimerization of benzaldehyde anion radical in aqueous ethanol were reported by Nadjo and Savéant.[92] The mechanism proposed involves the acid-base equilibrium (Equation 47) followed by rate-determining coupling reaction (Equation 48), and rapid protonation (Equation 49) with rate law (Equation 50).

$$-\dot{C}-O^- + H^+ \overset{K_{47}}{\rightleftharpoons} -\dot{C}-OH \qquad \text{(at equilibrium)} \tag{47}$$

$$-\dot{C}-OH + -\dot{C}-O^- \xrightarrow{k_{48}} -C(OH)-C-O^- \tag{48}$$

(rate determining)

$$-C(OH)-C-O^- + H^+ \rightleftharpoons -C(OH)-C-OH \tag{49}$$

$$\text{rate} = k_{app}[-\dot{C}-O^-]^2[H^+] \tag{50}$$

In rate law (Equation 50), $k_{app} = k_{48}K_{47}$. The reaction was also studied by ac polarography, and the data were found to be consistent with the second-order dimerization, but no attempt was made to verify the proton donor term in rate law (50).[93] Parker and Lerflaten[94] pointed out that it is highly improbable that reaction (48) can take place to the complete exclusion of the electron transfer reaction (Equation 51), which would lead to the formation of benzyl alcohol (Equation 52). They suggest that the

$$-\dot{C}-OH + -\dot{C}-O^- \xrightarrow{k_{51}} -\bar{C}-OH + \;\rangle C{=}O \tag{51}$$

$$-\overset{|}{\bar{C}}-OH + HA \longrightarrow -\overset{|}{C}-OH + A^- \quad (52)$$

mechanism should account for the conjugate base of the proton donor A^-, and this should be included in rate law (Equation 53).

$$\text{rate} = k_{app}[-\dot{C}-O^-]^2[HA]/[A^-] \quad (53)$$

Thus, the crucial test of Nadjo and Savéant's mechanism[92] is the kinetic behavior at different $[A^-]$. A detailed kinetic study using DPSC, DCV, and LSV indicated that the rate of the reaction is independent of $[Bu_4NOH]$ in aqueous ethanol, as long as the latter was 20 mM or greater. Under these conditions, a primary deuterium kinetic isotope effect was observed which implicates proton transfer in a rate-determining, rather than, equilibrium step.[93] These results were interpreted in terms of mechanisms (54) + (55)

$$2\,PhCHO^{\cdot-} \overset{K_{54}}{\rightleftharpoons} PhCH(O^-)CH(O^-)Ph \quad (54)$$

$$PhCH(O^-)CH(O^-)Ph + ROH \xrightarrow{k_{55}} PhCH(OH)CH(O^-)Ph + RO^- \quad (55)$$

$$\text{rate} = k_{55}K_{54}[PhCHO^{\cdot-}]^2[ROH] \quad (56)$$

and rate law (Equation 56). At lower $[Bu_4NOH]$ or in the presence of HOAc another mechanism competes which is first order in the anion radical. This was demonstrated by LSV and DCV reaction orders. The coulometric *n* value was observed to be 1 regardless of the conditions indicating that the overall reaction is dimerization in all cases. The most likely reaction pathway under these conditions was concluded to be rate-determining protonation (Equation 57) followed by reaction with substrate (Equation 58) and further

$$PhCHO^{\cdot-} + ROH \xrightarrow{k_{57}} Ph\dot{C}HOH + RO^- \quad (57)$$

$$Ph\dot{C}HOH + PhCHO \longrightarrow PhCH(OH)CH(O^{\cdot})Ph \quad (58)$$

reduction. An alternative mechanism involving coupling reaction (Equation 48) instead of (Equation 58) was deemed less likely, since electron transfer (Equation 51) would be expected to compete favorably, and this would lead to the formation of $PhCH_2OH$.

6.6. The Protonation of Anthracene Anion Radical by Phenol

The reduction of anthracene in DMF in the presence of phenol was studied by Amatore and Savéant using DPSC.[95] The objective of the study was to show that the simple *ECE* scheme is not followed, but rather that the process goes by the ECE_h mechanism. The basis for assuming the ECE_h mechanism was that $dE^p/d \log \nu$ during LSV analysis was observed to be close to 30 mV/decade, according to the theory for a first-order process.[95] In that study, the concentrations of both substrate and phenol were invariant at 10^{-3} and 10^{-2} *M*, respectively. A more detail study using both DPSC and DCV[38] supported the proposal of the ECE_h mechanism. The fit of experimental to theoretical DPSC data for the reaction is illustrated in Fig. 13. Thus, it appeared clear at this point that the reaction was of the ECE_h type and can be described by reactions (Equation 59)-(Equation 61) with step (Equation 59) rate determining and

$$AN^{\cdot-} + PhOH \underset{k_{-59}}{\overset{k_{59}}{\rightleftharpoons}} AN^{\cdot}{-}H + PhO^{-} \tag{59}$$

$$AN^{\cdot}{-}H + AN^{\cdot-} \underset{k_{-60}}{\overset{k_{60}}{\rightleftharpoons}} AN^{-}{-}H + AN \tag{60}$$

$$AN^{-}{-}H + PhOH \xrightarrow{k_{61}} AN(H_2) + PhO^{-} \tag{61}$$

essentially irreversible under the conditions of the studies.

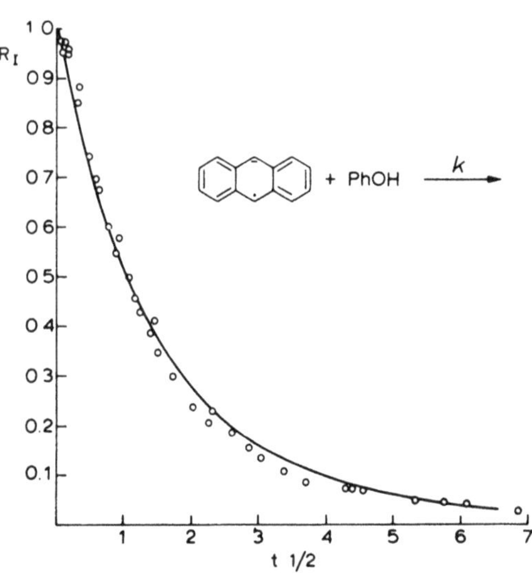

Figure 13. Double-potential step chronoamperometric data for the reduction of anthracene in DMF-Bu_4NBF_4 (0.10 *M*) in the presence of phenol. The solid line is the theoretical working curve for the ECE_h mechanism (reprinted from E. Ahlberg and V. D. Parker, *J. Electroanal. Chem.* **121**, 73 (1981) with permission).

However, a subsequent study of the reaction[96] resulted in the conclusion that the kinetics, under all conditions, are not adequately described by a simple ECE_h mechanism. In fact it was proposed that the reaction order in anthracene anion radical ($AN^{\overline{\cdot}}$) varies between 1 and 2, and that the reaction order in phenol is greater than 1. Besides the DCV evidence for a complex mechanism, LSV data for the reaction were clearly inconsistent with the simple scheme.[97] This study was sharply criticized by Amatore, Gareil, and Savéant,[98] who imply that the experimental data[96] are in error. They claimed that the mechanism is in fact the simple ECE_h scheme proposed earlier.[38,95] The viewpoints of the two groups have recently converged somewhat. Amatore, Gareil, and Savéant[99] have re-examined their data[98] and found that it requires a much more complex mechanism involving the initial formation of complexes containing one $A^{\overline{\cdot}}$ and either one or two molecules of phenol (PhOH).

The experimental aspect on which there has been the greatest disagreement is the value of $dE^p/d \log \nu$ at a phenol concentration of 10 mM, which has been claimed on the one hand[96] to be about 21 mV/decade with a substrate concentration of 1 mM and 29 mV/decade by others[98] under the same conditions. In both studies, it was assumed that the theoretical value for the ECE_h mechanism is 29.6 mV/decade. However, an important factor has been overlooked. Phenoxide ion formed in reactions (Equation 59) and (Equation 61) strongly complexes with phenol,[100] which results in an overall stoichiometry as indicated in Equation 62. Theoretical

$$2\,AN^{\overline{\cdot}} + 4\,PhOH \rightarrow AN(H_2) + AN + 2\,PhO^-/PhOH \qquad (62)$$

calculations taking the correct stoichiometry into account result in $dE^p/d \log \nu$ equal to 26.1 mV/decade,[101] i.e., about halfway between what is obtained in the two different laboratories. Thus, data from neither laboratory are consistent with the simple ECE_h mechanism.

The story regarding the mechanism of the protonation of anthracene anion radical by phenol is not yet complete. The conclusions of the earlier studies[96,97] that the kinetics of the reaction cannot be accounted for by the simple ECE_h mechanism under all experimental conditions still hold. There are still major inconsistencies with any explanation based on only the limiting cases. The results do show that one must be flexible in interpretation and be willing to change a conclusion when new data become available (or data are re-examined[99]). Sometimes, we have to remind ourselves of one of the fundamental rules of mechanism studies: Any mechanism can be considered to be valid as long as it is consistent with all known experimental facts and must be discarded when any new experimental evidence fails to comply.

7. CONCLUDING REMARKS

The methods for the investigation of electrode mechanisms have undergone major development over the past twenty years. At the present time these methods can be considered as precise physical organic kinetic tools.[102] The mechanistic discussions have become somewhat more sophisticated in recent years, and there is lively debate in the area. With the methods at their current high state of development, it is reasonable to predict that in the future the emphasis will be placed more on the chemistry of the reactive intermediates and less on methodology.

GLOSSARY OF TERMS

E^p	electrode peak potential
F	the Faraday
D	diffusion coefficient
τ	the potential step width in double-potential step chronoamperometry
ν	the voltage sweep rate used in cyclic voltammetry
LSV	linear sweep voltammetry
CV	cyclic voltammetry
DCV	derivative cyclic voltammetry
DPSC	double-potential step chronoamperometry
SHAC	phase sensitive second harmonic a.c. voltammetry
E	an electron transfer at the electrode
E_h	an electron transfer in homogeneous solution

REFERENCES

1. D. D. MacDonald, *Transient Techniques in Electrochemistry* (Plenum Press, New York, 1977).
2. R. S. Nicholson and M. L. Olmstead, *Computers in Chemistry and Instrumentation, Electrochemistry*, edited by J. S. Mattson, H. B. Mark, Jr., and H. C. MacDonald (Dekker, New York, 1972), Chapt. 5.
3. S. W. Feldberg, *Electroanalytical Chemistry*, edited by A. J. Bard (Dekker, New York, 1969), Vol. 3.
4. S. W. Feldberg, *Computers in Chemistry and Instrumentation, Electrochemistry*, edited by J. S. Mattson, H. B. Mark, Jr., and H. C. MacDonald (Dekker, New York, 1972), Chapt. 7.
5. K. B. Prater, *Computers in Chemistry and Instrumentation, Electrochemistry*, edited by J. S. Mattson, H. B. Mark, Jr., and H. C. MacDonald (Dekker, New York, 1972), Chapt. 8.

6. J. T. Maloy, *Computers in Chemistry and Instrumentation, Electrochemistry*, edited by J. S. Mattson, H. B. Mark, Jr., and H. C. MacDonald (Dekker, New York, 1972), Chapt. 9.
7. D. Britz, *Digital Simulation in Electrochemistry* (Springer-Verlag, Berlin, 1981).
8. J. Jaguar Grodzinksi, M. Feld, S. L. Yang, and M. Szwarc, *J. Phys. Chem.* **69**, 628 (1965).
9. B. S. Jensen and V. D. Parker, *J. Am. Chem. Soc.* **97**, 5211 (1975).
10. O. Hammerich and V. D. Parker, *Electrochim Acta* **18**, 537 (1973).
11. S. Hünig and H. Berneth, *Top. Curr. Chem.* **92**, 1 (1980).
12. T. G. McCord and D. E. Smith, *Anal. Chem.* **41**, 1423 (1969).
13. A. M. Bond and D. E. Smith, *Anal. Chem.* **46**, 1946 (1974).
14. E. Ahlberg and V. D. Parker, *Acta Chem. Scand.* **B34**, 97 (1980).
15. M. E. Peover and J. D. Davies, *J. Electroanal. Chem.* **6**, 46 (1963).
16. V. D. Parker, *Acta Chem. Scand.* **B38**, 189 (1984).
17. E. D. Becker, *Spectrochim. Acta* **17**, 46 (1963).
18. J. M. Savéant and D. Tessier, *J. Electroanal. Chem.* **61**, 251 (1975).
19. C. Amatore, J. Pinson and J. M. Savéant, *J. Electroanal. Chem.* **137**, 143 (1982).
20. V. D. Parker, *Acta Chem. Scand.* **B37**, 125 (1983).
21. G. R. Stevenson and M. Pourian, *J. Phys. Chem.* **86**, 1871 (1982).
22. C. L. Gardner, D. T. Fouchard and W. R. Fawcett, *J. Electrochem. Soc.* **128**, 2337 (1981).
23. V. D. Parker, *Acta Chem. Scand.* **A37**, 423 (1983).
24. V. D. Parker, *Acta Chem. Scand.* **B35**, 373 (1981).
25. A. C. Testa and W. H. Reinmuth, *Anal. Chem.* **33**, 1320 (1961).
26. P. Delahay and T. Berzins, *J. Am. Chem. Soc.* **75**, 2486 (1953).
27. J. M. Savéant and E. Vianello, *Electrochim. Acta* **8**, 905 (1963).
28. R. S. Nicholson and I. Shain, *Anal. Chem.* **36**, 706 (1964).
29. H. B. Herman and A. J. Bard, *J. Phys. Chem.* **70**, 396 (1966).
30. R. S. Nicholson, *Anal. Chem.* **37**, 667 (1965).
31. M. D. Hawley and S. W. Feldberg, *J. Phys. Chem.* **70**, 3459 (1966).
32. A. J. Bard and L. R. Faulkner, *Electrochemical Methods* (John Wiley, New York, 1980), Appendix B.
33. R. S. Nicholson, *Anal. Chem.* **37**, 1351 (1965).
34. S. P. Perone and T. R. Mueller, *Anal. Chem.* **37**, 3 (1965).
35. C. V. Evins and S. P. Perone, *Anal. Chem.* **39**, 309 (1967).
36. S. P. Perone, *Computers in Chemistry and Instrumentation, Electrochemistry*, edited by J. S. Mattson, H. B. Mark, Jr., and H. C. MacDonald, Jr. (Dekker, New York, 1972), Chapt. 13.
37. E. Ahlberg and V. D. Parker, *J. Electroanal. Chem.* **121**, 57 (1981).
38. E. Ahlberg and V. D. Parker, *J. Electroanal. Chem.* **121**, 73 (1981).
39. E. Ahlberg and V. D. Parker, *Acta Chem. Scand.* **B35**, 117 (1981).
40. E. Ahlberg, B. Svensmark, and V. D. Parker, *Acta Chem. Scand.* **B34**, 53 (1980).
41. W. M. Schwarz and I. Shain, *J. Phys. Chem.* **69**, 30 (1965).
42. W. V. Childs, J. T. Maloy, C. D. Keszthelyi, and A. G. Bard, *J. Electrochem. Soc.* **118**, 874 (1971).
43. T. Kuwana and N. Winograd, *Electroanalytical Chemistry*, **7**, 1 (1974).
44. H. N. Blount, N. Winograd, and T. Kuwana, *J. Phys. Chem.* **74**, 3231 (1970).
45. L. Nadjo and J. M. Savéant, *J. Electroanal. Chem.* **48**, 113 (1973).
46. L. Nadjo and J. M. Savéant, *J. Electroanal. Chem.* **44**, 327 (1973).
47. C. P. Andrieux and J. M. Savéant, *J. Electroanal. Chem.* **53**, 165 (1974).
48. N. Koizumi, T. Saji and S. Aoyagi, *J. Electroanal. Chem.* **81**, 403 (1977).
49. V. D. Parker, *Acta. Chem. Scand.* **B35**, 373 (1981).

50. B. Aalstad and V. D. Parker, *J. Electroanal. Chem.* **112**, 163 (1980).
51. J. C. Imbeaux and J. M. Savéant, *J. Electroanal. Chem.* **44**, 169 (1973).
52. C. P. Andrieux, J. M. Savéant and D. Tessier, *J. Electroanal. Chem.* **63**, 429 (1975).
53. J. M. Savéant and D. Tessier, *J. Electroanal. Chem.* **61**, 251 (1975).
54. L. Nadjo, J. M. Savéant, and D. Tessier, *J. Electroanal. Chem.* **52**, 403 (1974).
55. B. Aalstad and V. D. Parker, *J. Electroanal. Chem.* **122**, 183 (1981).
56. B. M. Bezilla, Jr. and J. T. Maloy, *J. Electrochem. Soc.* **126**, 579 (1979).
57. V. D. Parker, *Acta Chem. Scand.* **B35**, 233 (1981).
58. V. D. Parker, *Acta Chem. Scand.* **B35**, 259 (1981).
59. V. D. Parker, *Acta Chem. Scand.* **B37**, 165 (1983).
60. V. D. Parker, *Acta Chem. Scand.* **B37**, 243 (1983).
61. C. Amatore, J. Pinson, and J. M. Savéant, *J. Electroanal. Chem.* **137**, 143 (1982).
62. C. Amatore and J. M. Savéant, *J. Electroanal. Chem.* **144**, 59 (1983).
63. V. D. Parker, *Acta Chem. Scand.* **B35**, 51 (1981).
64. R. N. McDonald, J. R. January, K. J. Borhani, and M. D. Hawley, *J. Am. Chem. Soc.* **99**, 1268 (1977).
65. D. Bethell, P. J. Galsworthy, K. L. Handoo, and V. D. Parker, *J. Chem. Soc. Chem. Commun* **534** (1980).
66. V. D. Parker and D. Bethell, *Acta Chem. Scand.* **B34**, 617 (1980).
67. V. D. Parker and D. Bethell, *Acta Chem. Scand.* **B35**, 691 (1981).
68. A. Ronlán, K. Bechgaard and V. D. Parker, *Acta Chem. Scand.* **27**, 2375 (1973).
69. B. Aalstad, A. Ronlán, and V. D. Parker, *Acta Chem. Scand.* **B35**, 649 (1981).
70. B. Aalstad, A. Ronlán and V. D. Parker, *Acta Chem. Scand.* **B37**, 467 (1983).
71. E. Lamy, L. Nadjo, and J. M. Savéant, *J. Electroanal. Chem.* **50**, 141 (1974).
72. J. M. Savéant and D. Tessier, *J. Electroanal. Chem.* **61**, 251 (1975).
73. V. D. Parker, *Acta Chem. Scand.* **B35**, 147 (1981).
74. M. D. Ryan and D. H. Evans, *J. Electrochem. Soc.* **121**, 881 (1974).
75. M. J. Hazelrigg, Jr. and A. J. Bard, *J. Electrochem. Soc.* **122**, 211 (1975).
76. V. D. Parker, *Acta Chem. Scand.* **B37**, 393 (1983).
77. A. Yildiz and H. Baumgärtel, *Ber. Buns. Ges.* **81**, 1177 (1977).
78. O. Hammerich and V. D. Parker, *Acta Chem. Scand.* **B35**, 341 (1981).
79. O. Hammerich and V. D. Parker, *Acta Chem. Scand.* **B37**, 379 (1983).
80. P. Margaretha and V. D. Parker, *Acta Chem. Scand.* **B36**, 260 (1982).
81. O. Lerflaten and V. D. Parker, *Acta Chem. Scand.* **B36**, 225 (1982).
82. P. P. Levin, I. V. Khudyakov, and V. A. Kuzmin, *Int. J. Chem. Kinet.* **12**, 147 (1980).
83. D. J. Williams and R. Kreilick, *J. Am. Chem. Soc.* **90**, 2775 (1968).
84. L. R. Mahoney and M. A. DaRooge, *J. Am. Chem. Soc.* **97**, 4722 (1975).
85. D. F. Bowman, T. Gillan and K. U. Ingold, *J. Am. Chem. Soc.* **93**, 6555 (1971).
86. E. F. Caldin and K. J. Tortschanoff, *J. Chem. Soc., Faraday Trans* **1**, 74, 1804 (1978).
87. N. J. Turro, G. F. Leht, J. A. Butcher, Jr., R. A. Moss, and W. J. Guo, *J. Am. Chem. Soc.* **104**, 1754 (1982).
88. P. C. Wong, D. Griller, and J. C. Scaiano, *Chem. Phys. Lett.* **83**, 69 (1981).
89. U. Maharaj and M. A. Winnik, *J. Am. Chem. Soc.* **103**, 2328 (1981).
90. V. D. Kiselev and J. G. Miller, *J. Am. Chem. Soc.* **97**, 4036 (1975).
91. J. Jähme and C. Rüchardt, *Tetrahedron Lett.* **23**, 4011 (1982).
92. L. Nadjo and J. M. Savéant, *J. Electroanal. Chem.* **33**, 419 (1971).
93. J. W. Hayes, I. Ruzic, D. E. Smith, C. L. Booman, and J. R. Delmastro, *J. Electroanal. Chem.* **51**, 269 (1974).
94. V. D. Parker and O. Lerflaten, *Acta Chem. Scand.* **B37**, 403 (1983).
95. C. Amatore and J. M. Savéant, *J. Electroanal. Chem.* **107**, 353 (1980).

96. V. D. Parker, *Acta Chem. Scand.* **B35**, 349 (1981).
97. V. D. Parker, *Acta Chem. Scand.* **B35**, 373 (1981).
98. C. Amatore, M. Gareil, and J. M. Savéant, *J. Electroanal. Chem.* **147**, 1 (1983).
99. C. Amatore, M. Gareil, and J. M. Savéant, *J. Electroanal. Chem.* **176**, 377 (1984).
100. F. G. Bordwell, R. J. McCallum and W. N. Olmstead, *J. Org. Chem.* **49**, 1424 (1984).
101. M. Folmer Nielsen, O. Hammerich, and V. D. Parker, *Acta Chem. Scand.* **B39** (1985), in press.
102. V. D. Parker, *Adv. Phys. Org. Chem.* **19**, 131 (1983).

3

The Electrochemistry of Transition Metal Organometallic Compounds

John C. Kotz

The synthesis of ferrocene in 1952 led to an explosion of activity in organometallic chemistry, and *bis*-η^5-cyclopentadienyl compounds of other transition metals were soon synthesized.[1,2] Among the first properties of these compounds to be studied were their redox reactions. For example, Page and Wilkinson found that ferrocene could be oxidized to a stable Fe(III)-containing ferricenium ion,[3] and the ferricenium reduction potential was observed to be dependent on the nature of any ring substituents. The possibility of the isolation of organometallic compounds with the metal in unusual oxidation states, and the sensitivity of redox potentials to substituents, catalyzed the study of the electrochemistry of organometallic compounds, work which has led to the observation of a wide variety of interesting and unusual reactions.

The first major effort to examine the electrochemistry of organometallic compounds in a systematic manner was made by Dessy and his co-workers in a long series of papers.[4-21] With this impetus, the field has very rapidly developed. In addition to other systematic studies by several groups, there are now many casual references to the electrochemical behavior of new compounds in the literature. The vast majority of this work has described compounds based on transition metals. Thus, in an effort to prepare a thorough review of the field of reasonable length, we have limited the discussion to transition metal compounds.

John C. Kotz ● Chemistry Department, State University of New York, Oneonta, New York 13820.

This review covers the literature concerned with the electrochemistry of transition metal organometallic compounds to June 1983. Approximately 350 research publications have been surveyed and are included herein, but no attempt has been made at an exhaustive search of the literature. Rather, it was our aim to review the main areas of activity. We wish to review systematically those parts of the field where sufficient work has been done to allow us to arrive at useful generalizations, to point out areas still in need of systematic effort, and especially, to note chemical reactions induced by electron transfer.

The last topic is of particular interest, because the importance of electron transfer reactions in organometallic reaction chemistry has been recognized within the past decade, and their continuing study promises important developments. Such reactions are thoroughly discussed in Kochi's book,[22] and it is recommended that this book be used in conjunction with this review to obtain a complete picture of organometallic electron transfer reactions.

In 1970, Gubin briefly reviewed some of the early work on organometallic electrochemistry,[23] and Dessy reviewed his extensive research in 1972.[21] To the best of our knowledge, the present review is the first attempt to survey the electrochemical behavior of transition metal organometallic

TABLE 1
Redox Properties of the Metallocenes, $(\eta^5\text{-}C_5H_5)_2M$, and the Decamethylmetallocenes, $(\eta^5\text{-}C_5Me_5)_2M$. All potentials in Volts vs. SCE

Compound	$+2 \rightleftarrows +1$[a]	$+1 \rightleftarrows 0$	$0 \rightleftarrows -1$	Solvent/comments	Reference
Cp_2V	+0.59	−0.55	−2.74	THF	43
Cp_2Cr		−0.67	−2.30	MeCN	43
Cp_2Fe		+0.31		MeCN	43
Cp_2Ru				$2e^-$ oxidation; irreversible	54
Cp_2Co		−0.94	−1.88	MeCN	43
Cp_2Rh		−1.41	−2.2	MeCN	60
Cp_2Ni	+0.77	−0.09	−1.66	MeCN	43
$(C_5Me_5)_2Cr$		−1.04		MeCN	66
$(C_5Me_5)_2Mn$		−0.56	−2.50	Both reversible; MeCN	68
$(C_5Me_5)_2Fe$		−0.12		MeCN	66
$(C_5Me_5)_2Ru$		+0.55		CH_2Cl_2; reversible; 1 electron	69
$(C_5Me_5)_2Co$		−1.47		MeCN	66
$(C_5Me_5)_2Ni$	+0.31	−0.65		MeCN	66

[a] Charge on the molecule.

compounds since the early 1970s. However, a review of the closely related area of the "Electrochemistry of Metallaboron Cage Compounds" has very recently appeared,[24] as has a review of the electrochemistry of metal clusters.[25] Thus, work in these areas will be noted only where particularly appropriate. In addition, very little is included herein on electrochemical synthesis, since this area has also been reviewed.[26,27] The one area which might be construed as organometallic, but which is not covered herein, is that of the electrochemistry of compounds containing only isonitrile ligands, that is, compounds of the type $M(CNR)_x$, and related cyanide compounds. References are included in the bibliography for the reader wishing entry into the literature regarding the electrochemistry of this group of compounds.[28-40]

A word about abbreviations. The common electrochemical abbreviations will be routinely used. In addition, however, we note the use of R and Ar for sigma-bonded alkyl and aryl substituents (Me = methyl, Et = ethyl, and Ph = phenyl, etc.), Cp for the $[\eta^5\text{-}C_5H_5]^-$ ion, and Fc for the ferrocenyl group, $(\eta^5\text{-}C_5H_4)Fe(\eta^5\text{-}C_5H_5)$.

Finally, unless otherwise noted, all potentials quoted in the text are referenced to the saturated calomel electrode (SCE).

1. ELECTROCHEMISTRY OF THE METALLOCENES, $(\eta^5\text{-}C_5H_5)_2MX_n$ (n = 0, 1, 2)

The metallocenes represent one of the major classes of compounds in organometallic chemistry. To understand their properties, it is well to refer to one of the general principles of transition metal organometallic chemistry, the "effective atomic number" or "16/18 electron" rule.[2,41,42] In filling as completely as possible the bonding molecular orbitals of an organometallic complex, a total of 18- or sometimes 16-valence electrons are used. With certain exceptions, all transition metal organometallic compounds follow this rule. Thus, ferrocene, $(\eta^5\text{-}C_5H_5)_2Fe$, has 6 electrons contributed by each of the $C_5H_5^-$ ligands and 6 electrons provided by the Fe(II) ion.

As shown in Table 1, the electrochemical behavior of a number of the simple metallocenes of transition metals has been reported. We shall first describe the more straightforward behavior of the symmetrical *bis*-η^5-cyclopentadienyl compounds of the type $(C_5H_5)_2M$ found in Groups 5–8 before turning to the metallocenes of the earlier metals. As suggested by the "16/18 electron" rule, the earlier transition metals, with fewer valence electrons to contribute, must add additional ligands to their coordination sphere in order to fill as completely as possible their bonding molecular orbitals.[42] Thus, for example, d^0 Ti(IV) must add two-electron donor anionic ligands in addition to Cp^- in order to form the 16-valence electron molecules $(\eta^5\text{-}C_5H_5)_2TiX_2$ (X = halide, alkyl, etc.); the electrochemistry of such compounds will often involve the X substituents.

1.1. The Simple Metallocenes, $(\eta^5\text{-}C_5H_5)_2M$

Vanadium(II), a d^3 metal ion, can be isolated in the form of very air-sensitive, neutral vanadocene Cp_2V. The molecule does not obey the "16/18 electron rule," since only 15-valence electrons are available. This means that the compound is not only paramagnetic but quite reactive. Further, it behaves rather like many other vanadium compounds which typically display a range of oxidation states (0 to +5), and several of these states are electrochemically accessible for vanadocene (Table 1).[43,44] For example, vanadocene can be reduced reversibly at about −2.74 V (vs. SCE) to give the moderately stable vanadocene anion based on V(I), but little is known about this ion, a species isoelectronic with Cp_2Cr.

The oxidation of THF solutions of vanadocene, however, is better characterized.[43,44] The V(II) → V(III) transformation is reversible, but the second oxidation step [V(III) → V(IV)] is irreversible with an anodic peak potential highly dependent on scan rate, implying a slow electron transfer step. The difference in these two electron transfer steps is thought to reflect the fact that the V(III) → V(IV) step involves severe structural changes, and the reason for this goes back to the mode of bonding in the molecule. As noted previously, electron-poor, early transition metal ions form metallocenes in which ligands in addition to $C_5H_5^-$ must be bound to the metal. Owing to electronic repulsive effects, the complex can no longer be a symmetrical "sandwich," but rather it must adopt the conformation dictated by VSEPR. Thus, for example, vanadium(III) forms the triangular structure Cp_2VX, while vanadium(IV) produces Cp_2VX_2, with a tetrahedral structure. The electrochemistry of Cp_2VCl_2 is described below, but Equation 1 shows the connection between the redox chemistry of vanadocene and the dichloride derivative.

$$\begin{array}{ccccccc} [Cp_2V]^{2+} & \underset{}{\overset{+0.6\ V}{\rightleftharpoons}} & [Cp_2V]^{+} & \overset{-0.6\ V}{\rightleftharpoons} & Cp_2V & \overset{-2.7\ V}{\rightleftharpoons} & [Cp_2V]^{-} \\ & & \Big\updownarrow {\scriptstyle 2Cl^-} & & & & \Big\uparrow {\scriptstyle -2.7\ V} \\ [Cp_2VCl_2]^{+} & \overset{+0.2\ V}{\rightleftharpoons} & Cp_2VCl_2 & \overset{-0.3\ V}{\rightleftharpoons} & [Cp_2VCl_2]^{-} & \overset{-1.5\ V}{\rightleftharpoons} & Cp_2V + 2Cl^- \end{array} \quad (1)$$

The electrochemistry of chromocene, Cp_2Cr, a dark red, very air-sensitive solid, has been thoroughly examined.[44,45] The neutral compound, a 16-electron molecule based on d^4Cr^{2+}, can be both oxidized and reduced by one electron. Both reactions are reversible in THF, but in CH_3CN

follow-up reactions are observed.

$$[Cp_2Cr]^- \underset{-2.30\,V}{\rightleftharpoons} Cp_2Cr \underset{-0.67\,V}{\rightleftharpoons} [Cp_2\,Cr]^+ \qquad (2)$$

Manganocene (Cp_2Mn, 17-valence electrons) is absent from Table 1, as its electrochemistry has apparently not been reported. (As described below, the redox chemistry of the decamethyl compound is known.) However, as the bonding in this molecule is unique among the transition metal metallocenes, one would anticipate unusual behavior.

One of the first properties of ferrocene to be reported was its electrochemistry. Page and Wilkinson reported reversible oxidation of ferrocene to the ferricenium ion, $[(\eta^5\text{-}C_5H_5)_2Fe]^+$ at +0.31 V (vs. SCE) at the DME in 90 percent ethanol with 0.1 *M* $NaClO_4$.[3] Since that time, the reversibility of this redox reaction in various solvents, and in a molten salt mixture, has been confirmed many times.[44,46,47] (However, it has been noted that fast scan rates in cyclic voltammetry reveal quasi-reversible behavior.) Thus, with some restrictions, the ferricenium/ferrocene couple [$E^0 = 0.400$ V vs. NHE][48] has been suggested as an internal standard for electrochemical measurement.[46] As noted in the introduction, the ferricenium reduction potential is exceptionally sensitive to ring substituents, an aspect of ferrocene chemistry described in Section 1.5 below.

Unlike the metallocenes discussed thus far, the electrochemistry of all of the iron group metallocenes has been reported. However, there has been some confusion about the nature of the oxidation product, especially in the case of ruthenocene. There were early reports of a one-electron oxidation,[49] but more recently it has been firmly established that ruthenocene is oxidized in widely varying solvents (including a molten salt) by two electrons at a potential much higher than that of ferrocene.[50-54] Following electrolysis at a mercury electrode, the compound $[(C_5H_5)_2Ru]_2Hg^{2+}$ has been isolated.[51,52] At a Pt electrode, Diaz *et al.* observed the oxidation of ruthenocene to be electrochemically irreversible, but they claimed it is chemically reversible.[54] They suggested that the oxidized Ru(IV)-species may relax to a Ru(II)-containing compound having neutral cyclopentadiene ligands; this, in turn, may reduce back to Cp_2Ru at a much lower potential. This interesting possibility has precedent in the behavior of $[(\eta^6\text{-}C_6Me_6)_2Ru]^{2+}$ described in Section 2.[55,56] (See also Section 5.1.)

The electrochemical behavior of osmocene is somewhat similar to that of ruthenocene.[49-51] Two irreversible steps are observed at +0.75 V and +1.37 V at the Pt electrode. A salt, analyzed as $[(C_5H_5)_2Os]_2HgBF_4$, was isolated after electrolysis at a mercury electrode. However, the green salt $(C_5H_5)_2OsBF_4$, a one-electron oxidation product, was isolated following electrolysis in CH_3CN at a Pt electrode.

Like vanadium, cobaltocene (19-valence electrons) displays a range of oxidation states. The compound is usually isolated in the laboratory as the yellow air-stable, 18-valence electron, cobalticenium ion, $[Cp_2Co]^+$, an ion isoelectronic with ferrocene. This ion has been found to be capable of reduction to neutral cobaltocene, a very air-sensitive, purple compound based on $Co^{2+}(d^7)$.[57-59] Yet another one-electron, reversible reduction step provides the brown-orange anion, $[Cp_2Co]^-$ (20 valence electrons, isoelectronic with Cp_2Ni). This anion is long-lived in 1,2-dimethoxyethane, but in CH_3CN it decays to an electroactive product with a rate constant estimated to be about 0.1 sec^{-1}. On adding a proton donor, phenol, to the glyme solution, the red, ring-protonated compound $CpCo(C_5H_6)$ was formed.[58]

$$
\begin{aligned}
[Cp_2Co]^+ + e^- &\rightarrow Cp_2Co \qquad E_{1/2} = -0.81\ V \\
Cp_2Co + e^- &\rightarrow [Cp_2Co]^- \qquad E_{1/2} = -1.90\ V \\
[Cp_2Co]^- + C_6H_5OH &\rightarrow (\eta^5\text{-}C_5H_5)Co(\eta^4\text{-}C_5H_6) + C_6H_5O^- \\
Cp_2CoC_5H_6 + 2e^- &\rightarrow \text{products} \qquad E_{1/2} = -2.30\ V
\end{aligned}
\tag{3}
$$

Only in the iron and cobalt sub-groups is the opportunity provided for the study of the heavier congeners of the first row metallocenes. Although ferrocene and ruthenocene differ significantly, $[Cp_2Co]^+$ and $[Cp_2Rh]^+$ are rather similar.[60] Both reduce in two one-electron steps with the first step having a rapid electron transfer rate. This implies that in neither case is there any significant structural change in the molecule on electron transfer. In the case of cobalt, the second reduction is about 1 V more cathodic than the first, whereas it is 0.77 V more cathodic for Cp_2Rh. This latter fact has been taken as suggesting a similar metal vs. ligand charge distribution.

The major difference between the neutral, 19-electron cobalt and rhodium metallocenes is that Cp_2Co is stable in glyme, whereas Cp_2Rh is not. The neutral rhodocene forms a dimer which can be isolated in good yield following one-electron reduction of the rhodocenium ion. The electrochemistry of the ion is summarized as follows:

$$Cp_2Rh^+ + e^- \rightarrow Cp_2Rh \qquad E_{1/2} = -1.41\ V$$

$$Cp_2Rh \rightarrow \mathbf{1} \tag{4}$$

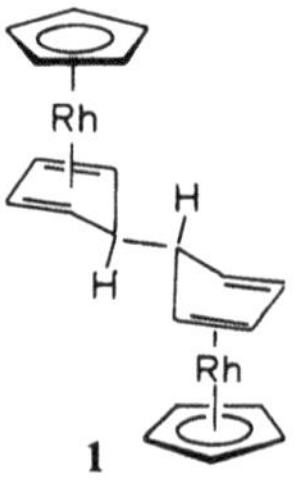

Rh H H Rh

1

$$Cp_2Rh + e^- \rightarrow Cp_2Rh^- \qquad E_{1/2} = -2.18\ V$$

and the dimer is oxidized at −0.4 V.

$$\text{dimer} \rightarrow [Cp_2Rh]^+ + e^- \qquad E_{pa} = -0.4\ V$$

It is interesting to compare the 19-electron, neutral rhodocene with the behavior of another 19-electron molecule, $(\eta^5\text{-}C_5H_5)Fe(\eta^6\text{-}C_6H_6)$.[61] This molecule is similarly unstable, dimerizing through the benzene ring to give $CpFe(\eta^5\text{-}C_6H_6\text{-}\eta^5\text{-}C_6H_6)FeCp$ (see Section 2).

The last of the stable metallocenes to be considered, nickelocene, has a very interesting behavior.[44,59,62–64] This 20-valence-electron molecule undergoes a highly reversible one-electron oxidation at −0.09 V in all solvents at the DME or Pt electrodes. A further one-electron oxidation to Ni(IV), that is, the d^6 metal electron configuration stable for so many other metallocenes, is also observed at +0.77 V; the process is reversible at −40°C is acetonitrile.

Although nickelocene is the most electron-rich metallocene, it is the easiest metallocene to reduce. Unfortunately, it is not a reversible reduction, the process being a rather complex one which involves more than one electron. Initially, the reduction does proceed by one electron to give the mono-anion. This ion is short-lived at room temperature, and, in a bulk electrolysis experiment, gave $C_5H_5NiC_5H_7$ as one product. (The latter compound was oxidized at $E_{1/2} = +0.6$ V and reduced at $E_{1/2} = -2.5$ V.)

Just as ferrocene[47] and ruthenocene[53] have been examined in a molten salt, so has nickelocene.[63] In a 1:1 $AlCl_3$-1-butylpyridinium mixture, Cp_2Ni undergoes a reversible one-electron process approximately −400 mV relative to ferrocene in the same solvent. This is approximately the same as the difference observed in various nonaqueous solvents. In a melt with a greater relative amount of aluminum chloride (1.1:1 $AlCl_3$:1-butylpyridinium chloride), reversible oxidation to the Ni(IV) dication is observed. The dication is reportedly stable under these conditions for at least several hours, apparently owing to the lack of water in these molten salts. Oxidized metallocenes are generally susceptible to nucleophilic attack,[65] and the nickelocene dication is no exception. It can be observed in acetonitrile, but only in a very rigorously dried solvent.

1.2. Decamethylmetallocenes, $(\eta^5\text{-}C_5Me_5)_2M$

A large number of reports have appeared regarding the electrochemical behavior of metallocenes substituted with one or more groups. These are described in Section 1.5. However, in connection with the simple metallocenes, Cp_2M, it is useful to note the behavior of decamethylmetallocenes, $(\eta^5\text{-}C_5Me_5)_2M$.

The methyl group is inductive and should lead to an increase in electron density at the metal in the decamethylmetallocenes relative to the simple metallocenes. This should lead to a lowering of the reduction potential, and this lowering is, indeed, observed. Although decamethyl derivatives are known for all of the first-row transition metals from V through Ni, as well as Ru, the electrochemistry of only the Cr, Mn, Fe, Ru, Co, and Ni complexes has been reported.[66-69] All undergo very reversible, one-electron oxidations to the mono-cation (it has been noted that the cyclic voltammogram of $(C_5Me_5)_2V$ is complex with no one-electron reversible waves), and all can be isolated in this form. As seen in Table 1, the oxidations of the permethylated complexes are generally about 500 mV less positive than the simple metallocenes.

Once again, the nickel derivative is interesting. The Ni(II) → Ni(III) → Ni(IV) sequence occurs for $(C_5Me_5)_2Ni$ just as for nickelocene. However, whereas the nickelocene dication was quite unstable, $[(C_5Me_5)_2Ni]^{2+}$ can be isolated as the PF_6^- salt.

1.3. General Conclusions Regarding Metallocene Electrochemistry

The basic electrochemical behavior of the simple first-row metallocenes and the decamethylmetallocenes is fairly clear. In general, the electron transfer series, where the metal is reduced in stepwise fashion from M(IV) to M(I),

$$[Cp_2M]^{2+} \rightleftharpoons [Cp_2M]^{+} \rightleftharpoons Cp_2M \rightleftharpoons [Cp_2M]^{-}$$

is possible. However, the complete series is known only for the metallocenes at the extremes. Nickelocene is the only metallocene for which the four separate ions are detected electrochemically, but the most reduced species is very reactive. Vanadocene also exhibits three redox steps, but the dication, $[Cp_2V]^{2+}$, is too unstable to detect by cyclic voltammetry.

Holloway and Geiger have collected the information available on the redox chemistry of the metallocenes and have classified them by electronic configuration.[43] The most interesting conclusion is that the electron configuration most commonly exhibited by the metallocenes is d^6. If the general order of molecular orbitals in metallocenes is that for ferrocene as pictured in Fig. 1, the essentially nonbonding e_{2g} and a_{1g} orbitals [which are, respectively, essentially the $(d_{x^2-y^2}, d_{xy})$ and d_{z^2} orbitals] are just filled with 6d electrons. In neutral cobaltocene and nickelocene, there are 1 or 2 electrons in the slightly antibonding e_{1g}^* level, and, not surprisingly, these molecules are more readily oxidized than ferrocene.

Because the molecular orbitals involved in the redox chemistry of the first-row transition metals are essentially metal atomic orbitals, the molecule

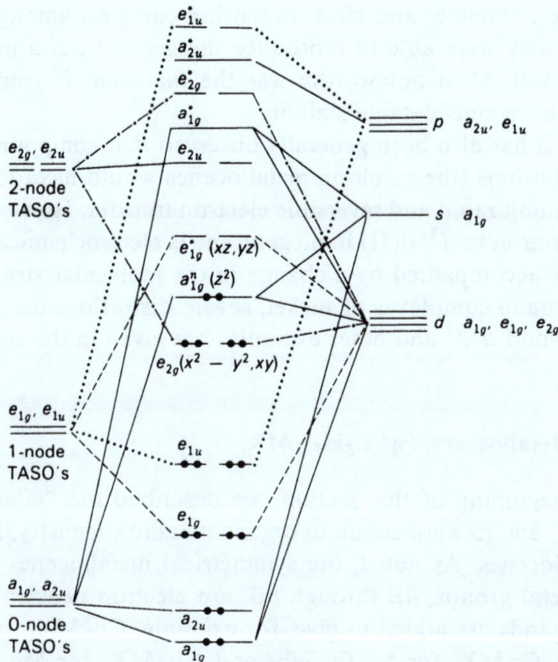

Figure 1. Molecular orbital diagram for ferrocene. Taken from K. F. Purcell and J. C. Kotz, *Inorganic Chemistry* (W. B. Saunders, Philadelphia, 1977).

can accommodate more than 18-valence electrons. This is apparently not as true for metallocenes of second- and third-row metals, since, as in the case of rhodocene, their reduction generally leads to dimerization.[61]

For the same reason that 19-electron species can exist, it is also not surprising that there is a reasonable correlation between the $E_{1/2}$ values of the $M(\text{II}) \rightarrow M(\text{I})$ metallocene reduction (where M = V, Cr, Co, and Ni) and the second ionization potential of the metal. The correlation predicts ferrocene reduction to $[Cp_2Fe]^-$ at about −2.3 V, and it was later observed at −2.93 V.[70] This again suggests the considerable stability of the d^6 configuration.

Another excellent correlation exists between the electrochemical potentials for metallocene oxidation and the gas phase values from UV photoelectron spectroscopy. The only exception to this is vanadocene oxidation, and it has been suggested that solvation energies may play a different role in this case than in the other metallocenes.[43]

Very recently van Gaal and van der Linden attempted to analyze the trends in redox potentials of transition metal complexes, including the metallocenes, on a more theoretical basis.[71] They considered the ligand field-splitting parameter and electron-pairing energies, among other parameters, and they were able to reproduce the general trend in $E_{1/2}$ values remarkably well. Most noteworthy was the fact that d^6 complexes were observed to have considerable stability.

Finally, it has also been generally observed that compounds based on d^3 to d^8 metal ions (the resulting metallocenes would have 15–20 valence electrons) exhibit rapid and reversible electron transfer. However, the addition of an electron to d^8Ni(II) in nickelocene is electrochemically slow and is apparently accompanied by a change in the molecular structure. In the related carborane complexes of nickel, severe distortions occur on adding electrons beyond d^8,[24] and other examples are given in the sections which follow.

1.4. Bent Metallocenes, $(\eta^5\text{-}C_5H_5)_2MX_n$

In the beginning of this section, we described the "effective atomic number rule" and its application to organometallic chemistry, in particular to the metallocenes. As noted, the symmetrical metallocenes of the early transition metal groups, 4B through 6B, are electron deficient. To rectify this, extra ligands are added to give, for example, Cp_2MX_2 for Ti Zr, Hf; Cp_2MX and $CpMX_3$ for V, Ta, Nb; and Cp_2MX_2 for Mo, W.[2,41,42] In attaining this stoichiometry, and a net of 16- or 18-valence electrons in each case, the molecules adopt the geometry dictated by VSEPR forces, the triangle in the case of Cp_2MX and the tetrahedron for Cp_2MX_2. Thus, this class of compounds is sometimes called the "bent metallocenes," and their electrochemistry offers much more variety than the simple metallocenes, since loss of X or even $C_5H_5^-$ can follow electron transfer.

The electrochemistry of the Group 4B compounds of the type Cp_2MX_2 has been described, but no systematic work has been done on an extended series. A large number of papers have been published regarding the electrochemistry of the dicyclopentadienyltitanium(IV) dihalides, Cp_2TiX_2,[8,72–83] largely because it was observed in the late 1960s that systems such as Cp_2TiCl_2 + Grignard reagent could be used to catalytically hydrogenate alkenes and alkynes and to convert dinitrogen to ammonia.[77] As expected for a metallocene based on Ti(IV), only reduction processes are observed, but the reported results vary widely.

In general, three reduction waves are observed for the dicyclopentadienyltitanium dihalides, the apparent reversibility depending on the solvent used. In THF, waves at $E_{1/2} = -0.80$, -2.10, and -2.38 V were found using a rotating platinum electrode. With one exception,[81] all investigators have concluded that the following series of reactions occurs:

$$
\begin{array}{lll}
\text{Electron Transfer:} & Cp_2TiX_2 + e^- \rightarrow [Cp_2TiX_2]^- & \\
\text{Rapid Loss of } X^-\text{:} & [Cp_2TiX_2]^- \rightarrow Cp_2TiX + X^- & \\
\text{Electron Transfer:} & Cp_2TiX + e^- \rightarrow [Cp_2TiX]^- & \\
\text{Loss of } X^-\text{:} & [Cp_2TiX]^- \rightarrow Cp_2Ti + X^- &
\end{array}
\qquad (5)
$$

Loss of halide following the first reduction was suggested by the results of Gubin and Smirnova who showed that the first reduction of Cp_2TiCl_2 (on mercury in dimethylformamide) was dependent on the Cl^- concentration in the solution,[73] and this was recently confirmed.[83] (El Murr *et al.*[81] have disagreed with this, suggesting instead that $[Cp_2TiCl_2]^-$ was stable in THF.) In the presence of a base, the product is Cp_2TiXL (L = Lewis base, for example, THF, DMF, phosphine). In addition, the reduced, coordinatively unsaturated species could dimerize to form $[Cp_2TiCl]_2$, a compound which is readily prepared by the reduction of Cp_2TiCl_2 with Al metal in THF.[84–86]

Considerable effort has been expended to define the product of the second one-electron reduction step. Although the product is written as "Cp_2Ti," and implies the formation of the symmetrical metallocene titanocene, this has proved to be quite elusive. No such compound has ever been isolated. Indeed, chemical reduction of Cp_2TiCl_2 has led to the isolation of at least two different compounds, neither of which is simple Cp_2Ti.[87] The nature of the ultimate product in the electrochemical studies has not been established. However, it is known that Cp_2TiCl_2 can be cathodically reduced in dimethoxyethane of THF to give a blue solution.[8,77,78] There was considerable early interest in this observation, since the complex implicated in nitrogen fixation, $[Cp_2Ti]_2N_2$, is intensely blue.[77] However, the cathodically produced blue solutions can be generated in an argon atmosphere as well. The current explanation of this blue color is that it is apparently a secondary product. There is evidence, though, that "Cp_2Ti" can indeed be a transient intermediate in the cathodic reduction. Chemical or cathodic reduction in a CO atmosphere leads to a high yield of the stable dicarbonyl $Cp_2Ti(CO)_2$.[77,82] However, once the blue solution noted above has been produced, the product will not react with CO.

Changing the solvent to aqueous alcohol as a medium for the cathodic reduction of Cp_2TiX_2 leads to quite different behavior. Under these conditions, Gubin and Smirnova found that the first reduction was reversible for

Cp_2TiX_2 (X = Cl, Br, I) and occurred at the same potential for all three halides.[73] They suggested prior hydrolysis of the dihalide to give $Cp_2Ti(OH)^+$ which is then reduced to Cp_2Ti^+, a presumably solvated cation.

$$\begin{aligned} Cp_2TiX_2 + H_2O &\rightarrow [Cp_2TiOH]^+ + H^+ + 2X^- \\ [Cp_2TiOH]^+ + e^- + H^+ &\rightarrow Cp_2Ti^+ + H_2O \end{aligned} \tag{6}$$

The electrochemistry of a series of aryloxy cyclopentadienyltitanium compounds, Cp_2TiRR' (R and R' = Cl or OAr, where Ar is a substituted aryl group) is illuminating. In DMF at the dropping mercury electrode, Laviron *et al.*[80] observed three irreversible waves for Cp_2TiCl_2. Replacing one halide substituent with an OAr group also gives a compound having three reduction waves, the middle wave being at the same potential as the middle wave for Cp_2TiCl_2. If both halides are replaced by OAr, then two irreversible waves, with substituent-sensitive potentials, are seen, the second occurring at about the same potential as the third in $Cp_2TiCl(OAr)$. These observations can be rationalized by the following schemes:

For $Cp_2TiCl(OAr)$ (7)

Wave 1: $2Cp_2TiCl(OAr) + 2e^- \rightarrow Cp_2TiCl + Cp_2TiOAr + Cl^- + OAr^-$

Wave 2: same as the second wave for Cp_2TiCl_2

$$Cp_2TiCl + e^- \rightarrow \text{products}$$

Wave 3: $Cp_2TiOAr + e^- \rightarrow$ products

For $Cp_2Ti(OAr)(OAr')$ (8)

Wave 1: $Cp_2Ti(OAr)_2 + e^- \rightarrow Cp_2Ti(OAr) + OAr^-$

Wave 2: same as the third wave for $Cp_2TiCl(OAr)$

$$Cp_2TiOAr + e^- \rightarrow \text{products}$$

In general, the stability of the initially formed anion, Cp_2TiXY^-, increases as the number of aryloxy groups is increased.

Using platinum or glassy carbon electrodes, El Murr and co-workers have obtained results for the reductive electrochemistry of $Cp_2Ti(OR)_2$

which contrast with the work described above.[88] That is, they have suggested that the d^1 Ti(III) dicyclopentadienyl compounds are unstable with respect to loss of $C_5H_5^-$, rather than OR^-.

Only a few observations of dicyclopentadienyltitanium dialkyls and diaryls have appeared. El Murr has found that, just as the compounds $Cp_2Ti(OR)_2$ lose $C_5H_5^-$ on reduction, so do alkyls.[88] This finding suggests that an earlier report of the reductive behavior of Cp_2TiR_2 (R = Me, Ph, *o*-carboranyl) should be re-interpreted.[89] Kotz *et al.*[90] have also recently reported the diferrocenyl derivative, Cp_2TiFc_2 (Fc = ferrocenyl group). In contrast with analogous Cp_2TiPh_2, the diferrocenyl compound is unstable to oxidation; the green color of the molecule is rapidly discharged in solutions exposed to the air or oxidizing agents such as Br_2. The molecule is reduced in quasi-reversible fashion at −1.66 V; this can be compared with the diphenyl analog which reduces at −1.53 V, in agreement with the relative inductive abilities of the two substituents. The molecule also undergoes totally irreversible oxidation at +0.25 V and apparently decays rapidly to form ferrocene and biferrocene. This very unusual behavior must be studied further. Indeed, it is suggested that a new and systematic study of a series of compounds of the type Cp_2MX_2 (M = Ti, Zr, Hf) would be highly desirable.

In contrast with titanium compounds, only a few brief reports have appeared of the electrochemistry of analogous Zr(IV) compounds, and there is one brief mention of a hafnium(IV) compound.[91,92] Work on the zirconium compounds bears directly on the studies of titanium compounds described above. Lappert and co-workers have synthesized a series of Zr(IV) compounds of the type $(\eta^5\text{-}C_5H_4R)_2ZrXY$. In general, when $X = Y$ = Cl, a reversible reduction, about 1 V more cathodic than that of Cp_2TiCl_2, is observed in THF at a platinum electrode. (This very great difference in reduction potentials indicates a 80–100 kJ difference in the LUMO in the Ti(IV) and Zr(IV) compounds.) However, when one or both of the Cl substituents were replaced by $R = CH_2Ph$, $CHPh_2$, or $CH(SiMe_3)_2$, the one-electron reduction was totally irreversible. In agreement with El Murr,[88] Lappert *et al.* suggest that the compounds are unstable with respect to loss of $C_5H_5^-$; indeed, an anodic wave, at a potential appropriate for oxidation of $C_5H_5^-$, is seen following the one-electron reduction.[91,92]

In view of the more recent work, the early report of the electrochemistry of Cp_2ZrCl_2 in dimethoxyethane (Hg electrode) is interesting and probably should be re-investigated.[49] Two reduction waves are reported, but no esr signal was observed after either step. A white, insoluble powder was the product of the second reduction.

In comparison with the widely varying electrochemical behavior of the Group 4B Cp_2MX_2 compounds, that of the Group 5B metallocenes seems better defined. Vanadocene dichloride, Cp_2VCl_2, has been studied in ethers or acetone by Dessy *et al.*,[8] Bond *et al.*[93] and most recently by Holloway and Geiger[43] and others.[94] Holloway and Geiger observed three polarographic waves at −0.29 V, −1.54 V, and −2.74 V; the first and third waves were reversible, and the third wave represented a product of the second reduction. The following scheme was proposed to rationalize their observations:

$$\begin{aligned} Cp_2VCl_2 + e^- &\rightleftharpoons [Cp_2VCl_2]^- && \text{first wave, reversible} \\ [Cp_2VCl_2]^- + e^- &\rightarrow [Cp_2VCl_2]^{2-} && \text{second wave, irreversible} \\ [Cp_2VCl_2]^{2-} &\rightarrow Cp_2V + 2Cl^- && \\ Cp_2V + e^- &\rightarrow [Cp_2V]^- && \text{third wave, reversible} \end{aligned} \tag{9}$$

The connection between the electrochemistry of vanadocene dichloride and that of vanadocene itself was shown in Equation 1.

Bond and co-workers have described the electrochemistry of a very extensive series of vanadocene derivatives, (**2**), where chelating ligands such as those illustrated were used.[95,96]

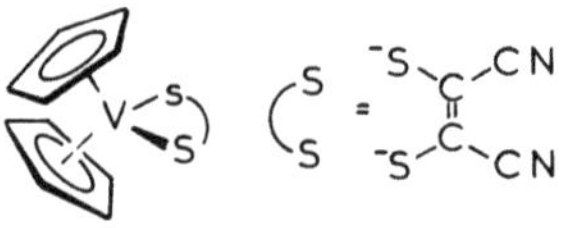

2

Approximately 25 compounds were studied, and, in general, all reduced at a mercury electrode according to the ECE mechanism

$$\begin{aligned} [Cp_2V^{IV}(SS)]^+ + e^- &\rightarrow Cp_2V^{III}(SS) \\ Cp_2V^{III}(SS) &\rightarrow [Cp_2V^{III}] + SS^- \\ SS^- + Hg &\rightarrow \text{mercury dithio chelates} \end{aligned} \tag{10}$$

where the final "E step" is associated with complexes formed between the free ligands and mercury of the electrode. (This step is of course not observed at Pt electrodes.) The stability of the vanadium(III) complex, $Cp_2V(SS)$, depends on the nature of the thio-chelate and is critically affected by oxygen. The formal reduction potential of $[Cp_2V^{IV}SS]^+$ is also dependent on the nature of the ligand. For example, mono-anionic ligands, such as the dithiocarbamates, led to more easily reduced complexes than di-anionic ligands such as maleonitrile dithiolate.

The last periodic group in which "bent metallocenes" exist is 6B. Molybdenum(IV) and tungsten(IV) commonly form Cp_2MX_2, and we have recently described some aspects of the anodic electrochemistry of compounds where X is a halide or a thiolate (in acetonitrile).[90] We observed that, when X is a halide, the Mo(IV) compounds all oxidized quasi-reversibly at about 0.6 V. Since the Mo(V) → Mo(IV) reduction potential changes very little with change in halide electronegativity, this implies that the HOMO is relatively little influenced by the halide and is probably largely metal in character. This conclusion is supported by esr spectra, as well. When the metal is tungsten, the reduction potential of $[Cp_2WX_2]^+$ was more cathodic than those of the molybdenum compounds by about 0.1 V. (Just as in the case of Cp_2MX_2 (M = Ti, Zr), the reduction potential became more cathodic on descending the periodic group.) The metal thiolates, $Cp_2M(SR)_2$, were oxidized in the range 0 to +0.4 V for M = Mo(IV) or W(IV), with the tungsten compounds again less anodic; there were no clear trends in redox potential with R. For both the dihalides and the dithiolates, controlled potential electrolysis in MeCN led ultimately to replacement of at least one substituent with solvent to give $[Cp_2M(NCMe)X]^+$. The nature of this substitution reaction, and those of Cp_2MH_2,[97] are discussed in more detail in Section 5 below.

1.5. Substituent Effects in Metallocene Derivatives

One of the earliest discoveries in organometallic electrochemistry was the sensitivity of metallocenes oxidations to ring substituents. This effect has been thoroughly and widely studied for the insight which can be provided into the electronic structure of the metallocenes, especially that of ferrocene.[50,98–106]

The molecular orbital diagram of ferrocene in Fig. 1 suggests that the oxidation potential of ferrocene should be affected, if the cyclopentadienyl rings bear one or more substituents. It has been of great interest to many to see if this is indeed correct and to uncover the nature of the influence. At least five fairly extensive studies have been done on the relationship between substituent type and redox potential for ferrocene derivatives,[99–103] and one of the first of these was a study of the effect of substitution on the oxidation of phenylferrocene.[98] Using a limited series of para-substituted phenylferrocenes, the difference between the oxidation potential for ferrocene and the substituted phenylferrocene in aqueous acetic acid was found to vary linearly with the Hammett σ_p constant.

Paralleling the phenylferrocene study above, a much more extensive study of the same effect was done by Little *et al.*[101] Chronopotentiometric quarter-wave potentials for the oxidation of forty-nine different *o*-, *m*-, and *p*-substituted phenylferrocenes were measured as well as those for nine-*p*-ferrocenylazobenzenes. Using twelve *m*- and *p*-substituted derivatives whose primary substituent constants were known, it was found that the difference between the ferrocene oxidation potential and that of the substituted phenylferrocene varied according to the equation $E_{1/4} = 0.128\sigma + 0.024$ V. From this line, secondary substituent constants were derived for fifteen other groups, and the values obtained were in good agreement with those obtained by other methods. For *ortho*-substituted phenylferrocenes, Taft *o*-constants were used, and an excellent correlation was again obtained with the ferrocene group redox potential ($E_{1/4} = 0.126\sigma_o^* + 0.031$ V). The lack of a steric factor in this equation demonstrates that the steric effects of ortho-substituents play no role in determining the ferrocenyl group redox potential; only electronic effects are operative.

Little *et al.*[101] have also demonstrated that substituent effects are transmitted over a considerable distance with some predictability. A series of nine substituted *p*-ferrocenylazobenzenes was examined. Although the ferrocenyl group redox potential in acetonitrile varied by only about 18 mV, a reasonable correlation was obtained with Hammett sigma-constants.

At least three reasonably systematic studies have been done on the effect of substituents directly bound to the ferrocenyl group.[100,102,103] All showed that the ferrocenyl oxidation potential was excellently correlated with standard substituent parameters. However, the most thorough analysis was done by Hall and Russell who showed that the substituent effects for a large number of mono- and di-substituted ferrocenes in acetonitrile were best correlated using a blending of Hammett σ_m and σ_p constants,[103] specifically, with a "calculated constant" $(\sigma_m + 2\sigma_p)/3$. Hall and Russell argue that the electronic effects of the substituents are transmitted by a mechanism that is primarily inductive, and that σ_p overestimates group resonance effects on the oxidation of the ferrocenyl iron(II) ion.[103]

An interesting sidelight to the correlation studies was that the oxidation potentials of acetamido- and urethano-substituted ferrocenes did not fit the correlations. They were much more readily oxidized than expected, an effect ascribed to the possibility that the substituent carbonyl could interact with the iron(III) ion in the oxidized molecule. That is, an "internal solvation" effect is operative. Such substituent–iron intramolecular interactions are not unknown in ferrocene chemistry; hydrogen bonding between a substituent hydroxyl group and the ferrocene iron has been observed in ferrocenyl alcohols.[107]

In a study closely related to the electrochemical work cited above, Hall, Hill, and Richards examined the solvolysis of methylferrocenylcarbinyl acetates which had a substituent on the other C_5 ring.[106] The solvolysis rate correlated well with the same "calculated constant" noted above, and thus gave an excellent correlation of solvolysis rate with the ferrocenyl oxidation potential. This suggested again that resonance interactions play a minor role in such reactions; that is, inductive effects predominate.

The studies cited above gathered information only on ferrocenes having one or two substituents. The ferrocenyl oxidation potential has also been examined (in 90 percent ethanol) for a series of compounds having at least three substituents distributed between the two rings.[105,108] The results show that the substituent effects are additive and do not depend on ring position. For example, 1,1′,2-triethylferrocene and the 1,1′,3 isomer have the same oxidation potential (+0.19 V, a value 150 mV less anodic than the potential of ferrocene under the same conditions). In general, introduction of an alkyl group decreases the ferrocene oxidation potential by about 47 mV relative to ferrocene in 90 percent ethanol (or 57 mV in acetonitrile),[108] whereas a phenyl substituent increases the oxidation potential by about 23 mV. Thus, 1,1′,3,3′-tetraphenylferrocene is oxidized at 0.45 V (compared with ferrocene at +0.34 V).

The "ferrocenophanes" (**3**) are a very interesting class of ferrocene-based compounds.

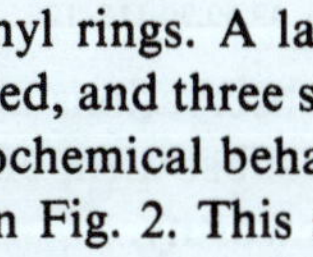

CH_2 Fe CH_2 CH_2

[3](1,1′)ferrocenophane = [3]Fcp

3

These compounds have a bridge of varying length joining the two cyclopentadienyl rings. A large number of compounds in this class have been synthesized, and three studies, two of them extensive, have been made of their electrochemical behavior.[108–110] Perhaps the most interesting result is illustrated in Fig. 2. This shows, not unexpectedly, that, as the number of methylene groups of the bridge is increased, and as the number of bridges is increased, the oxidation potential declines, the relationship between potential and number of methylene groups, *n*, being

$$E_{1/2}(\text{V vs. SCE}) = -0.0221n - 0.3654$$

What is unexpected, however, is that the [3][3]- and [3][3][3]-ferrocenophanes (where [3] is the number of methylene carbons in the bridge)

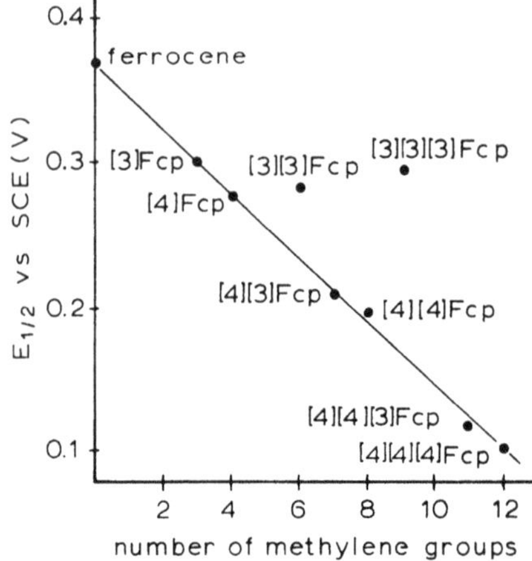

Figure 2. The potential for the oxidation of ferrocenophanes (for example, **3**) vs. the number of bridging CH_2 groups, n. (On the figure, [3] and [4] indicate bridges of 3 or 4 CH_2 groups, respectively.) T. Ogata *et al.*, *Bull. Chem. Soc. Japan*, **54**, 3723 (1981).

do not fit this correlation. The reason for this apparently lies in a change in electron density about the iron brought on by molecular compression. As stated by Ogata *et al.*,[109] "on the basis of molecular models of [these ferrocenophanes], the compression of the metal-ring bond needed to accommodate the bridge was found to be remarkable." Fujita *et al.*[108] have also observed this effect, and have further found that there is a rough correlation between the iron oxidation potential and the *change* in iron-ring distance on going from ferrocene to the ferrocenophane. That is, compression of the molecule along the ring-metal axis increases the electron density about the iron. This in turn attenuates the expected electronic effect of adding methylene groups, and the observed potential is higher than expected.

In addition to the electrochemistry of the ferrocene-based mixed valence compounds described in Section 1.6, a large number of other ferrocene derivatives has also been reported.[49,59,90,111-126] Most of these reports are incidental to other work, and the observations are those anticipated on the basis of the now well-established substituent effects described above. There are, however, some that are noteworthy.

An especially intriguing observation was reported by McCleverty *et al.*[115-117] regarding ferrocenyl-based carbene complexes of chromium and tungsten carbonyl, $Fc(X)C:Cr(CO)_5$ (X = OEt, OMe, pyrrolidin-1-yl). The reversible ferrocenyl redox wave was observed at 0.70 V in CH_2Cl_2 for X = OEt and M = Cr, for example. As ferrocene oxidation is generally observed at about 0.4 V in CH_2Cl_2, the increase of about 200 mV represents a relatively large decrease in ease of oxidation. The more interesting aspect of these compounds, however, was the fact that the compounds do not show the expected two separate oxidation waves [one for $Cr(0) \rightarrow Cr(I)$

and another for Fe(II) → Fe(III)] or even perhaps one two-electron wave. Rather, only one electron is removed within the accessible potential range, an observation which was explained by stating that "the redox orbital of the Cr(CO)5[FcC(*X*)] complex encompasses both potential redox centers (and the single redox wave) corresponds to the generation of $[A\text{-}B]^+$." [117] This is most unusual behavior, and it should be examined further in a broader, systematic study.

We have published a study of the electrochemical behavior of the ferrocenyl phosphines (**4**) and some of their complexes of the type L-BH_3, L-$M(CO)_5$, and LMe^+I^-.[121]

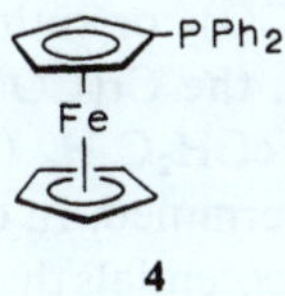

4

All of the compounds were electrochemically well-behaved, the parent phosphines having one, two, or three reversible waves for ferrocenyl oxidation, depending on the number of ferrocenyl groups. In general, the first of these waves was 70–150 mV more anodic than ferrocene under similar conditions. As expected, the ferrocenyl oxidation for complexes of $FcPh_2P$ was even more anodic (generally about 200 mV for neutral acids). Surprisingly, the potentials for oxidation of Fc_2PhP and Fc_3P were little changed on complexation. This suggests that, on attachment of the second and third ferrocenyl group, there is little change in the net electron density at the iron after complexation to most Lewis acids. Finally, it was also observed that the stretching frequencies of the carbonyl groups in the $LM(CO)_5$ complexes were changed very little (*ca.* 5 cm^{-1}), suggesting that the coordinating ability of the ferrocenylphosphines changes very little upon oxidation.

The electrochemistry of a bidentate ferrocene-based amine-phosphine (**5**) has also been described.[120] The uncomplexed base has a complex electrochemistry, but complexing the amine and phosphine sites with BH_3 leads to a molecule with a single, reversible redox wave for ferrocene at $E_{1/2} = 0.87$ V (in CH_2Cl_2). When the ligand functions in a bidentate fashion with $M(CO)_4$(M = Cr, Mo, W), the ferrocenyl group is again reversibly oxidized at 0.90–0.93 V. In the case of the chromium compound, two reversible waves were observed, and it was proved that the less anodic was for Cr(O) → Cr(I), while the more anodic was for ferrocenyl oxidation. What is intriguing in this simple series is that the ferrocenyl group oxidizes at approximately the same potential in both the BH_3 and $M(CO)_4$ complexes, in spite of the fact that the M atom is oxidized before the ferrocenyl iron, and thereby places a positive charge on the complex. One would have expected that M oxidation would have strongly affected the iron oxidation by a field effect.

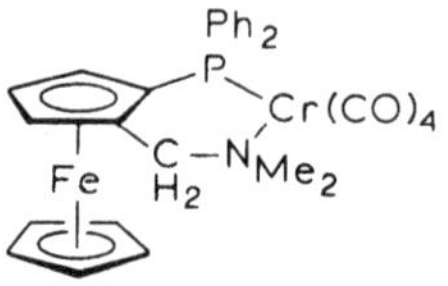

5 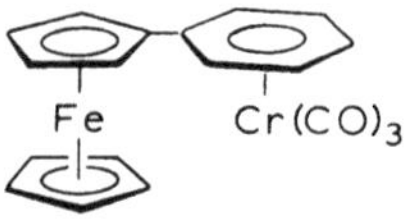 **6**

Gubin and co-worker have also used the ferrocenyl group as a probe to determine the effect of $Cr(CO)_3$ or ruthenocene as a substituent.[122] In the case of the metal carbonyl, the $Cr(CO)_3$ group was complexed to the phenyl group of Fc-C_6H_6 or $FcCH_2C_6H_6$ (for example, **6**) and the redox potentials of the metal sites determined. In contrast with (**5**), the ferrocenyl group is oxidized at less anodic potentials than the chromium(0). An increase of 160 mV in $E_{1/2}$ for ferrocenyl group oxidation was found for (**6**), indicating that $C_6H_5Cr(CO)_3$ is strongly withdrawing. (This shift in potential is larger by about 30 mV than the largest shift induced by any "ordinary" substituents on phenylferrocene.[101]) Even when a methylene group intervened between the ferrocenyl and phenyl groups, a potential shift of 100 mV was observed. The strong inductive effect of $Cr(CO)_3$ is indicated by the Hammett σ_p^0 constant (+0.26), derived from these electrochemical results.

Aside from the decamethylmetallocenes described above, substituent effects in derivatives of other metallocenes of the type Cp_2M have been briefly examined only in the case of some cobaltocenium derivatives.[127] In general, the ease of reduction of the ion to the neutral, Co(I) metallocene is affected as expected by electron donating or withdrawing groups.

1.6. Organometallic Mixed Valence Compounds: The Electrochemistry of Multi-Metal Compounds

A mixed valence compound is one in which two or more metals, usually of the same kind, are in adjacent or nearly adjacent oxidation states.[128-130] If the centers interact, an electron can be effectively transferred from the lower to the higher valent site,

$$M_1^+ - M_2 \rightarrow M_1 - M_2^+ \tag{11}$$

the transfer being accomplished both thermally and photochemically. Mixed valence compounds can be highly colored and have interesting magnetic and electrical properties. For this reason, and for the insight they provide into the electronic structure of compounds and the factors affecting electron transfer between interacting sites, mixed valence compounds have been of continuing interest for some years. Organometallic compounds can of course be synthesized with several metals, and electrochemical methods can be used to produce metals in different oxidation states. Therefore, there has been considerable activity in the study of mixed valence organometallic compounds.[128,131]

Some years ago, Robin and Day proposed a classification of mixed valence compounds according to the extent of delocalization of the valence electrons.[129] Class I compounds have no interaction between sites of different formal oxidation number; the valences are said to be trapped, and the properties of the mixed valence compound are simply the sum of the properties of the two different sites. In contrast, in class III ions, there is extensive delocalization between sites, and two different oxidation states are no longer distinguishable by techniques having an observation time greater than the rate of electron transfer. Class II compounds are intermediate between those of classes I and III. There is a weak but nonnegligible interaction between the adjacent oxidation states. To a first approximation, different integral oxidation states can be ascribed to the two sites. However, there is a valency interchange in class II compounds by two different mechanisms: (i) photon-driven electron transfer (this usually gives rise to a broad band of low energy called an intervalent transfer or IT band); and (ii) an associated, thermally activated process.

In organometallic chemistry, the most commonly studied mixed valence compounds have been based on ferrocene,[54,131–158] because the ferrocenyl group is usually stable when the iron is in the formal oxidation states of 2+ and 3+. It has been possible to examine the mixed valence behavior of such compounds as a function of the distance between two or more interacting ferrocenyl groups, as a function of the nature of the bridge between them, and as a function of their relative orientation.

In general, the compounds which have been examined are biferrocene and its derivatives (**7**),[132,139] ferrocenyl groups bridged by one or more atoms of carbon or other elements (**8, 9**),[133,136,137,140–144] or [1.1]ferrocenophanes (**10–12**).[134–136,141,145,146] Electrochemical data for a number of these

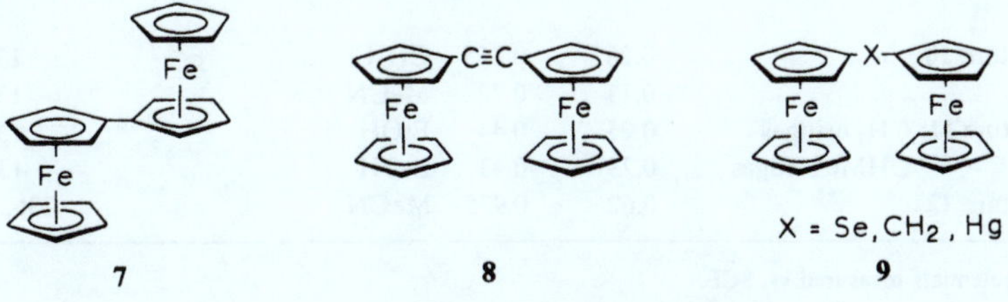

10 **11** R = H, Me **12**

compounds has been collected in Table 2; all of the potentials given are for one-electron, reversible processes.

Biferrocene has been one of the most widely studied compounds. As indicated in Table 2, it undergoes two, successive one-electron oxidations separated by about 350 mV. The first gives the mixed valent Fe(II)/Fe(III)

TABLE 2
Redox Potentials for some Compounds Containing Two or More Ferrocenyl Groups (Fc)

Compound	Reduction potentials[a]		Solvent/comments	Ref.
	$+1 \rightleftharpoons 0$	$+2 \rightleftharpoons +1$		
Fc—Fc (**7**)	0.435	0.785	CH_2Cl_2	136
	0.31	0.64	MeCN	134
Fc—Fc—Fc	0.22	0.44	Oxid. to +3 at 0.82 V; CH_2Cl_2/MeCN	147
Fc—Fc—Fc—Fc	0.16	0.36	Oxid. to +3 at 0.61 and to +4 at 0.89 V; CH_2CL_2/MeCN	147
Fc—CH_2—Fc	0.39	0.56	MeCN/EtOH	143, 133
Fc—Se—Fc	0.46	0.68	MeCN	143
Fc—C≡C—Fc (**8**)	0.625	0.753	CH_2Cl_2	136, 140, 141
Fc—C≡C—C≡C—Fc	0.58	0.68	CH_2Cl_2	136
1′,6′-DiiododiFc (**7**)	0.42	0.70	MeCN	135
2-AcetyldiFc (**7**)	0.470	0.975	CH_2Cl_2	139
1′-AcetyldiFc (**7**)	0.52	0.975	CH_2Cl_2	139
1′,1‴-DiacetyldiFc (**7**)	0.715	1.025	CH_2Cl_2	139
2-MethoxycarbonyldiFc (**7**)	0.475	0.965	CH_2Cl_2	139
1′-MethoxycarbonyldiFc (**7**)	0.490	0.960	CH_2Cl_2	139
1′,1‴-Dimethoxycarbonyl-diFc (**7**)	0.690	1.020	CH_2Cl_2	139
Structure **10**	0.265	0.855	CH_2Cl_2	136
	0.13	0.72	MeCN	134
Structure **11**: CH_2 bridges	0.25	0.44	EtOH	135
CHMe bridges	0.23	0.43	EtOH	136
Structure **12**	0.62	0.975	MeCN	136, 141

[a] All potentials measured vs. SCE.

species—the focus of interest. This mixed valent species is stable in solution, the equilibrium constant for the disproportionation reaction being about 10^{-6}.

$$2\text{Fc} - \text{Fc}^+ \rightarrow \text{Fc} - \text{Fc} + \text{Fc}^+ - \text{Fc}^+,$$

$$\Delta G = 350\,\text{mV}, \qquad K = 10^{-6} \tag{12}$$

The electronic spectrum of the cation in acetonitrile has bands at 545 nm, 840 nm, and, most importantly, at 1800 nm ($\varepsilon = 750$) in the near infrared. The 1800 nm band, which is solvent dependent, is not observed in either the Fe(II)/Fe(II) or the Fe(III)/Fe(III) species. Thus, it is assigned as an intervalent transfer or IT band arising from optical excitation of electron transfer between the two valence sites in the mixed valence ion. The observation of this band strongly suggests that the mixed valence ion is a class II ion, and this is further confirmed using the theoretical treatment of Hush for weakly interacting mixed valence species.[130] (This treatment allows one to calculate the expected band width, as well as a parameter related to the extent of electron delocalization. The latter is directly proportional to the intensity and width of the IT band and inversely proportional to the energy of the band and the distance between the interacting centers.) Finally, Mössbauer spectra of $\text{Fc} - \text{Fc}^+$ clearly show that the cation contains Fe(II) and Fe(III) and gives a thermal electron exchange rate constant of less than $10^7\,\text{sec}^{-1}$.

Adding one or two more ferrocenyl groups to biferrocene gives 1,1′-terferrocene (Fc-Fc-Fc) and 1,1′-quaterferrocene (Fc-Fc-Fc-Fc), two compounds with interesting electrochemical behavior. It has been observed that biferrocene is more readily oxidized than ferrocene by about 90 mV [Table 2].[147] Taking this as the ferrocenyl group substituent effect, it is expected, and, indeed, observed that terferrocene is more easily oxidized than biferrocene by approximately 90 mV, and quaterferrocene is in turn more readily oxidized than terferrocene. More interesting, however, is the possibility of "oxidation state isomerism" in terferrocene and quaterferrocene. That is, the ion $[\text{Fc-Fc-Fc}]^+$ can exist in two energetically equivalent isomers, Fc-Fc-Fc^+ or $\text{Fc}^+\text{-Fc-Fc}$, or the energetically nonequivalent isomer $\text{Fc-Fc}^+\text{-Fc}$. The distribution of these isomers will depend on their free energy differences, and, at least for the following case,

$$\text{Fc}^+\text{-Fc-Fc}^+ \rightarrow \text{Fc}^+\text{-Fc}^+\text{-Fc} \tag{13}$$

it was estimated that the "symmetrical" form is favored by 0.12 V (ΔG = 11.6 kJ/mol).[147]

Cowan and co-workers have investigated a number of substituted biferrocenes (7) in order to test a fundamental aspect of the Hush theory of class II mixed valent compounds.[139] An important by-product of this work was further information on substituent parameters. It is clear from the discussion in Section 1.5 that substituent effects are generally additive, and so it was suggested that the reduction potential of a 1,1′-disubstituted ferrocene. $(C_5H_4X)Fe(C_5H_4Y)$, could be calculated from

$$E_{\text{calc}} = E_{(\text{Fc}^{+,0})} + \delta_x + \delta_y$$

where the first term is the potential for the parent compound, and the second and third terms are substituent parameters. For example, $\delta_{\text{Fc}} = -0.105$ V can be derived on comparing ferrocene and biferrocene, $\delta_{\text{CO}_2\text{Me}} = +0.240$ V by comparing ferrocene and methoxycarbonylferrocene, and $\delta_{\text{FcCO}_2\text{Me}} = -0.05$ V by comparing ferrocene and methoxycarbonylbiferrocene. These parameters allow us to estimate that the reduction potential of 1′,1‴-dimethoxycarbonylbiferrocene should be about +0.73 V, in reasonable agreement with observation (+0.690 V). A very useful result of such calculations is that the difference in energy between oxidation state isomers can be calculated. For example, oxidation of 1′-acetylbiferrocene should occur at the group bearing the electron-withdrawing acetyl group to give $\text{MeCOFc}^+\text{-Fc}$. However, thermal or photon-driven electron transfer will give MeCOFc-Fc^+. Using substituent parameters, it can be estimated that ΔG for this process, for example, is of the order of 15–20 kJ/mol. (Estimates for other biferrocenes in Table 2 are in the same general range.)

A notion which arose some years ago among those studying mixed valence compounds was that the difference in the reduction potentials of two metal sites was a measure of the extent of electron delocalization between the sites. This does not appear to be true, except perhaps in very closely related series.[137] Flanagan *et al.* state that "the differences in potential between the half-reactions of . . . successive electron transfers . . . can depend on the extent of interaction between the sites, solvation changes, ion pairing, and structural changes of the molecule . . . ."[142] This view is shared by other groups, and the results outlined below underscore this.[137,148]

When one or two atoms are placed between two ferrocenyl groups, there can still be considerable difference between the reduction potentials for the two steps

$$\begin{aligned} \text{Fc}^+\text{-}X\text{-Fc}^+ + e^- &\rightarrow \text{Fc-}X\text{-Fc}^+ \quad E_1 \\ \text{Fc-}X\text{-Fc}^+ + e^- &\rightarrow \text{Fc-}X\text{-Fc} \quad E_2 \end{aligned} \tag{14}$$

Unfortunately, when X = Se,[143,144] Se-Se,[143] CH_2, or P,[149] no intervalent transfer band was observed. However, when X is an acetylenic linkage, or even two acetylenic linkages, an intense invervalent transfer band was observed.[136,137,140,141] (In the series Fc-Fc, Fc-C≡C-Fc, Fc-C≡C-C≡C-Fc, it is true that the differences in reduction potentials get smaller as the IT band intensity declines and the band energy increases, both effects indicating a smaller electron delocalization between sites.) One would infer that, in order to observe an IT band, there must be a direct metal-metal interaction, or the two ferrocenyl groups must be linked through a bridge that can allow the pi systems of the two ferrocenyl groups to interact. Given this point of view, the mixed valence ion ferricenyl(III)*tris*(ferrocenyl(II))borate,$[Fc^+BFc_3]^-$ is of interest.[150] This molecule is formally a tetraferrocenylborate anion; however, the anionic charge is internally compensated by an iron(III)-containing ferricenyl group. A distinct intervalent transfer band is observed, in spite of the fact that one would consider a tetrahedral boron atom far less capable of allowing electron delocalization than a selenium atom in Fc-Se-Fc, where no IT band was observed. Kramer *et al.* argue that this observation suggests that "electronic coupling probably depends on a direct interaction between the two iron ions . . .", and the crystal structure of Fc_4B suggests this is possible.[144]

When two ferrocenyl groups are separated by a saturated ethyl linkage, both groups oxidize at the same potential, because the ferrocenyl groups interact weakly, if at all.[133] Not surprisingly, this is also true for poly(vinylferrocene) [PVF].[142] Flanagan *et al.* have carefully examined the electrochemistry of PVF in order to prove that for a system with multiple, noninteracting redox centers, the current but not the shape of the electrochemical wave is affected.[142] (That is, the anodic and cathodic peak potentials and the peak and half-peak potentials should be separated by 58 mV at 25°.) They also observed that, while the $E_{1/2}$ for vinyl ferrocene is 0.56 V, that for PVF with various degrees of polymerization ranges from 0.45 to 0.48 V.

It was mentioned above that there is not necessarily any connection between the difference in reduction potentials of multimetal molecules and the existence of an intervalence transfer band. Therefore, even if all of the metal sites in a molecule oxidize at the same potential, or nearly so, the system should still be examined for an IT band for some indication of electronic interaction. In his review of the synthesis of mixed valence species, Brown has spoken to this point.[128] He suggested that, if a fully reduced, two-site system is oxidized by one electron, the equilibrium constant for disproportionation of the mixed valence compound is 0.25. This value comes about because the mixed

$$2M^+\text{-}M \rightarrow M\text{-}M + M^+\text{-}M^+ \quad (15)$$

valent species has twice the probability of either the fully reduced or fully oxidized species. Therefore, an equilibrium mixture should contain the mixed valent species as 50 percent of the species in solution. Given a sufficiently intense IT band, a mixed valent species with electronically interacting metal centers should be capable of observation.

The point made above can be applied to a number of papers in the literature. We synthesized *B*-triferrocenyl borazine, $(FcBNH)_3$, in the hope of generating, by one- or two-electron oxidation at the ferrocenyl group, a mixed valence compound.[119] The pi electrons of the borazine ring system can potentially be delocalized and boron-bound ferrocenyl groups could thereby interact. We found that the molecule undergoes a three-electron oxidation at +0.36 V in DMF. Unfortunately, even when the molecule was oxidized by just one electron, no IT band was observed. In a somewhat related study, Wollmann and Hendrickson synthesized meso-tetraferrocenylporphyrin and its copper complex.[151] The oxidation of the four ferrocenyl groups was observed as a broad oxidation "wave" running from about 0.25 V to 1.05 V, but no IT band was observed for a partially oxidized material. Finally, in another case, that of (**5**) (Section 1.5), we stated that there was no evidence from electrochemistry for Fe(II)-Cr(O) or Fe(II)-Cr(I) interaction.[120] We now know that such interaction is impossible to rule out on the basis of electrochemical data alone.

The point made in the previous paragraph also applies to a very interesting series of compounds (**13**) studied by Fry and co-workers.[152,153]

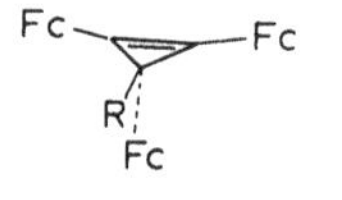

13 Fc = ferrocenyl

Three redox waves were observed by cyclic voltammetry for all compounds; however, as the waves were relatively closely spaced, a computer method was devised to determine more accurate $E_{1/2}$ values. It is interesting to note that the spacing between the first and second waves is in the same range as that observed for diferrocenylacetylene (**8**), a molecule having some degree of electron delocalization between the two iron sites. One would predict, therefore, that an IT band would be observed. Furthermore, these molecules are of interest owing to the fact that "oxidation state isomerism" should be observed.

The cyclopropenes above were synthesized from the parent triferrocenylcyclopropenium ion, an ion which has two irreversible reductions at −1.49 and −2.10 V (by CV in MeCN, vs. $Ag/AgNO_3$).[152] Not surprisingly, this ion is more difficult to reduce than the triphenyl analog. Controlled potential reduction of triferrocenylcyclopropenium ion at a mercury pool electrode at 0°C gave dark orange hexaferrocenylbicyclopropenyl, a molecule unstable with respect to decomposition to a compound having the formula Fc_6C_6.

If two ferrocenyl groups are bound through two rings, the result is bisfulvalenediiron (BFDFe)(**10**). The two iron sites of this molecule are oxidized at very different potentials [Table 2].[134–138,141,145,146] The Fe(II)/Fe(III) species gives rise to a near infrared band, but its energy is higher, and its width narrower, than predicted.[137] Most especially, a Mössbauer spectrum shows identical iron sites in the mixed valence cation, thus strongly suggesting a delocalized ground state and a class III mixed valence ion.

For comparison with bisfulvalenediiron [BFDFe], Smart and coworkers have synthesized the analogs with V, Cr, Co, and Ni.[154] These compounds all have very interesting magnetic, spectroscopic, and electrochemical properties. Just as BFDFe can be oxidized to +1 and +2 ions, so can the other metal analogs. Notice that the data in Table 3 show that, just as in the symmetrical metallocenes, the iron compound is by far the most resistant to oxidation, again illustrating the stability of the d^6 configuration. Furthermore, the BFD compounds follow the same general trend in reduction potentials as the Cp_2M analogs.[71] And finally, just as BFDFe is much more readily oxidized than Cp_2Fe, so are the bisfulvalene compounds more readily oxidized (by about 0.2 V) than the corresponding metallocene. The one exception to this trend is the vanadium compound, but it should be noted that the (+3, +2) ↔ (+2, +2) redox process is electrochemically irreversible. This may indicate a structural change on oxidation. Indeed, if

TABLE 3
Redox Properties of *Bis*(fulvalene)dimetal Compounds Compared with the Simple Metallocenes, $(\eta^5\text{-}C_5H_5)_2M^{a,b}$

Metal	Metallocene reduction[c] potential, +1 ⇄ 0	*Bis*(fulvalene)dimetal reduction potentials[d]	
		+1 ⇄ 0	+2 ⇄ +1
V	−0.55	−0.09	+0.14
Cr	−0.67	−0.84	−0.18
Fe	+0.31	+0.13	+0.72
Co	−0.94	−0.95	−0.07
Ni	−0.09	−0.29	+0.12

[a] All potentials are in volts vs. SCE, and all were measured in MeCN.
[b] Charge given is that on the complex.
[c] See reference 43.
[d] See reference 154.

BFDV is chemically oxidized in acetonitrile to the +2 ion, each V(III) is also bound to an acetonitrile, and a crystallographic structure determination shows that the two vanadium ions have slipped toward one another, perhaps to promote a metal–metal bond.

Several papers have also appeared regarding [1.1]ferrocenophanes having bridging groups between the ferrocenyl groups (**11**) [Table 2].[110,135,136,141] All exhibit two redox waves and the mono- and di-cations have been isolated for all but $R = R' =$ Me. The most interesting aspect of the series is that all (except $R = R' =$ Me which has not been examined) exhibit some degree of delocalization between Fe(II) and Fe(III) in the +1 cation. When the bridge is —C≡C— (**12**), the compound is a class III mixed valence species, class II when $R =$ Me, $R' =$ H, and when $R = R' =$ H, the ion approaches class III behavior.

The Fe(II)/Ru(II) and Ru(II)/Ru(II) analogs of (**11**) (R = R' = H) have also been isolated, and the electrochemistry of the di-ruthenium molecule is intriguing.[54] Ruthenocene is oxidized by two electrons in an electrochemically irreversible manner (but chemically reversible) (E_{pa} = 0.920 V in PhCN). However, the di-ruthenium molecule undergoes a reversible, two-electron redox process at only +0.380 V to give a diamagnetic di-cation. Diaz *et al.*[54] speculate that this may indicate oxidation of only one ruthenium from +2 to +4. (However, it must be noted that the diiron species (**11**) ($R =$ Me, $R' =$ H) also gives a diamagnetic dication, and it seems clear that each iron site is d^5 iron(III) in this case.) The di-ruthenium system clearly deserves more attention.

In the field of organometallic-mixed valence studies, perhaps the most interesting recent development is the observation of intervalent transfer between two different metals. Thus far, four papers have appeared,[155-158] but many more are promised. The first such species was a ferrocenylcyanide-ruthenium complex, $[FcCNRu(NH_3)_5]^{2+}$, wherein both the Ru(II) and the ferrocenyl group are oxidized reversibly (at +0.44 V and +0.85 V, respectively; the reduction potential of $[FcCN]^+$ is +0.77 V).[155] An intervalent transfer band was observed for the Fe(II)/Re(III) species, indicating some degree of electronic interaction between the metal sites. Similar results were obtained for Ru(III) complexes of 1,1'-dicyanoferrocene.[156]

The other two heteronuclear-mixed valence compounds both involve Fe(III) and low valent Co. We found that the redox behavior of (**14**) and its trans isomer are almost identical.[157] For (**14**) the ferrocenyl groups oxidize reversibly at +0.33 V and +0.52 V and the cobalt irreversibly at 1.38 V. A near infrared band was observed for *both* the mono- and di-cations (1250 nm and 1120 nm, respectively), and it was argued that the band arises from photon-driven electron transfer between Co(I) and Fe(III). The ferrocenyl-cobalt cluster (**15**) was found to have reversible redox waves for ferrocenyl group oxidation (+0.80 V) and cluster reduction (−0.59 V, vs. Ag/AgCl in CH_2Cl_2).[158] If the basal CO groups are substituted by phosphite ligands, a second cluster oxidation wave was observed for Co_3C oxidation. The monocation was observed to exhibit an intervalent transfer band (1545 nm), presumably owing to electron transfer from the cluster to Fe(III).

2. OTHER "SANDWICH" COMPOUNDS OF PI DONOR LIGANDS

In this section, we shall discuss "sandwich" compounds wherein the metal is held between two pi donor ligands, only one of which may be the cyclopentadienyl ion. Included in this class are complexes having two arene ligands, one arene and one η^5-cyclopentadienyl ligand, a η^5-cyclopentadienyl ligand and an olefin, two heterocyclic rings, and other miscellaneous complexes.

2.1. Symmetrical *Bis*-Arene Complexes: (Arene)$_2$*M*

In comparison with the metallocenes, relatively little has been reported on the electrochemistry of *bis*-arene metal "sandwich" compounds,[2,159] in large part because few such compounds have been synthesized. Symmetrical diarene complexes are known for all of the first row transition metals, vanadium through cobalt, with most of the work having been done with chromium. Just as d^6 Fe^{2+} forms the most stable metallocene, d^6 Cr(O) has so far been found to form the greatest range of *bis*-arene complexes.

The best known symmetrical metal arene complex is dibenzenechromium, and its electrochemistry, and that of a few of its derivatives, has been studied by a number of groups. In general, a single redox process, reversible both chemically and electrochemically, is observed at −0.74 V (in DMF) for[160]

$$[(\eta^6\text{-benzene})_2\text{Cr}]^+ + e^- \rightarrow (\eta^6\text{-benzene})_2\text{Cr} \quad (16)$$

Substituent effects have not been studied to the same extent as in metallocenes, but the same trends, in general, are observed.[161-166] For example, a wide range of compounds $(C_6H_4RR')_2Cr$ was examined in MeCN by

Treichel *et al.*[166] They observed that the potentials for oxidation fell over a range from −0.767 V (for R = OMe, R' = H) to +0.385 V (for $R = R'$ = CF_3), a much greater range than was observed for ferrocene derivatives.

Some very interesting results were published recently concerning $(arene)_2Cr$ compounds where the arene was benzene, fluorene, naphthalene, phenanthrene, chrysene, pyrene, anthracene, and fluoranthene.[160] As Dessy *et al.* had previously observed for a very limited range of compounds,[7] all of these compounds undergo reversible oxidation-reduction at nearly identical potentials. This is in spite of the fact that the reduction potentials of the free arenes are spread over about 1.7 V. This result was taken as evidence in support of an MO bonding scheme developed by Anderson and Drago for dibenzenechromium.[167] In this scheme the HOMO is an orbital composed of the Cr $3d_{z^2}$ orbital and the sigma frameworks of the benzene rings. This feature explains the fact that substituents can affect the redox potential—they can influence the ring sigma framework—while changes in the pi system by fusing another ring to benzene will have little effect on the HOMO.

The d^6 metal ions Fe^{2+} and Ru^{2+} both form symmetrical arene complexes with hexamethylbenzene. The iron complex $[(\eta^6\text{-}C_6Me_6)_2Fe]^{2+}$ can be reversibly reduced by one electron at −0.48 V and irreversibly by another electron at −1.46 V (in aqueous ethanol). (As expected, an analogous *bis*-mesitylene complex is reduced at less cathodic potentials: −0.28 V and −1.16 V.[168,169])

The symmetrical *bis*-hexamethylbenzene complex of ruthenium(II), $[(\eta^6\text{-}C_6Me_6)_2Ru]^{2+}$ (**16**), is much more interesting than the iron analog.[170,171] The ion was reduced by two electrons at −1.02 V (E_p = 35 mV; i_{pa}/i_{pc} = 0.36 at 100 mV/s; MeCN) to give (η^6-hexamethylbenzene) (η^4-hexamethylbenzene)ruthenium(0) (**17**), a fully characterized complex.

16 $\xrightarrow{2e^-}$ **17** (17)

Similarly reduced are (η^6-hexamethylbenzene)(η^6-durene)ruthenium(2+) (−0.93 V) and (η^6-hexamethylbenzene)(η^6-*p*-xylene)-ruthenium(2+) (−0.85 V). Note that the reduction potentials were less cathodic, as expected, but the reductions were also considerably less reversible than that of the *bis*-hexamethylbenzene complex. In this regard, it is noteworthy that $(\eta^6\text{-}C_6H_6)_2Ru(0)$ is unstable with respect to benzene and ruthenium metal.

The most interesting aspects of the electrochemistry of the arene ruthenium complexes are the two-electron change and the structural change accompanying (or following) electron transfer (see also Section 5). The structural change, as illustrated in Equation 17, is the transformation of one C_6 ring from η^6-planar to η^4-bent. This change was seen more clearly in the electrochemistry of cyclophane complexes such as (**18**) and (**19**). For example, (**18**) was reduced at −0.89 V while reduction of (**19**) occurred at −0.50 V, and the reversibility increased from only about 38 percent to almost 100 percent. This was apparently the result of the ease of attaining the η^4 geometry. Compound (**18**) is already partly preformed toward the geometry of the Ru(0) product, while (**19**) is nearly fully preformed.

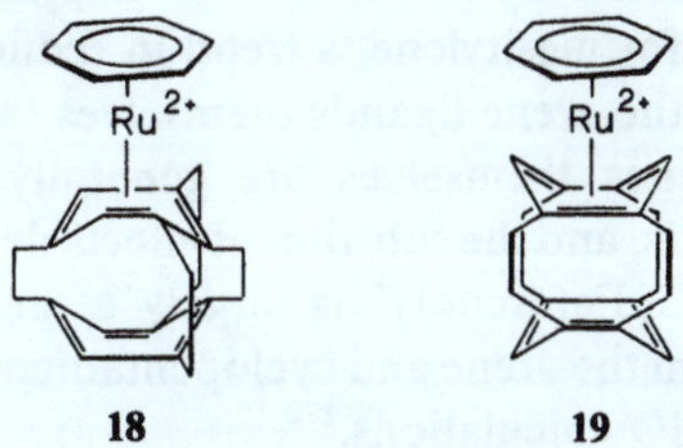

In two-electron changes, the second reduction is usually separated from the first by coulombic effects. (For example, as we have so often seen in Section 1.6 on mixed valence compounds, coulombic effects usually lead to the first and second reduction potentials of nondirectly bonded metals being separated by 50–500 mV.) Observation of a single, two-electron wave for the ruthenium complexes in MeCN was taken to indicate that the second reduction occurs at a more positive potential than the first, $E_2^0 \geq E_1^0$. Although in MeCN the reduction waves overlapped, a change in solvent to CH_2Cl_2 decreased the solvation energy of the Ru^{2+} complex relative to Ru^+ and increased the ease of the first reduction step. Thus, in CH_2Cl_2, (**16**), (**18**), and (**19**) were all reduced in two separate, one-electron steps. This fact, and the observation that the electron transfer rate of the second step was slow, suggested to Finke *et al.* that the overall mechanism was EE_C, at least in CH_2Cl_2.[171] Structural changes of this type are further discussed in Section 5.1.

2.2. $(\eta^6\text{-Arene})(\eta^5\text{-}C_5H_5)M$ Complexes

Complexes such as (**20**) and (**21**) can be synthesized by ligand exchange between ferrocenes and arenes, the yields often being nearly quantitative. All can be reduced reversibly to d^7 Fe(I)-centered, 19-electron complexes,[172–175] which Astruc has called "electron reservoirs," since they can be used in the activation of O_2 to O_2^- and in the catalytic reduction of NO_3^-

to NH_3.[172] All compounds exhibit a second, more cathodic, irreversible step described below.

Fe $\xrightarrow{e^-}$ Fe $\xrightarrow[e^-]{H^+}$ Fe (18)

20

The benzene ligand in (**20**) has been replaced with naphthalene, phenanthrene, and other similar molecules.[173] The first reduction potential of these molecules varied from −1.07 V for naphthalene to −1.45 V for benzene and −1.55 V for mesitylene, a trend in reduction potentials which is the same as that of the arene ligands themselves (although the reduction potentials of the arenes themselves are generally more cathodic than −2.5 V).[173] These results, and the substituent effects described below, suggest that the LUMO of $[CpFe(arene)]^+$ is largely a metal-based orbital with some contribution from the arene and cyclopentadienyl pi orbitals, a supposition supported by MO calculations.[172]

Several additional points are interesting and bear on the relative energy of the LUMO in these d^6 iron(II) "sandwich" complexes. While the reduction of ferrocene can apparently occur only at a very negative potential,[43,70] the reduction of $[CpFe(arene)]^+$ occurs at approximately −1 to −1.5 V, and that of $[(arene)_2Fe]^{2+}$ at about −0.5 V. Further, it has been observed that the $[CpFe(arene)]^+$ complexes are extremely resistant to oxidation, in great contrast with the ease of oxidation of ferrocene and its derivatives. (Apparently the electron-donor and -acceptor molecular orbitals have been raised in energy by at least 1 eV.) The effect on the first redox process of substituting either the C_5 or the C_6 ring has been briefly examined with the expected results.[173] The effect of substituents on the C_5 ring is given by $E_{1/2} = -1.39 + 0.520\sigma_p^0$, while the effect of C_6 ring substituents is described by $E_{1/2} = -1.42 + 0.515\sigma_p^0$. It is interesting, especially in view of the fact that one is comparing a Fe(III) → Fe(II) reduction for ferricenium ions and a formal Fe(II) → Fe(I) reduction for $[CpFe(arene)]^+$, that a nearly identical slope of the $E_{1/2}$ vs. σ_p^0 plot is obtained for these compounds and for ferrocene.[103]

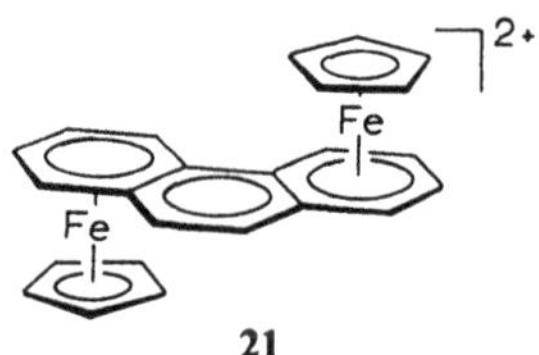

21

Hendrickson and co-workers have synthesized fifteen complexes, such as (**21**), with a view to observing an Fe(I)-Fe(II) interaction in the half-reduced species.[175] In general, the complexes have two reversible waves in the region −0.9 to −1.5 V for the sequential reduction of the two iron centers

$$Fe^{II}Fe^{II} \rightarrow Fe^{I}Fe^{II} \rightarrow Fe^{I}Fe^{I}$$

Just as in the sequential oxidation of the iron centers of various bridged biferrocenes, the separation of the redox waves for $[CpFe(arene)FeCp]^{2+}$ depends on three effects: "(1) direct iron-iron electron exchange interactions, (2) iron-iron dipolar interactions, and (3) electron-exchange interactions propagated through the rings."[175] Thus, only one, two-electron wave is seen for $[CpFe(bibenzyl)FeCp]^{2+}$, whereas the separation is about 150 mV for the anthracene complex, even though the CpFe groups are probably on opposite sides of the anthracene molecular plane. In this regard, it is interesting to compare $[CpFe(biphenyl)FeCp]^{2+}$ and $[(benzene)Fe(C_5H_4\text{-}C_5H_4)Fe(benzene)]^{2+}$. The former has a wave separation of only about 100 mV, while the latter is about 400 mV. This may reflect closer iron-iron approach in the Cp-Cp bridged species, as well as greater ring propagated interactions.

The second, irreversible, one-electron reduction of the $[CpFe(arene)]^+$ series gives an η^5-hexadienyl complex in the presence of an electrophile (for example, CO_2, H^+, or RX) (Equation 18).[174] (In the absence of an electrophile, the reduced species decomposes to $C_5H_5^-$, arene, and iron metal.) The Fe(II)-hexadienyl complex is reversibly oxidized to Fe(III) form.

2.3. (η^5-Cyclopentadienyl)(Olefin)Metal Complexes

Combining $CpCo(CO)_2$ with acetylenes leads to cyclopentadienone complexes such as (**22**). In general, such compounds can be reversibly reduced (in the range −1 to −2 V) to the monoanion wherein the metal is formally M(0); most such complexes also have a second, generally irreversible reduction.[176] Where R = OBu, oxidation has been successfully used to prepare croconic acid (Equation 19).[177]

22 $\xrightarrow[R\,=\,OBu]{-e^-}$ (BuO)₄-cyclopentadienone ⇢ croconic acid (19)

Demetallation was also the major reaction observed when (η^4-cyclooctatetraene)CoCp or (η^4-cyclooctadiene)CoCp was oxidized (Equation 20), the reaction being much more rapid in MeCN than in CH_2Cl_2. In either case, the free olefin was isolated in quantitative yield along with the cobalticenium ion.[178,179]

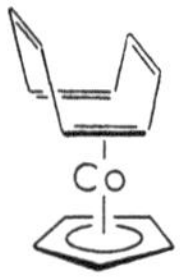

23

$$(\mathbf{23}) \rightarrow [(COD)CoCp]^+ + e^-, \qquad E^0 = +0.24\ V \tag{20}$$

$$[(COD)CoCp]^+ \rightarrow 1,5\text{-}COD + 0.5[Cp_2Co]^+ + \text{other Co products}$$

Reduction of these complexes leads to a rearrangement of the COT or COD bonding mode as described in Section 5.1.

El Murr and co-workers have extensively studied the electrochemistry of complexes such as (**24**) and find that they are readily oxidized in aqueous alcohol to give the cobalticenium ion.[180–182]

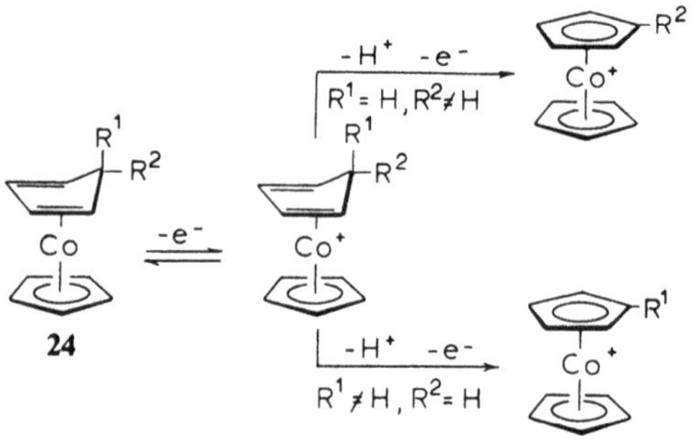

(21)

24

The chemistry of this system is further described in Section 5.3.

2.4. Complexes of Metals with Heterocyclic Rings

There are few complexes known with pi-bonded heterocyclic rings, but some interesting results have been obtained for several such systems.

The ligand $[\eta^6\text{-}C_5H_5BR]^-$ is isoelectronic with $\eta^5\text{-}C_5H_5^-$, and analogs of vanadocene, chromocene, ferrocene, and cobaltocene (**26**) have been reported, as well as $(\eta^5\text{-}C_5H_5)Fe(\eta^6\text{-}C_5H_5BR)$ (**25**).[183] The ligand is

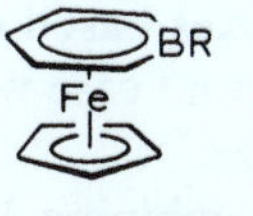

R = Me, Ph M^{2+} = Fe, Co, V

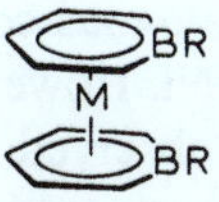

25   **26**

apparently a stronger electron acceptor but a weaker electron donor than the cyclopentadienyl anion, as the potentials for the reversible reduction of the Fe(III) complexes are in the order

Complex (*R* = Ph):	$[Fe(C_5H_5BR)_2]^+$	$[CpFe(C_5H_5BR)]^+$	$[Cp_2Fe]^+$
$E_{1/2}(CH_2Cl_2)$:	1.13 V	0.88 V	0.48 V

For this same reason, $Fe(C_5H_5BPh)_2$ can be reduced reversibly at −1.77 V, a much less cathodic potential than ferrocene.[70]

The same effects are seen in the V(II), Cr(II) and Co(II) analogs of (**26**). For example, the chromium compound (*R* = Ph) is oxidized over 1 V more anodically than chromocene (0.53 V vs. −0.61 V), and, unlike Cp_2Cr, it can be reduced reversibly (−1.34 V).

For the sake of completeness, we note heterocyclic complexes such as (**27**).

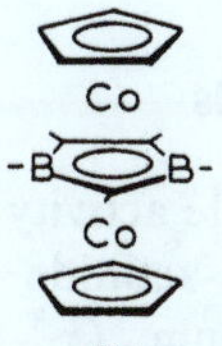

27

An extensive paper describing their electrochemistry has very recently appeared.[184] However, since Geiger reviewed their electrochemistry in 1982,[24] we shall note here only a few salient features of these compounds. The borolenyl ligand in (**27**) is considered a three-electron donor for electron counting purposes. Thus, neutral (**27**) has a total of 31 valence electrons (12 for the two Cp^- ligands, 16 for the two Co(I) ions, and 3 for the borolenyl). Five species are electrochemically accessible, from the 29-electron dication to the very transient 33-electron dianion.

$$\underset{29}{[CoCo]^{2+}} \rightleftharpoons \underset{30}{[CoCo]^{+}} \rightleftharpoons$$

$$\underset{31}{CoCo} \rightleftharpoons \underset{32}{[CoCo]^{-}} \longrightarrow \underset{33}{[CoCo]^{-}} \quad \text{valence electrons}$$

Similar series exist for analogous complexes based on the metal combinations CoFe, CoFe, and NiNi. However, in no case was a species with more than 34-valence electrons observed.

Compound (**28**) was prepared by combining $FeCl_2$ and tetramethylthiophene in the presence of $AlCl_3$. It is reversibly reduced by one electron at −0.27 V, a reduction potential almost identical with that of *bis*(η^6-mesitylene)iron(2+).[168]

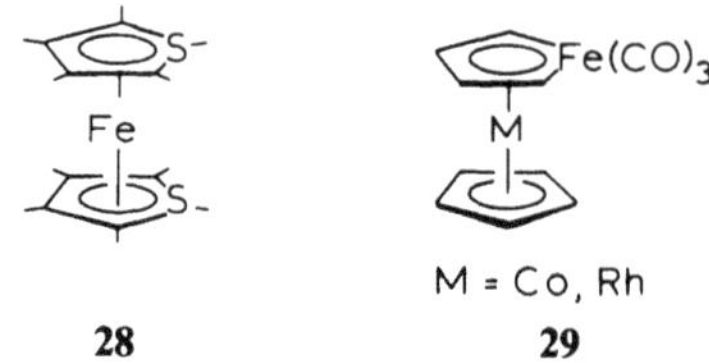

28 **29**

Some years ago Hoffmann suggested the "concept of isolobal species," wherein isoelectronic systems could be obtained by replacing CH^+ with $Fe(CO)_3$.[185] This has been realized experimentally with the synthesis of ferracobaltocene and -rhodocene (**29**), compounds isoelectronic with $[Cp_2Co]^+$ and $[Cp_2Rh]^+$. Both compounds are reversibly reduced, the cobalt compound at −1.39 V and the rhodium compound at −1.76 V (vs. Ag/AgCl).[186]

2.5. Other Sandwich Compounds

Another area of considerable activity in organometallic chemistry has been in organolanthanide and -actinide chemistry, and the synthesis of *bis*(η^8-cyclo-octatetraenyl)uranium, (η^8-$C_8H_8)_2U$, was one of the first important discoveries. The molecule is extremely air-sensitive, and so it is not surprising that the redox potential is only +0.06 V. However the monocation is very unstable, giving as at least one product, a green solid, thought to be a tri-uranium cluster.[187,188]

Finally, several interesting phosphate complexes (**30**) have been synthesized recently, and their electrochemistry reported.[189] The free phosphonate ligand was oxidized at +0.54 V, while the complex with M = Co(II) was oxidized reversibly at +1.02 V. In contrast, the Fe(II) complex oxidation potential was much less anodic (−0.46 V). These results suggests that the site of oxidation is the coordinated metal and not the Co(I) of the ligand.

M = Co, Fe, Hg, Cu

O=P = O=P(OR)$_2$

30

3. THE BINARY METAL CARBONYLS

All of the first-row transition metals, from vanadium through nickel, form simple binary carbonyls containing only one or two metals and carbon monoxide. Examples include $Cr(CO)_6$, $Mn_2(CO)_{10}$, and $Fe(CO)_5$.[2,41] In addition, some form molecular polyhedra or clusters, a network of metal atoms wherein each metal atom is bound to at least two others of the same type; tetrahedral $Co_4(CO)_{12}$ and triangular $Fe_3(CO)_{12}$ are excellent examples of binary metal carbonyl clusters.[41,190] In this section, we shall describe only the electrochemistry of the binary mono- and bi-metallic carbonyls, and, in the section which follows, we shall describe some aspects of the derivatives of these compounds. We shall not detail the electrochemical properties of metal clusters, since an excellent review of this area has very recently appeared.[25] (A related review of the redox chemistry of compounds with metal-metal bonds is also available.[191])

To satisfy the "16/18 electron rule,"[41] vanadium carbonyl is most stable as the 18-valence electron anion $[V(CO)_6]^-$. The couple $V(CO)_6/[V(CO)_6]^-$ is quasireversible in MeCN with E_{pc} reported to be -0.21 V ($\Delta E_p = 270$ mV).[192]

Chromium hexacarbonyl is reversibly oxidized by one electron at +1.50 V in MeCN to give the cation $[Cr(CO)_6]^+$.[192–195] This ion is not stable, however, and the esr spectrum of the cation shows that it decays according to second-order kinetics ($k = 2.5$ liter mol^{-1} s^{-1} at 25°C). This observation, and the fact that controlled potential electrolysis at +1.6 V showed that two electrons are involved, is best explained by the following reaction sequence.

$$Cr(CO)_6 \rightarrow [Cr(CO)_6]^+ + e^-$$

$$2[Cr(CO)_6]^+ \rightarrow Cr(CO)_6 + [Cr(CO)_6]^{2+} \tag{22}$$

$$[Cr(CO)_6]^{2+} \rightarrow Cr^{2+} + 6CO$$

The cation $[Cr(CO)_6]^+$ was found to be considerably more stable when formed in trifluoroacetic acid as the solvent ($E_{1/2} = 1.06$ V). Indeed, this low nucleophilicity solvent has been found to be generally useful for the stabilization of cations, which are otherwise unstable in common organic solvents such as MeCN or THF. (See Section 4.3.)

The molybdenum and tungsten analogs of $Cr(CO)_6$ were both irreversibly oxidized.[192–195] There was no evidence for a stable cationic species. This is not surprising, given the fact that it is commonly observed that $Mo(CO)_6$ in particular is much more rapidly substituted than $Cr(CO)_6$.[196] (Oxidatively induced substitutions will be discussed below in Section 5.)

The Group 6B carbonyls were also examined in a molten salt mixture (2:1 $AlCl_3$ and ethylpyridinium bromide + benzene).[195] The behavior in this medium was in general the same as in MeCN. The one notable exception, however, was that the oxidation peak potentials in MeCN were all 1.53 V, whereas they varied from 1.61 (M = Cr) to 1.74 (M = W) in the molten salt. This suggests the very interesting possibility that there is an interaction between the $AlCl_3$ of the melt and $M(CO)_6$. Such an interaction could be the formation of a bond between a carbonyl oxygen and aluminum, that is, $M\text{-}C{\equiv}O{:}AlCl_3$.[197]

Cathodic reduction of the Group 6B hexacarbonyls also occurs by one electron, but the anion $[M(CO)_6]^-$ rapidly decays by loss of CO. This coordinatively unsaturated anion dimerizes, or, after oxidation, gives the solvent substituted pentacarbonyl.[192-195] (The reactions are illustrated using MeCN, but the same results are found in THF or DMF.)

$$\begin{aligned} M(CO)_6 + e^- &\rightarrow [M(CO)_6]^- \\ [M(CO)_6^- &\rightarrow [M(CO)_5]^- + CO \quad \text{(fast)} \\ 2[M(CO)_5]^- &\rightarrow [M_2(CO)_{10}]^{2-} \quad \text{(slow)} \qquad (23) \\ [M(CO)_5]^- - e^- &\rightarrow [M(CO)_5] \\ [M(CO)_5] + MeCN &\rightarrow M(CO)_5(NCMe) \end{aligned}$$

The dimeric dianions of Mo and W were not electroactive in THF, but $[Cr_2(CO)_{10}]^{2-}$ was oxidized in MeCN to give $Cr(CO)_5(NCMe)$.

The simple binary carbonyls of the Group 7B metals are bimetallic. In order to satisfy the requirement of 18-valence electrons, a metal–metal bond is formed to give $(OC)_5M\text{-}M(CO)_5$. The Mn_2 and Re_2 compounds are both reduced by two electrons to give the well-known pentacarbonyl anions, the di-rhenium compound having a considerably more cathodic potential.[7,192,198,199] Both can be re-oxidized to the neutral dimer. Such reactions are typical of all symmetrical compounds having single metal–metal bonds.[9,191]

$$Mn_2(CO)_{10} + 2e^- \underset{E_p=-0.02\text{ V}}{\overset{E_p=-1.65\text{ V}}{\rightleftharpoons}} 2[Mn(CO)_5]^- \qquad (24)$$

$$Re_2(CO)_{10} + 2e^- \underset{E_p=-0.1\text{ V}}{\overset{E_p=-2.36\text{ V}}{\rightleftharpoons}} 2[Re(CO)_5]^- \qquad (25)$$

(The potentials above were observed for MeCN solutions; less cathodic values were observed in DMF.[198] When electrolysis was carried out at a mercury electrode, $(OC)_5M$-Hg-$M(CO)_5$ was formed.[7]) In DMF, the unsymmetrical $(OC)_5$Re-Mn$(CO)_5$ was reduced to $[Re(CO)_5]^-$ and $[Mn(CO)_5]^-$ at a potential intermediate between those of Mn_2 and Re_2. Lemoine and Gross found that the force constant for the M-M bond stretch and the energy of the $\sigma \rightarrow \sigma^*$ electronic transition correlated well with the reduction potential of the Mn_2, MnRe, and Re_2 decacarbonyls.[198] Thus, it was concluded that the strength of the metal-metal bond decreased in the sequence Re-Re > Mn-Re > Mn-Mn. On substituting one CO ligand on each metal with PPh_3, a better sigma base but poorer pi acid, the reduction potentials of $(Ph_3P)(OC)_4M$-$M'(CO)_4(PPh_3)$ are all shifted cathodically by about 0.2 V.[198]

As expected from the requirement for 18-valence electrons, the simplest binary carbonyl of iron is trigonal bipyramidal $Fe(CO)_5$. There is, however, some disagreement in the literature as to the nature of the cathodic process. Dessy *et al.* reported that a one-electron reduction occurred at a mercury electrode in DME (at −2.4 V vs. $Ag/AgClO_4$) to give a very unstable anion.[7] Pickett and Pletcher also observed an irreversible reduction on Pt (at −2.48 V) in THF.[192] Controlled potential electrolysis indicated it was a one-electron process, and the final product was the bimetallic anion $[Fe_2(CO)_8]^{2-}$. Similarly, El Murr and Chaloyard reported irreversible reduction in THF on mercury (−1.77 V), glassy carbon (−2.0 V) or platinum; electrolysis on mercury in THF required one faraday/mol and gave the di-iron di-anion.[200] A possible sequence of reactions leading to this product is

$$\begin{aligned} Fe(CO)_5 + e^- &\rightarrow [Fe(CO)_5]^- \\ [Fe(CO)_5]^- &\rightarrow [Fe(CO)_4]^- + CO \\ 2[Fe(CO)_4]^- &\rightarrow [Fe_2(CO)_8]^{2-} \end{aligned} \qquad (26)$$

This is in agreement with Krusic *et al.* who found that chemical reduction of $Fe(CO)_5$ with one equivalent of sodium naphthalenide gave the di-iron product, whereas two equivalents of reducing agent gave $[Fe(CO)_4]^{2-}$.[201,202]

In contrast with the results above, Bond *et al.* could obtain no evidence (by DME polarography or CV) to support a one-electron reduction.[203] Rather, they reported a two-electron reduction in acetone at mercury. (They also noted that no reduction was observed at a Pt electrode.) In this case, the product was presumably $[Fe(CO)_4]^{2-}$, which can combine with $Fe(CO)_5$

to give the usual di-iron species.

$$Fe(CO)_5 + [Fe(CO)_4]^{2-} \rightarrow [Fe_2(CO)_8]^{2-} + CO \quad (27)$$

However, no product analysis was done.

Bearing on the results outlined above, El Murr and Chaloyard found that, if water was present in the THF or ethanol solutions of $Fe(CO)_5$ solution, a two-electron reduction was indeed observed.[200] The following reaction sequence was suggested.

$$\begin{aligned} Fe(CO)_5 + e^- &\rightarrow [Fe(CO)_5]^- \\ [Fe(CO)_5]^- &\rightarrow CO + [Fe(CO)_4]^- \\ [Fe(CO)_4]^- + H^+ &\rightarrow [HFe(CO)_4] \\ [HFe(CO)_4] + e^- &\rightarrow [HFe(CO)_4]^- \end{aligned} \quad (28)$$

Iron pentacarbonyl is also oxidized in THF by one electron (+1.5 V) to give a very unstable cation.[192] In a molten salt medium, however, the cation is more stable (+1.02 V vs. an Al reference), and there is evidence of reversibility.[195] More interesting, however, is the fact that, while $Fe(CO)_5$ and $M(CO)_6$ (M = Cr, Mo, W) are oxidized at almost identical potentials in MeCN, the oxidation of $Fe(CO)_5$ is considerably less difficult than the Group 6B carbonyls in the molten salt. Again, a specific interaction between $AlCl_3$ and CO was suggested.

Iron also forms the binary carbonyls $Fe_2(CO)_9$ and $Fe_3(CO)_{12}$. The former is quite insoluble in organic solvents, but Dessy *et al.*[7] reported that it was reduced by two electrons in DMF at the same potential as $Fe(CO)_5$. Bearing on this is an observation by Dawson *et al.*,[202] that when $Fe_2(CO)_9$ "dissolved" in THF, the solution was found to contain $Fe(CO)_5$ and $Fe_3(CO)_{12}$.

The electrochemistry of the triangular cluster $Fe_3(CO)_{12}$ has been reviewed very recently and was re-examined in a paper appearing at that same time.[25,200] In brief, the cathodic behavior of the molecule is complex, but the main processes appear to be

$$\begin{aligned} Fe_3(CO)_{12} &\rightarrow [Fe_3(CO)_{12}]^- \rightarrow [Fe_3(CO)_{12}]^{2-} \\ [Fe_3(CO)_{12}]^- &\rightarrow 2Fe(CO)_5 + \text{other carbonyls} \end{aligned} \quad (29)$$

Dicobalt octacarbonyl is the simplest binary carbonyl of cobalt(0). Again, as Co(0) has an odd number of electrons, the requirement for

18-valence electrons dictates a metal-metal bonded dimer. Its electrochemistry closely resembles that of the bimetallic Group 7B carbonyls. That is, reduction of $Co_2(CO)_8$ leads to cleavage of the metal-metal bond and formation of $[Co(CO)_4]^-$.[204]

The last element forming a simple binary carbonyl is nickel, but the electrochemistry of $Ni(CO)_4$ has only been briefly examined. Reduction occurred at −2.71 V to give an unstable purple species.[7,192]

4. METAL CARBONYL DERIVATIVES

The CO ligands of all of the simple binary metal carbonyls can be substituted by a wide variety of other donor ligands. These alternate ligands can be sigma donor ligands such as halide ions, amines, isonitriles (CNR), and especially phosphines (PR_3) and phosphites $[P(OR)_3]$. Carbonyls can also be replaced by pi donor ligands, such as olefins, $[\eta^5\text{-}C_5H_5]^-$, and arenes. Electrochemical studies have been done on all of these classes of compounds, especially those with phosphine substituents. We shall survey each of these types, with some emphasis on the correlation of trends in redox behavior as a function of ligand type. By and large, the organization of this section will be according to periodic group.

Again, we note that the electrochemistry of derivatives of metal carbonyl clusters has been recently reviewed.[25] Information on this subject will not be repeated herein unless particularly germane.

4.1. Metal Carbonyls with Sigma Donor Ligands $M(CO)_xL_y$ (L = Halide, Phosphine, Phosphite, Isonitrile)

The vast majority of the derivatives studied have been those based on phosphine substitution. Phosphine is apparently a better sigma base than CO, but a poorer pi acid, and this leads invariably to lower oxidation potentials for phosphine metal carbonyls than for the parent binary metal carbonyls. (This will be elaborated for specific cases below.) Further, it is generally observed that 17-valence electron phosphine-substituted metal carbonyls (that is, oxidized compounds) are kinetically more stable than oxidized parent binary carbonyls. This is especially true with compounds bearing bidentate phosphines, as indicated in the compounds described in the following paragraph. The reasons for this improved stability are not clear, but, as esr spectra of 17-valence electron phosphine metal carbonyls usually indicate substantial phosphorus involvement in the HOMO, it has been suggested that such ligands promote stability by allowing considerable delocalization of the unpaired electron onto the ligand phosphorus atoms.[205]

The simple binary carbonyls of tantalum and niobium are not known, but seven-coordinate phosphine carbonyls have been synthesized (**31**) and their electrochemistry examined.[205]

31

Both Ta(I) and Nb(I) derivatives are readily oxidized, reversibly, to stable 17-valence electron cations, the tantalum derivatives being slightly more readily oxidized than their niobium analogs. Both also can be oxidized to their 16-valence electron cations, but these are not stable in solution. On varying the X substituent in $Ta(CO)_2(DMPE)_2X$, it was observed that the $X = CH_3$ complex was more readily oxidized, as expected, than X = halide.

By far the greatest number of carbonyl derivatives studied have been Group 6B compounds, and a recent paper nicely illustrates the effect of substitution of CO in $Cr(CO)_6$ by a variety of sigma donor groups (Fig. 3).[206]

Owing to the fact that phosphines, amines, and nitriles are poorer pi acceptor ligands than CO, the reduction potentials of the complexes decline as CO is progressively replaced by these other ligands.[207–210] As Fig. 3 clearly shows, phosphines retain some pi acceptor ability, since E^0 declines less on substitution with PR_3 than on substitution with MeCN or pyridine. Fig. 3 also shows, as is generally observed, that for comparable substitution the reduction potentials of $[L_xM(CO)_{6-x}]^+$ are in the order Cr ≪ W ≤ Mo. As will be seen in Section 5.2 on oxidatively induced ligand substitution, Fig.

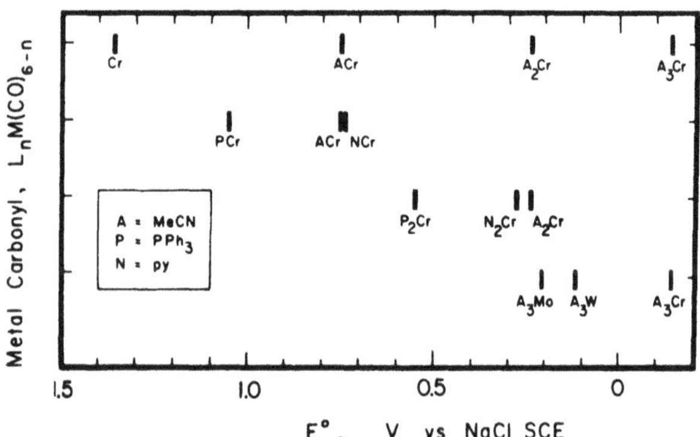

Figure 3. The effect of carbonyl group substitution on the reduction potential of $[M(CO)_{6-n}L_n]^+$ (M = Cr, Mo, W). J. W. Hershberger, R. J. Klingler, and J. K. Kochi, *J. Am. Chem. Soc.* **104**, 3034 (1982).

Fig. 3 provides a great deal of useful information. For example, the cation of a phosphine is predicted to act as an oxidizing agent toward a nitrile complex of the same type. That is,

$$[(PPh_3)_2W(CO)_4]^+ + (MeCN)_2W(CO)_4 \rightarrow (PPh_3)_2W(CO)_4 + [(MeCN)_2W(CO)_4]^+ \quad (30)$$

Figure 3 clearly shows the effect of CO substitution on the reduction potential and shows that it is an approximately linear effect. As more CO ligands are exchanged for nitrile, for example, the potential for oxidation declines by about 0.5 V. Such an effect has often been found, and it has been perhaps most thoroughly studied in the 18-valence electron isonitrile complexes of the type $M(CO)_{6-n}(CNR)_n(M = Cr, Mo)$ and $[M(CO)_{6-n}(CNR)_n]^+$ (M = Mn, Re).[211-219] For the series based on Mn(I), the change in $E_{1/2}$ is +0.33 to +0.53 V as CNMe is progressively exchanged for CO, that is, the ease of oxidation decreases as the number of CNMe ligands is increased. Molecular orbital calculations support the notion that this effect arises from the fact that CO is better able to stabilize the HOMO of the molecule, largely owing to the fact that CO is a better pi acceptor than CNMe.[215,216] Thus, $E_{1/2}$ is proportional to the HOMO energy as shown by the fact there is an excellent linear correlation between these two quantities for the Mn(I) complexes: the more negative the HOMO energy (lower energy), the harder the complex is to oxidize.

There have been several attempts to develop empirical equations to rationalize and predict redox potentials of simple organometallics of the type $[M(CO)_{6-x}L_x]^{y+}$. The first of these was [217]

$$E^0 = A + x(dE^0/dx) + 1.48y$$

where A is a constant dependent on the solvent and reference potential, (dE^0/dx) is a parameter defining the change in E^0 on substituting CO with L, and y is the charge on the molecule. The parameter A was found to be +1.12 V using MeCN as solvent and the ferrocene/ferricenium ion couple as reference, and ligand parameters were determined for a number of ligands. Values of (dE^0/dx) for some ligands are: +0.42 V for Br^-; −0.36 V for PPh_3, −0.44 V for CNMe, −0.54 V for MeCN, and −0.80 V for NH_3. In general, these parameters correlate with our current view of bonding in transition metal compounds. That is, electronegative Br^- should decrease the ease of oxidation of the complex; NH_3, a ligand not capable of functioning as a pi acceptor, leads to the most easily oxidized molecules, while molecules having some pi acceptor ability will be intermediate.

Another attempt at a correlation was recently published by Bursten who has developed a remarkably successful three-parameter equation.[216]

$$E_{1/2} = A + Bn + Cx_{\text{HOMO}}$$

where A, B, and C are empirically determined parameters, A depending on solvent and B and C depending on the ligand. The term x_{HOMO} is the number of ligands L that interact with the d_{pi} orbital that comprises the HOMO of the complex (the values range from 0 to 4). With this equation, Bursten showed that the general trend in $E_{1/2}$ for $ML_nI'_{6-n}$ complexes (where L is a better pi acceptor ligand than L') should be as illustrated in Fig. 4. The most interesting aspect of this plot is that the correlation is not quite linear: isomers are usually predicted to have different $E_{1/2}$ values owing to the x_{HOMO} term. Linear correlations are limited to the series $n =$ 0, 1, *trans*-2, and to *trans*-2, *mer*-3, *cis* or *trans*-4, 5, and 6. After determining the values of A, B, and C, Bursten calculated $E_{1/2}$ values for $[Mn(CO)_n(CNR)_{6-n}]^+$ (R = Ph, Me) and $Cr(CO)_n(CNMe)_{6-n}$ with remarkably good agreement with experiment. It is especially interesting that isomers are predicted to have different $E_{1/2}$ values, except in the case of the *cis* and *trans* isomers of $M(CO)_4L_2$. As discussed in more detail in Section 5.1, this prediction is essentially correct: the $E_{1/2}$ values for *cis*- and *trans*-$Mo(CO)_4[P(n\text{-}C_3H_7)_3]_2$ are nearly identical.[218]

Bond *et al.* have stated that "comparisons of electrochemical data . . . with almost every conceivable spectroscopic or other parameter . . . relevant to electrochemistry have been reported . . . ."[219] Indeed, these authors have reported a modest correlation between $E_{1/2}$ and CO ^{13}C nmr chemical shifts for a series of Cr(O) complexes, $Cr(CO)_5L$. No correlation was observed,

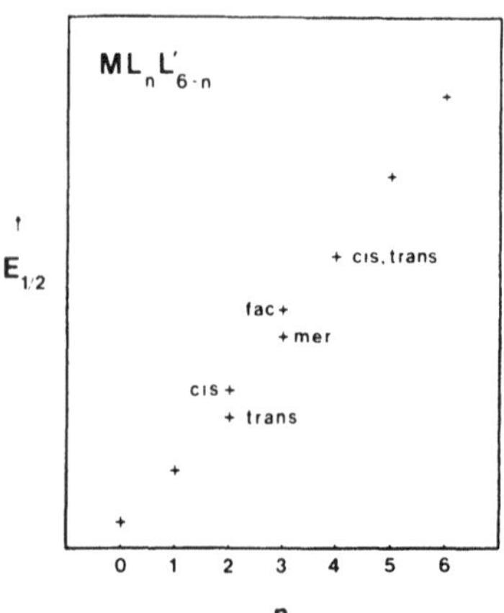

Figure 4. The predicted trend in the reduction potential for $[ML_nL'_{6-n}]^+$. B. E. Bursten, *J. Am. Chem. Soc.* **104**, 1299 (1982).

however, with ^{31}P chemical shifts. An excellent correlation was reported between $E_{1/2}$ and ν_{CO} in (**32**).[220] Other such correlations are described in Sections 4.2.2 and 4.2.3.

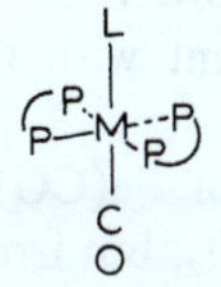

32 M = Mo

The oxidation of the series of complexes $Cr(CO)_5L$ is generally reversible at low temperatures or on a short time scale at room temperature. In the case of L = halide, the complexes decayed following one-electron oxidation to $Cr(CO)_5X$. It was proposed, however, that this Cr(I) species disproportionated to $[Cr(CO)_5X]^-$ and $[Cr(CO)_5X]^+$, the latter ion further decaying to other products. Such a disproportionation was of course also suggested for $[Cr(CO)_6]^+$ (Equation 22).[221]

Dessy and co-workers have examined the consequences of electrochemical reduction of an extensive series of Group 6B compounds of the type $LM(CO)_5$ and $L_2M(CO)_4$ where L was a mono- or bidentate amine (for example, pyridine and 2,2′-bipyridine).[17] Among other findings, they reported that the force constants for CO stretching were all lowered, but more for the CO groups trans to the ligands L than for those cis to it. This they construed as suggesting a "directional sigma [charge distribution] effect."

At the beginning of this section we described a seven-coordinate tantalum complex (**31**). The electrochemistry of the seven-coordinate Mo(II) and W(II) complexes $M(CO)_3(YPh_3)_2X_2$ (M = Mo, W; Y = P, As, Sb) has also been described.[222] The compounds all underwent irreversible two-electron reductions at Pt or Hg electrodes, but reaction 31 occurred in the presence of CO.

$$M(CO)_3(YPh_3)_2X_2 + 2CO + 2e^- \rightarrow [M(CO)_5X]^- + 2YPh_3 + X^- \quad (31)$$

Anodic oxidation of phosphine iron carbonyl complexes of the type LFe(CO)$_4$ and *trans*-L_2Fe(CO)$_3$, and the bidentate phosphine complex $(Ph_2PCH_2CH_2PPh_2)Fe(CO)_3$, have been examined in several solvents at both mercury and platinum electrodes.[203,223,224] Cations of the disubstituted complexes were found to be stable in CH_2Cl_2 (although not in acetone or MeCN) and could also be generated chemically using $AgPF_6$. (Using halogens as the oxidizing agent led to an "oxidative-elimination reaction" to give $FeI_2(CO)_3(PPh_3)$.[224]) (When L = $AsPh_3$ or $SbPh_3$, however, the

17-electron cations are unstable in all solvents tried.) In the case of *trans*-$L_2Fe(CO)_3$, the CO stretching frequency was linearly related to the oxidation potential, the CO stretch moving to lower frequencies as the complex became more easily oxidized in the order $L = P(OPh)_3 > PPh_3 > PPh_2Me > P(NMe_2)_3$.[223] This is in agreement with the expected increase in sigma basicity of the ligands.

As noted above, oxidation of $Fe(CO)_3(PPh_3)_2$ was reversible if performed at a Pt electrode in CH_2Cl_2, but irreversible behavior was found in acetone and MeCN.[225] At the mercury electrode, however, oxidation in MeCN was substantially more reversible, and a specific interaction with mercury was suggested. This is not an unreasonable suggestion, since addition of MeCN to a deep green solution of $[Fe(CO)_2(PPh_3)_3]^+$ in CH_2Cl_2 led to immediate decomposition of the cation with evolution of CO.

The bidentate ligand $Ph_2PCH_2CH_2PPh_2$ (often abbreviated diphos or dpe) often leads to interesting complexes, and several have already been mentioned above (for example, (**32**)). The electrochemistry of an extensive series of such complexes based on Fe(II), Co(II), and Ni(II) (**33**) has been

33 M = Fe^{2+}

reported. Zotti and co-workers have found that *trans*-$[Fe(dpe)_2(MeCN)_2]^{2+}$ can be electrochemically synthesized by anodic dissolution of an iron foil electrode in MeCN containing dpe.[226] This complex has a fascinating electrochemistry itself, but in the presence of CO, iron carbonyl complexes are formed which themselves have an extensive electrochemistry as outlined in the following scheme. (The symbol dpe* indicates that one of the dpe ligands in the complex is monodentate.)

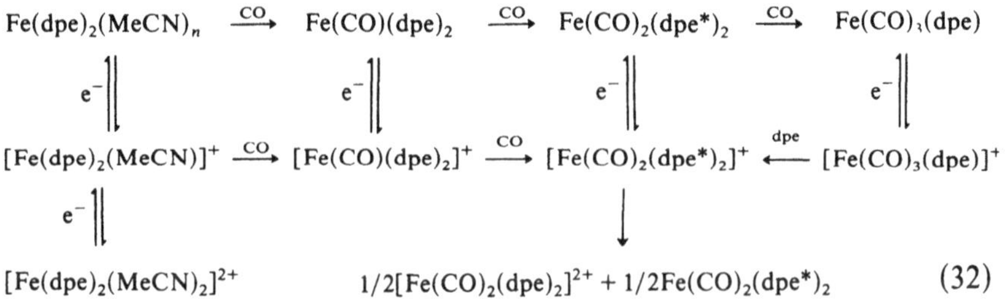

$$
\begin{array}{ccccccc}
Fe(dpe)_2(MeCN)_n & \xrightarrow{CO} & Fe(CO)(dpe)_2 & \xrightarrow{CO} & Fe(CO)_2(dpe^*)_2 & \xrightarrow{CO} & Fe(CO)_3(dpe) \\
e^- \Updownarrow & & e^- \Updownarrow & & e^- \Updownarrow & & e^- \Updownarrow \\
[Fe(dpe)_2(MeCN)]^+ & \xrightarrow{CO} & [Fe(CO)(dpe)_2]^+ & \xrightarrow{CO} & [Fe(CO)_2(dpe^*)_2]^+ & \xleftarrow{dpe} & [Fe(CO)_3(dpe)]^+ \\
e^- \Updownarrow & & & & \downarrow & & \\
[Fe(dpe)_2(MeCN)_2]^{2+} & & & & 1/2[Fe(CO)_2(dpe)_2]^{2+} + 1/2Fe(CO)_2(dpe^*)_2 & &
\end{array}
\qquad (32)
$$

Compounds based on cobalt, rhodium, and iridium have been widely used as catalysts or catalyst precursors for organic reactions. As these compounds function by what is thought to be a sequence of oxidation-reduction reactions, it is not surprising that some electrochemical studies have been published on such compounds.

A very interesting study has been done on the reduction of $[Co(CO)_3L)]_2$ and $Hg[Co(CO)_3L]_2$ as $L(= CO, PR_3$ or $P(OR)_3)$ was systematically varied.[204] The compounds were found to reduce by two electrons at a mercury electrode. (Potentials measured in THF vs. Ag/$AgClO_4$.)

$$[Co(CO)_3L]_2 + 2e^- \rightarrow 2[Co(CO)_3L]^- \quad E_{1/2} = -0.75 \text{ V for } L = CO \tag{33}$$

$$Hg[Co(CO)_3L]_2 + 2e^- \rightarrow 2[Co(CO)_3L]^- + Hg$$

$$E_{1/2} = -0.96 \text{ V for } L = CO$$

The reduction potential was observed to be remarkably well correlated with the basicity of L as measured by the method of Streuli.[227] The potentials were always more cathodic than when $L = CO$, and the most cathodic potentials were observed for $L = PEt_3$ and PMe_3 and the least cathodic for $L = P(OPh)_3$. (Reduction of the mercury compound was consistently about 0.2 V more cathodic than reduction of $[Co(CO)_3L]_2$.) Since the method of basicity determination measures only the sigma donor capacity of the phosphine, the $E_{1/2}$ vs. basicity correlation suggests that the reduction potentials are determined solely by ligand sigma donor effects. In contrast, plots of the ν_{CO} vs. $E_{1/2}$ were slightly curved, suggesting that the stretching frequency is influenced, not only by ligand sigma donor effect, but also by the ligand pi acceptor ability.

Reduction of the Co(I) cation $[Co(CO)_2(PMe_3)_3]^+$ at a mercury electrode occurred by two electrons and gave, not a mercury derivatives, but the Co-Co bonded dimer $Co_2(CO)_4(PMe_3)_4$ which itself was reducible at more cathodic potentials.[228] The sequence of reactions to give this product was thought to be

$$[Co(CO)_2L_3]^+ + 2e^- \rightarrow [Co(CO)_2L_2]^- + L$$

$$[Co(CO)_2L_2]^- + [Co(CO)_2L_3]^+ \rightarrow Co_2(CO)_4L_4 + L \tag{34}$$

$$Co_2(CO)_4L_4 + 2e^- \rightarrow 2[Co(CO)_2L_2]^-$$

An Ir(I) compound which has played a key role in the development of our ideas of the reactivity of low-valent metal centers is $Ir(CO)Cl(PPh_3)_2$, commonly known as Vaska's compound.[2,41] As the compound readily undergoes so-called "oxidative addition" reactions with a variety of electrophiles, the electrochemistry of compounds of this class based both on Ir(I) and Rh(I) has been studied by electrochemistry.

Compounds of the type $Ir(CO)XL_2$ were observed to undergo irreversible anodic oxidation in the range of 1 V.[229] Some variation in potential was observed, as expected, with different halides X and phosphines L. Reduction in CH_2Cl_2 of an oxidative addition product of Vaska's compound with mercuric halides, $IrCl(X)(CO)L_2HgX$ was thought to occur by initial one-electron reduction followed by a disproportion of the initial product.

$$\begin{aligned} IrCl(X)(CO)L_2HgX + e^- &\rightarrow IrCl(CO)L_2HgX + X^- \\ 2IrCl(CO)L_2HgX &\rightarrow IrCl(CO)L_2 + Hg + IrCl(X)(CO)L_2HgX \end{aligned} \quad (35)$$

Reductions of Vaska's compound and its rhodium(I) analog were found to be two-electron processes in THF, and there was evidence that the final product was a carbonyl bridged rhodium dimer.[228]

$$\begin{aligned} RhCl(CO)L_2 + 2e^- &\rightarrow [Rh(S)(CO)L_2]^- + Cl^- \quad (S = \text{solvent} = \text{THF}) \\ [Rh(S)(CO)L_2]^- + RhCl(CO)L_2 &\rightarrow [Rh(S)(CO)L_2]_2 + Cl^- \end{aligned} \quad (36)$$

As is so often the case, the compounds based on the third row transition metal, iridium(I), were reduced at a more negative potential (about 200 mV) than those of the second row element rhodium(I).

The cathodic electrochemistry of $[M(CO)(PPh_3)_3]^+$ (M = Rh, Ir) is closely related to that of $MCl(CO)L_2$.[230] Both are reduced in two, one-electron, reversible steps in dimethoxyethane (DME), the potentials for both complexes being about −1.35 V and −1.5 V. Reduction at a potential between the two potentials gave $Rh(CO)(PPh_3)_3$ which led, by a route similar to Equation 36, to the carbonyl bridged dimer where S was DME. Electrolysis at potentials more cathodic than the second wave, in the presence of excess PPh_3, gave the d^{10}, Rh(-I) complex, $[Rh(CO)(PPh_3)_3]^-$, an ion isoelectronic with $Ni(CO)(PPh_3)_3$. The overall electrochemistry is as follows:

$$[Rh(CO)(PPh_3)_3]^+ \xrightarrow{e^-} Rh(CO)(PPh_3)_3 \xrightarrow{e^-} [Rh(CO)(PPh_3)_3]^- \quad (37)$$
$$Rh(CO)(PPh_3)_3 \xrightarrow[-\,PPh_3]{+\,DME} 1/2[Rh(CO)(PPh_3)_2(DME)]_2$$

The Ir(I) complex was slightly different from the Rh(I) complex in that no dimer formation was observed. Further, the one-electron reduction product, $Ir(CO)(PPh_3)_3$, was extremely reactive, giving $Ir(OH)(CO)(PPh_3)_2$ and $HIr(CO)(PPh_3)_3$, if traces of water were present.

A very thorough study of the electrochemistry of $HIr(CO)(PPh_3)_3$ and its Rh(I) analog have also been made in the mixed solvent MeCN/toluene.[231] Both complexes were reversibly oxidized by one electron [$E_{1/2} = -0.30$ V vs. $Ag/AgClO_4$ for Ir(I)] and irreversibly by a second ($E_{1/2} = -0.05$ V). For the Ir(I) complex the electrochemistry could be described as in Equation 38.

$$HIr(CO)(PPh_3)_3 \xleftarrow[-MeCN]{+2e^-} [HIr(CO)(PPh_3)_3(MeCN)]^{2+}$$

$$HIr(CO)(PPh_3)_3 \underset{+e^-}{\overset{-e^-}{\rightleftharpoons}} [HIr(CO)(PPh_3)_3]^+ \xrightarrow[+MeCN]{-e^-} [HIr(CO)(PPh_3)_3(MeCN)]^{2+}$$

$$[HIr(CO)(PPh_3)_3]^+ \xrightarrow{+MeCN} 1/2HIr(CO)(PPh_3)_3 + 1/2[HIr(CO)(PPh_3)_3(MeCN)]^{2+} \quad (38)$$

The electrochemistry of the Rh(I) analog is basically the same. The main difference between the two metals comes in the fact that the oxidized Rh(I) species both lose CO.

Finally, in the cobalt sub-group, the very interesting five-coordinate complexes $[M(CO)(dpe)_2]^+$ [M = Co(I), Rh(I), Ir(I)] have been examined.[232] In CO-saturated acetonitrile, all undergo a two-electron reduction as follows.

$$[(L\text{-}L)_2M(CO)]^{\oplus} \xrightarrow{+2e^-} (L\text{-}L)M(CO)(L\text{-}L^{\ominus}) \xrightarrow{+CO} [(L\text{-}L)M(CO)_2]^{\ominus} + L-L \quad (39)$$

The last element forming simple binary carbonyls is nickel, and the electrochemistry of the complete series of complexes $Ni(CO)_{4-n}L_n$ [L = PPh_3 or $P(OPh)_3$] has been studied.[204] The parent nickel tetracarbonyl was irreversibly reduced by one electron in THF at -2.80 V (vs. $Ag/AgClO_4$). As expected the reduction potential became more cathodic, and the ν_{CO} values moved to lower energy, on adding phosphite ligands.

4.2. Compounds with Pi Donor Substituents

The carbon monoxide ligands of binary metal carbonyls may be replaced with pi donor ligands, such as ethylene or acetylene (two-electron donor), conjugated and nonconjugated dienes or the allyl ion (four-electron donors), or the η^5-cyclopentadienyl ion or benzene (six-electron donors). Such complexes are widely observed in organometallic chemistry, both as stable species and as intermediates in metal-catalyzed processes. To organize the material on the electrochemistry of such compounds, we shall divide it into sections on (i) olefin and allyl complexes, (ii) cyclopentadienyl complexes, (iii) arene complexes, and (iv) other six-electron pi donors, such as the cyclobutadiene dianion and the cycloheptatrienyl cation.

4.2.1. Compounds with Olefin and Allyl Ligands

The main feature of the electrochemistry of complexes of this type is their reactivity after electron transfer. A variety of reactions occur, including structural rearrangements discussed in Section 5.1.

Complexes of η^3-allyl and its derivatives are known with most transition metals, and the polarographic behavior of at least complexes with Co(I) and Fe(II) has been examined.[233,234] In general, complexes such as $\eta^3\text{-}C_3H_5Co(CO)_2L$ (L = CO or PR_3) and $\eta^3\text{-}C_3H_5Fe(CO)_2NO$ are reduced in a single, irreversible, two-electron step at quite cathodic potentials (−1.60 V for Co(I) with L = CO). Coulometric analysis, however, gave a net of one mole of electrons per mole of compound, and propene and $[Co(CO)_4]^-$ or $[Fe(CO)_3NO]^-$, among others, were observed as products. These results and others were interpreted in terms of a reaction sequence, such as

$$(\eta^3\text{-}C_3H_5)Co(CO)_2L \xrightarrow[+H^+]{+2e^-} [Co(CO)_2L]^- + C_3H_6 \qquad (40)$$

$$[Co(CO)_2L]^- + (\eta^3\text{-}C_3H_5)Co(CO)_2L \begin{cases} \nearrow [Co(CO)_3L]^- + \text{propene} + ? \\ \searrow [Co(CO)_4]^- + \text{propene} + ? \end{cases}$$

Electrochemical reduction of benzylideneacetonetricarbonyliron gave rise to a rich series of reactions.[235] Two reversible, one-electron reductions in THF produced first a green radical anion and second a yellow dianion. The radical anion slowly decomposed only in solution to yield the free ligand enone and $[Fe(CO)_3(\text{solvent})]^-$. The latter was oxidized by tropylium cation to yield an iron carbonyl complex of coupled cycloheptatriene rings.

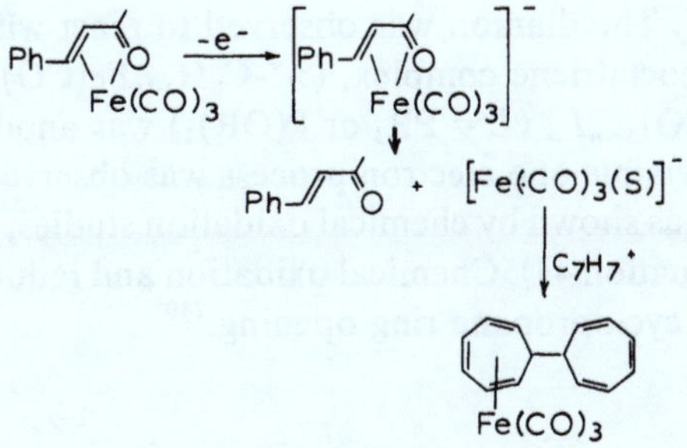

(41)

If the enone iron tricarbonyl was reduced in the presence of crotyl bromide, a quantitative yield of the bisallyl iron dicarbonyl complex was obtained.

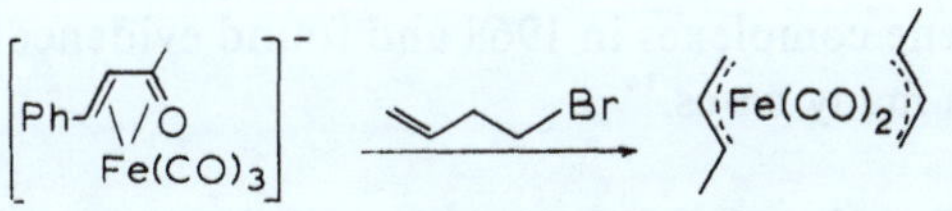

(42)

A preliminary report of the electrochemistry of the acetylene complex (**34**) suggests that it will have a rich chemistry.[236] Four molecular oxidation states were reversibly accessible, and electrolysis after E_{pa} of the first wave gave the mono-cation, which reacted readily with a wide variety of sigma or pi donor molecules (for example, dpe, PPh_3, cyclooctadiene) to give $[(\eta^5\text{-}C_5Ph_5)PdL_2]^+$.

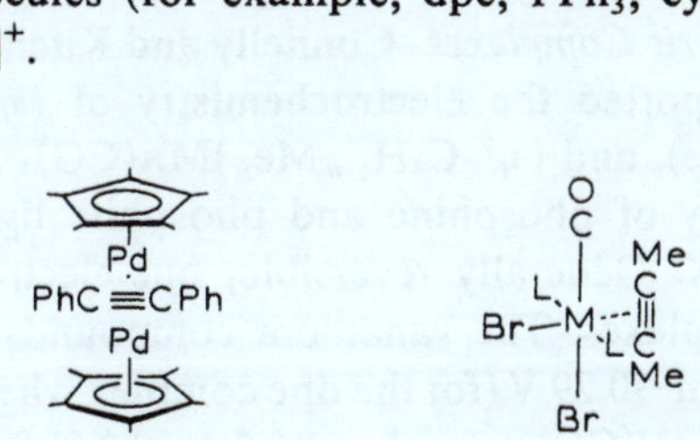

34 **35** M = Mo L = P(Et)$_3$

In yet another preliminary report, compound (**35**) was reported to undergo reduction at −0.99 V and to show the onset of oxidative processes at +0.87 V.[237]

Several groups have examined the electrochemistry of cyclooctatetraene (COT) metal carbonyl complexes and the ligand reactivity following electron transfer.[238–240] (See Section 5.1 for olefinic ligand isomerizations induced by electron-electron transfer.) For example, $COTFe(CO)_3$ was reduced in DMF by two, one-electron steps to give $[COTFe(CO)_3]^-$ and $[COTFe(CO)_3]^{2-}$. The dianion was observed to react with protons to produce an η^6-cyclooctatriene complex, $(\eta^6\text{-}C_8H_{10})Fe(CO)_3$.

If $COTFe(CO)_{3-n}L_n$ ($L = PR_3$ or $P(OR)_3$) was anodically oxidized in CH_2Cl_2, an irreversible one-electron process was observed.[239,240] The result of this oxidation, as shown by chemical oxidation studies, was stereospecific dimerization (Equation 43). Chemical oxidation and reduction of this dimer led ultimately to cyclopropane ring opening.[240]

2 FeL3 —(-2 e⁻)→ [FeL3 FeL3]2+ (43)

Finally, we note that Dessy and Pohl surveyed the electrochemistry of 35 metal-acetylene complexes in 1968 and found evidence for stable mono- and dianions in many cases.[14]

4.2.2. Cyclopentadienyl Metal Carbonyls and Related Compounds

Electrochemistry of a variety of compounds of the type $CpM(CO)_xL_y$, where L can be a ligand, such as a phosphine, olefin, acetylene, NO, etc., have been reported. In addition, extensive work has been done on various sulfur-bridged cyclopentadienyl metal compounds.[7,8]

4.2.2a. Monomeric Complexes. Connelly and Kitchen,[241,242] as well as others,[243-245] have reported the electrochemistry of $(\eta^5\text{-}C_5H_5)V(CO)_3L$, $(\eta^5\text{-}C_5H_5)V(CO)_2(\text{dpe})$, and $(\eta^5\text{-}C_5H_{5-n}Me_n)Mn(CO)_{3-x}L_x$ (where L represents a wide variety of phosphine and phosphite ligands), as well as $(\eta^5\text{-}C_5H_5)Fe(CNR)_2X$. Generally reversible, one-electron oxidation was observed for all complexes. The vanadium compounds were oxidized at potentials ranging from -0.29 V (for the dpe complex where two CO ligands were substituted in $CpV(CO)_4$) to about 0.2 to 0.3 V for complexes with only one CO substituted by a phosphine. The Mn(I) compounds were oxidized over a greater range: from -0.28 V where $L = PMePh_2$, $x = 2$, $n = 1$ to 0.92 V for $L = P(OPh)_3$ and $x = 1$, $n = 0$. In general, the trend in redox potentials was as expected on the basis of the known sigma basicity/pi acidity properties of the various ligands L. Of course, the potential was most dependent on x, the number of L ligands attached, changing to less anodic potentials by about 800 mV on going from one PPh_3 to two, for example, in the V(I) or Mn(II) complexes. Replacing a cyclopentadienyl H atom by a methyl group shifted the potential only about -40 mV. Finally, the Mn(I) series allowed a very extensive effort to correlate redox potentials with carbonyl stretching frequencies. It was found that, although there was a reasonable correlation between ν_{CO} and E^0 for a limited series of compounds, a much better correlation existed for E^0 vs. the CO-stretching force constant.[242]

In the series of Mn(I) compounds described above, the parent compound $CpMn(CO)_3$ was only irreversibly oxidized in CH_2Cl_2, but reversible behavior was reported in trifluoroacetic acid.[242,244] It was noted in Section 3 that this solvent, of very low nucleophilicity, allowed the observation of otherwise reactive binary metal carbonyl cations.[194] In Section 4.2.3 we describe the use of this solvent to stabilize $[(arene)Cr(CO)_3]^+$ cations.[246]

Closely related to the Mn(I) carbonyls is an extensive series of cyclopentadienylmanganese(I) nitrosyl complexes such as (**36-38**) and related (**39**).[247,248] complexes such as (**36-38**) are all reversibly oxidized. Positively charged complexes (for example, **36**) have potentials from +0.84 to +1.60 V, neutral complexes (for example, **37**) oxidize in the range +0.41 to +0.54 V, and anionic complexes (**38**) are oxidized at −0.51 to +0.20 V. A crude relationship between ν_{NO} and $E_{1/2}$, reminiscent of that for $[Co(CO)_2L(NO)]$,[249] was also observed.

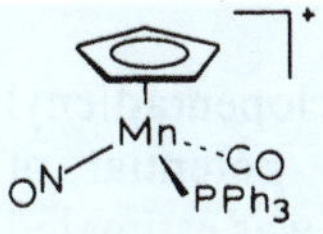

36 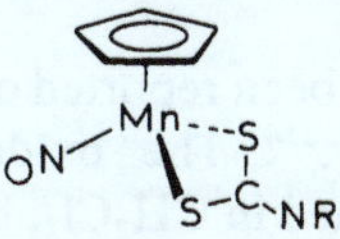 **37** 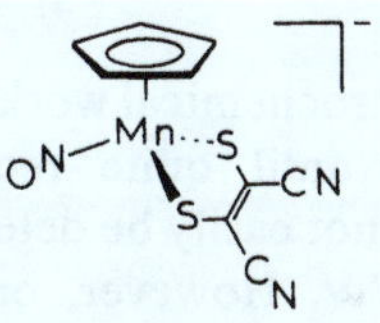 **38** 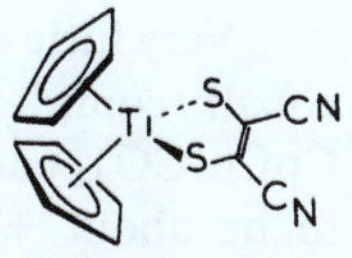 **39**

Cyclopentadienyliron carbonyl complexes are readily prepared, and some electrochemical work has been reported on them. On such series of compounds, $(\eta^5\text{-}C_5H_5)Fe(CO)LR$ has been used to study one of the most important reactions in organometallic chemistry: the cleavage of metal-carbon bonds. Baird and co-workers, among others, have thoroughly studied oxidatively induced Fe-C bond cleavage in this class of compounds, and some electrochemical measurements have been made.[250,251] The reaction is described in Section 5.3 on the general aspects of electron-transfer-induced bond cleavage reactions.

Quasi-reversible, one-electron oxidations have also been observed in CH_2Cl_2 for a lengthy series of complexes based on CpFe(dpe)X (where X = Cl, Br, I, H, Me, $SnMe_3$, CN, SCN, and SPh).[252] There was no apparent correlation of $E_{1/2}$ with the type of ligands involved, but it was observed that the oxidation potential was greatest for the good pi acceptor ligands CN^- (+0.535 V) and $SnCl_3^-$ (+0.90 V). Most of the complexes were oxidized in the range from −0.2 to +0.2 V. As a consequence it was found that $AgPF_6$ was an appropriate chemical oxidant, and many of the mono-cations could be isolated. Magnetic measurements showed the presence of one unpaired electron.

In Section 2.3, we noted a complex that arose from Hoffmann's suggestion that CH^+ could be replaced by an "isolobal" $Fe(CO)_3$ (Structure **29**). Thus, it has been possible to synthesize compounds similar to cyclopentadienyliron carbonyls, that is, the ferroles (**40–42**).[253] Ferrole (**40**) was reduced at about −1.30 V (vs. Ag/AgCl) by one electron in an electrode reversible process. In contrast, both of the benzoferroles underwent two-electron reductions. The unsymmetrical complex (**41**) was reduced in THF in two steps (−0.89 V and −1.30 V), the first one reversible, while (**42**) was reduced in a single, two-electron step (−1.14 V). The di-anion of the unsymmetrical complex (**41**) was unstable, while that of the symmetrical ferrole (**42**) was stable in solution. Molecular orbital arguments have been advanced to rationalize the stability differences of the reduced complexes.

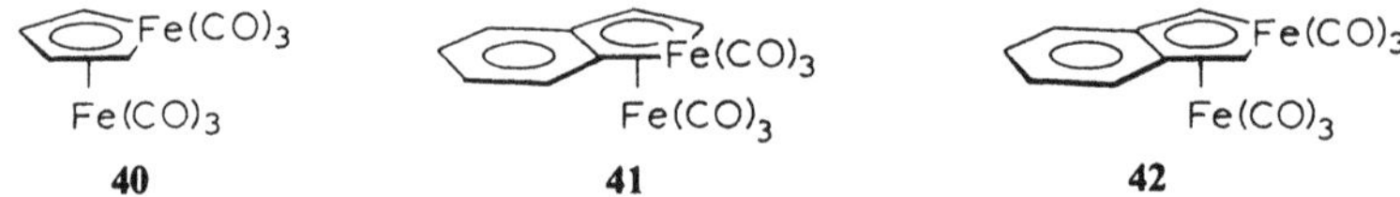

Very little electrochemical work had been reported on cyclopentadienyl cobalt complexes until quite recently.[254] The oxidation potential of $CpCo(CO)_2$ could not easily be determined in CH_2Cl_2, but it was estimated to be about +0.97 V. However, on substitution of one CO by a phosphine, the oxidation potential dropped drastically; E^0 for reduction of $[CpCo(CO)L]^+$ was about 0 V, the exact value depending on L. Replacement of both CO ligands with PPh_3 led to a complex with $E^0 = -0.71$ V (in CH_2Cl_2). Both electrochemical steps were highly reversible. The most interesting aspect of these cobalt complexes was the observation that, even though the CO ligand in $CpCo(CO)L$ could not be replaced thermally or photolytically, it was readily replaced in $[CpCo(CO)L]^+$ by PR_3. Apparently oxidation of cobalt to Co(II) weakens the Co-CO bond, and substitution is facilitated. This was also illustrated by the observation that, although $CpCo(CO)_2$ was not oxidized by the ferricenium ion, as expected from their relative redox potentials, rapid formation of $[CpCoL_2]^+$ occurred when PPh_3 was added to the mixture of $CpCo(CO)_2$ and $[Cp_2Fe]^+$. Apparently enough $CpCo(CO)L$ was formed thermally, and then oxidized by ferricenium, that the equilibria involved were forced to favor product formation.

There have been two reports of the electrochemistry of Ni(II) complexes of the type $[CpNiL_2]^+$ (L = phosphines and phosphites; L_2 = cyclooctadiene or norbornadiene).[255,256] Cyclic voltammetry of $[CpNi(PPh_3)_2]^+$ showed a cathodic wave at $E_{pc} = -1.25$ V and an anodic wave for a product of reduction at -0.35 V (vs. Ag/AgCl), the latter corresponding to nickelocene. A solution of Cp_2Ni and $Ni(PPh_3)_4$ exhibited the same behavior, showing that the 17-electron Ni(I) species $CpNi(PPh_3)_2$ must be disproportionate to nickelocene and NiL_4. Not surprisingly, if a large excess of PPh_3 were added to $[CpNi(PPh_3)_2]^+$, the reduction of the cation became reversible.

4.2.2b. Dimeric Complexes. The remaining cyclopentadienylmetal complexes to be described in this section are dimers. In his early series of papers on organometallic electrochemistry, Dessy and his co-workers surveyed the behavior of 26 different bridged bimetallic complexes.[15] A number of these have since been studied in more detail.

Among the best characterized in terms of their electrochemistry are $Cp(CO)Fe(\mu\text{-}CO)_2Fe(CO)Cp$ (**43**) and the closely related compounds having bridging phosphines (**44**).[257–259] Compound (**43**) was irreversibly oxidized by two electrons in all solvent media; using acetonitrile, the complex $[CpFe(CO)_2(NCMe)]^+$ was isolated following electrolysis at +0.7 V.

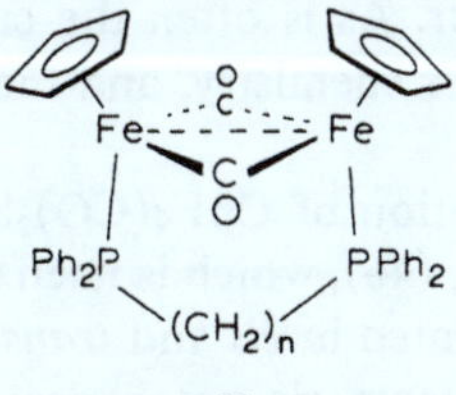

Fe C O C O Fe Ph2P PPh2 (CH2)n

44

In contrast with (**43**), oxidation in a variety of solvents of the phosphine bridged di-iron compound (**44**) occurred in two steps, the first one reversible. ($E_{1/2}$ was in the range -0.05 to $+0.10$ V, depending on the solvent.) The second one-electron oxidation (at ca. +0.9 V) was irreversible, and the dication decomposed by both symmetrical and unsymmetrical routes. Reduction of the product of the unsymmetrical route gave free diphosphine and the di-iron species $[CpFe(CO)_2]_2$ (**43**).

$$\mathbf{44} \xrightarrow{e-} [\mathbf{44}]^+ \xrightarrow{e-} [\mathbf{44}]^{2+} \qquad (44)$$

$$[44]^{2+} \rightarrow [CpFe(CO)_2(P\text{-}P)]^+$$

$$+ [Cp(CO)(MeCN)Fe(P\text{-}P)Fe(NCMe)(CO)Cp]^{2+} + Fe^{2+}$$

$$[CpFe(CO)_2(P\text{-}P)]^+ + e^- \rightarrow P\text{-}P + 1/2[CpFe(CO)_2]_2$$

It is worth noting that MO calculations have been done to probe the electronic structure of (**44**).[260] They show that the HOMO is not an Fe-Fe bond; rather, it is an Fe-Fe nonbonding MO. It is thought that oxidation results in weakening the iron-bridge group (Fe-CO or Fe-P) bonds.

Compounds based on structure (**45**) can exist in a range of oxidation states, which can be electrochemically interconverted.[261] When both bridges are NO, molecular oxidation states, z, from -1 to $+2$ are electrochemically accessible for the dimer with pentamethylcyclopentadienyl groups. The $+1$ and 0 states are electrochemically interconverted for the dimers with two CO bridges or with one NO and one CO.

X and Y = CO, NO

45

The remainder of the dimeric complexes to be described are bridged by thiolate anions or sulfur. As is often the case, metal-sulfur compunds have a rich electron transfer chemistry, and these complexes illustrate this point.[15]

Electrochemical oxidation of $CpFe(CO)_2SR$ results in the formation of a thiolate bridged dimer, (**46**), which is itself electroactive.[262,263] Interestingly, the dimer can be isolated in *cis* and *trans* forms which have different E^0s. (See Section 5.1 for isomeric octahedral complexes.) For example, *cis*-$[Cp(CO)Fe(\mu\text{-SMe})]_2^+$ is reversibly reduced to the zero state at -0.40 V, while the *trans* form is reduced at -0.35 V (potentials vs. $Ag/AgClO_4$). Furthermore, the electron transfer reactions of *cis*- or *trans*-$[Cp(CO)Fe(\mu\text{-SR})]_2^z$ ($z = 0, +1, +2$) are stereoselective; no isomerization was observed on changing the oxidation state. A second, quasi-reversible oxidation of (**46**) was observed at about $+0.5$ to $+0.7$ V; in MeCN, the carbonyl ligands were replaced in the dication to yield $[CpFe(NCMe)(\mu\text{-SR})]_2^{2+}$.

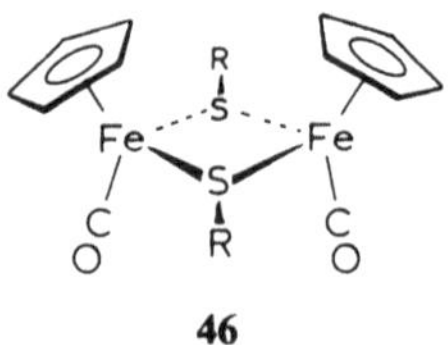

46

An important adjunct to the electrochemical study of (**46**) is the structural work on the neutral and mono-cationic complexes. The most striking finding was that the Fe-Fe distance decreased from a nonbonding distance of 339 pm in (**46**) to 292.5 pm in $[\mathbf{46}]^+$, implying that the electron had been removed from a molecular orbital that is antibonding with respect to the iron atoms.[264]

A number of thiolate-bridged nickel dimers, $CpNi(\mu\text{-}SR)_2NiCp$, have been studied.[15,263] All underwent oxidation to the +1 form at about 0 V and to the +2 form at about +1.0 V. As is often observed, the first redox process was reversible, while the second was irreversible. Only the $+1 \leftrightarrow 0$ redox process was observed for the analogous dicobalt dimers, but very few compounds as such were surveyed.

In an attempt to synthesize a Mn(I) compound similar to (**46**), $[CpMn(CO)_2(SEt)]^-$ was chemically oxidized.[265] However, the product was $[CpMn(CO)_2]_2(\mu\text{-}SEt)^+$, a cation formally based on Mn(I) and having a bridging thiolate and an Mn-Mn bond. The cation could be cathodically reduced in two, well-behaved, one-electron steps to an anion in which an Mn-Mn bond presumably no longer exists.

Yet another type of thiolate bridged metal dimer is $CpMo(SC_2H_4S)_2MoCp$.[266] The sulfur atoms of each dithiolate are both associated with both molybdenum centers, and the S-C-C-S plane is thus perpendicular to the Mo-Mo axis. Three molecular oxidation states are electrochemically accessible, several compounds of this type being all anodically oxidized at +0.13 V and +0.79 V in a reversible manner. The chemistry of these dimers is also quite noteworthy; for example, acetylene abstracts hydrogen from the bridging dithiolate to give ethylene and leave an unsaturated dithiolate bridge.

A very thorough structural and electrochemical study of the dimer $Cp_2Fe_2(S_2)(SEt)_2$ and the related tetranuclear Fe-S "cubane" $Cp_4Fe_4S_6$ (Fig. 5) has been reported.[267,268] As is often the case with multi-metal complexes, several molecular oxidation states are readily accessible. The dimer electrochemistry in MeCN is outlined in Equation 45.

$$Cp_2Fe_2(S_2)(SEt)_2 \xrightarrow[-1.2\text{ V}]{e^-} \text{irreversible reduction product} \quad (45)$$

$$Cp_2Fe_2(S_2)(SEt)_2 \xrightarrow{+0.21\text{ V}} [Cp_2Fe_2(S_2)(SEt)_2]^+ \xrightarrow{+0.90} [Cp_2Fe_2(S_2)(SEt)_2]^{2+}$$

$$\xrightarrow[+MeCN]{-S_2} [CPFe(NCMe)(SEt)]_2^{2+}$$

Notice that the dication loses S_2 (the fate of which is unknown) to give a

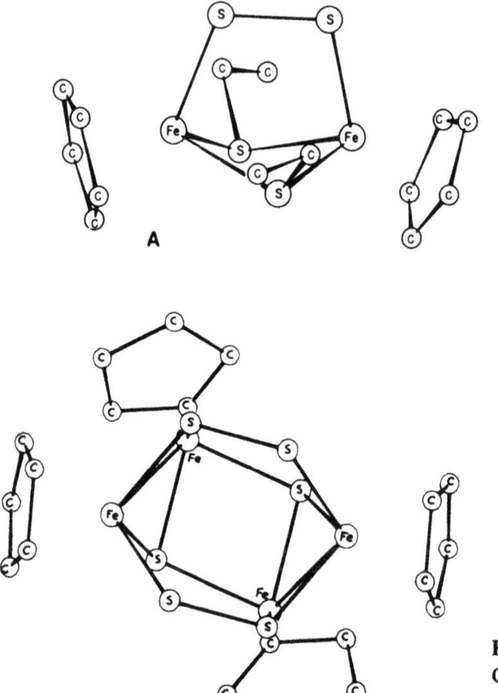

Figure 5. Molecular structures of $Cp_2Fe_2(S_2)(SEt)_2$ (**A**) and $Cp_4Fe_4S_6$ (**B**) G. J. Kubas and P. J. Vergamini, *Inorg. Chem.* **20**, 2667 (1981).

thiolate bridged dimer with MeCN ligands, just as in the case of $[Cp(CO)Fe(SR)]_2^{2+}$.

The tetra-iron species (Fig. 5) also has a rich electrochemistry as outlined in Equation 46. Many of the electron trasfer reactions are electrochemically reversible.

$$[Cp_4Fe_4S_6]^- \xrightarrow{-1.325\text{ V}} Cp_4Fe_4S_6 \xrightarrow{+0.06\text{ V}} [Cp_4Fe_4S_6]^+ \xrightarrow{+0.367\text{ V}} [Cp_4Fe_4S_6]^{2+} \xrightarrow{-S} [Cp_4Fe_4S_5]^{2+} \qquad (46)$$

The ultimate product of this redox sequence, $[Cp_4Fe_4S_5]^{2+}$, also has an extensive electrochemistry.

$$Cp_4Fe_4S_5 \xrightarrow{-0.28\text{ V}} [Cp_4Fe_4S_5]^+ \xrightarrow{0.02\text{ V}} [Cp_4Fe_4S_5]^{2+} \xrightarrow{1.14\text{ V}} [Cp_4Fe_4S_5]^{3+}$$

This very brief survey of sulfur-containing bimetallic complexes (and a tetra-iron complex) suggests that such compounds will generally have a very extensive electrochemistry, and future work in this area should be quite profitable.

4.2.3. Arene Metal Carbonyls

As was largely true in the case of *bis*(arene)metal compounds, all of the arenemetal carbonyl compounds studied have been based on Group 6B metals, that is, $(\eta^6\text{-arene})M(CO)_{3-n}L_n$ where M = Cr, Mo, or W, and L is a sigma donor ligand, such as PR_3, $P(OR)_3$, or an acetylene.

Rieke and his co-workers have thoroughly examined a number of arenechromium tricarbonyl complexes and have found that their electrochemical behavior depends critically on the nature of the arene and its substituents and on the solvent.[269-271] For example, reduction of $(\eta^6\text{-toluene})Cr(CO)_3$ in MeCN occurred by two electrons in a totally irreversible manner ($E_{1/2} = -2.25$ V).[269] In contrast, the two-electron reduction of (naphthalene)$Cr(CO)_3$ was at least quasi-reversible ($E_{1/2} = 1.66$ V). It is usually thought that, in two-electron reductions, electron transfer occurs in two separate steps, E^0 of the second reduction step being anodic of E^0 of the first. This requires an ECE mechanism, and Rieke suggested that the initial one-electron reduction led to an electronic rearrangement of the ligand (for example, naphthalene) in such a way that the second reduction step occurred more readily than the first.

Controlled potential electrolysis of (toluene)$Cr(CO)_3$ and similar compounds on the plateau of the reduction wave gave a solution having an oxidation wave anodic of the wave for the original complex.[269] Oxidation at this potential regenerated the original complex. It was thought that this behavior arose from protonation of the di-anion, the original complex being regenerated on oxidation. For (naphthalene)$Cr(CO)_3$ and analogous compounds, the di-anion is persistent, and the starting material can be recovered on oxidation.

If an unsymmetrical naphthalene, 2,3-dimethylnaphthalene, is used as a ligand, two isomers are possible for (arene)$Cr(CO)_3$. Rieke *et al.* isolated both isomers and found that no rearrangement occurred on exhaustive reduction.[269]

In sharp contrast with the two-electron reduction observed for (arene)$Cr(CO)_3$, compounds, such as (biphenyl)$[Cr(CO)_3]_2$, where a $Cr(CO)_3$ moiety is coordinated to each ring, are reduced by one electron per Cr atom (at potentials ca. 300 mV less cathodic than their mono-$Cr(CO)_3$ analogs) to give a very persistent dianion.[270] (For (biphenyl)$[Cr(CO)_3]_2$ $E_{1/2} = -1.606$ V and $E_{3/4} - E_{1/4} = 30$ mV.) The CO stretching modes moved to considerably lower frequencies, suggesting that the chromium atoms are approaching at formal −1 oxidation state. Again, an ECE mechansim (Equation 47) was proposed in order that E_2^0 could be less cathodic than E_1^0.

$$[C_6H_5Cr(CO)_3]_2 \xrightarrow[E_1]{+e^-} [C_6H_5Cr(CO)_3]_2^- \longrightarrow \xrightarrow[E_2]{+e^-} \quad (47)$$

The oxidative electrochemistry of arene$Cr(CO)_{3-n}L_n$ has been examined by several groups.[272–274] When $L = CO$, quasi-reversible behavior for the one-electron oxidation was observed in propylene carbonate, $E_{1/2}$ for a wide range of compounds being of the order of +0.8 V (vs. Ag/AgCl). Cation stability could be improved by adding methyl groups to the arene, presumably owing to hindrance of nucleophilic attack at the Cr(I) by solvent. The greatest improvement in cation stability came, however, from a change of solvent to trifluoroacetic acid, a solvent of notably low nucleophilicity (see Section 3).[194] Seventeen-electron cations of arene$Cr(CO)_3$ had half-lives on the order of hours in this solvent.[273]

One reason Rieke and co-workers examined arene$Cr(CO)_3$ complexes was to look at those of the type $Me_2E[C_6H_5Cr(CO)_3]_2$ where E was a Group 4A element.[271] It was their intention to search for interactions between chromium sites in a mixed valence Cr(0)-Cr(I) mono-cation. Although no such interactions were apparent from electrochemistry, they did find that the apparent stability of the cation increased, and that there was a general trend to less anodic potentials, on descending the Group 4A series.

Improvement in the stability of $[(arene)Cr(CO)_{3-n}L_n]^+$ was also observed when $L = PR_3$ or $P(OR)_3$ and the arene was C_6Me_6.[273] Further, as expected from the stronger sigma donor/weaker pi acceptor character of phosphines as compared with CO, the oxidation potentials were roughly 800 mV less anodic for the comparable $n = 0$ complexes. When L was an acetylene such as PhC≡CPh, oxidation occurred at still less anodic potentials, and a further irreversible oxidative process was seen, possibly leading to loss of the acetylene. Finally, an excellent correlation between E_{pa} and ν_{CO} was also observed for L = phosphine or phosphite, just as in the case of the isoelectronic $CpMn(CO)_{3-n}L_n$ series in Section 4.2.2.

Bond *et al.* have found that the very unusual molecule $M(CO)_2DAM$ (M = Cr, Mo) (**47**) could be reversibly oxidized at low temperatures (for M = Cr but not Mo), but reduction on a mercury electrode led to a mercury-containing cation $[HgM(CO)_2DAM])^+$.[275]

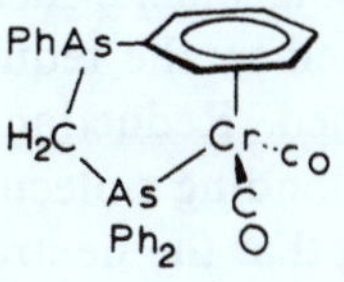

47

Finally, also noteworthy is the finding that the anodic oxidation of (arene)$Cr(CO)_3$ is catalyzed if ferrocene is placed in solution,[276] and examination of the electrochemistry of (arene)$Cr(CO)_3$ with a ferrocenyl substituent (see (**6**) in Section 1.5) showed that the ferrocenyl group was oxidized prior to the Cr(0) site.[122] Three reports have appeared regarding the reduction of acetophenone and benzophenone coordinated with $Cr(CO)_3$;[277–279] as these represent reactions at a coordinated ligand, they will be described in Section 5.4.

4.2.4. Other Pi Donor Ligands

Cyclobutadiene complexes of iron and ruthenium, such as $(\eta^4\text{-}C_4Ph_4)Fe(CO)_2L$, can be oxidized in CH_2Cl_2 by one electron in an electrochemically reversible process to give the radical cation.[280] For $L = PPh_3$, $E_{1/2} = 0.57$ V; $E_{1/2}$ for phosphite ligands was somewhat greater, reflecting the weaker sigma donor ability of $P(OR)_3$. Comparable Ru(II) complexes were oxidized at potentials about 200 mV more anodic than their Fe(II) analogs. In contrast to a report of the one-electron reduction of such complexes at mercury electrodes,[14] there was no evidence for reduction at a Pt electrode. Chemical oxidation of the cyclobutadiene complexes was complex, but halogens, for example, gave $[(\eta^4\text{-}C_4Ph_4)MX(CO)_2L]^+$.

Several interesting studies have been done of cycloheptatriene and of cycloheptatrienyl metal carbonyls.[272,281–283] Both are donors of six-pi electrons, and complexes of the type $(\eta^6\text{-}C_7H_8)W(CO)_3$ are known. This particular molecule exhibited two reduction waves in THF ($E_{1/2} = -1.52$ V and -2.55 V) and one oxidation wave ($E_{1/2} = +0.74$ V). It was suggested that, on reduction, the di-anion was formed, but this reacted rapidly with protons from trace water to give $[C_7H_9W(CO)_3]^-$, which could be isolated as $C_7H_9W(CO)_3Me$ after addition of methyl iodide.

Organometallic mixed valence compounds are most intriguing as already described in Section 1.6. A series of bimetallic, mixed valence compounds based on the cycloheptatrienyl ligand can be synthesized by several approaches as illustrated in Equation 48 below.[282,283] The paramagnetic dimers could be oxidized reversibly in CH_2Cl_2 to diamagnetic cations at potentials in the range of −0.1 V. The Mo-Mo distance in the dimeric cation (X = Cl) is 307.6 pm, suggesting a metal-metal bond. If this is correct, each Mo(I) metal site is surrounded by the requisite 18-valence electrons, and the diamagnetism is explained. Reduction would presumably place an electron in a metal-metal antibonding molecular orbital, and give a paramagnetic species. It is interesting that the neutral dimer exhibits an electronic transition in the near infrared region (1040 nm) in a range typical of intervalent transfer bands for class II-mixed valence species (see Section 1.6).

+ halide or OR^- ; e^- (48)

5. ELECTROCHEMICALLY INDUCED REACTIONS

One of the most active areas of investigation in organometrallic electrochemistry is the inducement of chemical change by electron transfer. A number of papers have appeared on structural changes which occur as a result of electron transfer, on electrochemically induced substitution reactions at metal centers, and on ligand-based reactions. Some examples have already been mentioned. For instance, in Section 2 (Equation 17) we noted that *bis*(η^6-hexamethylbenzene)ruthenium(2+) was reduced by two electrons to give the (η^6-C_6Me_6)(η^4-C_6Me_6)Ru(O) wherein one ring has only four carbon atoms in intimate contact with the ruthenium center. Other examples include Equations 37–39, 44, 45, and 48.

In this section, we shall describe electrochemically induced isomerizations, substitution reactions, and ligand-based addition and elimination reactions.

5.1. Electrochemically Induced Structural Changes

Although most 18-electron organometallic complexes are quite stable to rearrangement or substitution, on oxidation to 17-electron molecules their lability increases greatly. This is nicely illustrated by two forms of structural change which have been induced by electron transfer: (i) the *cis-trans* isomerization of six-coordinate complexes and (ii) the change in bonding mode of a pi donor ligand.

5.1.1. Isomerization of Six-Coordinate Complexes

Electrochemically induced *cis-trans* isomerization has been found on oxidation of several types of octahedral complexes of Groups 6B and 7B metals.[284-293]

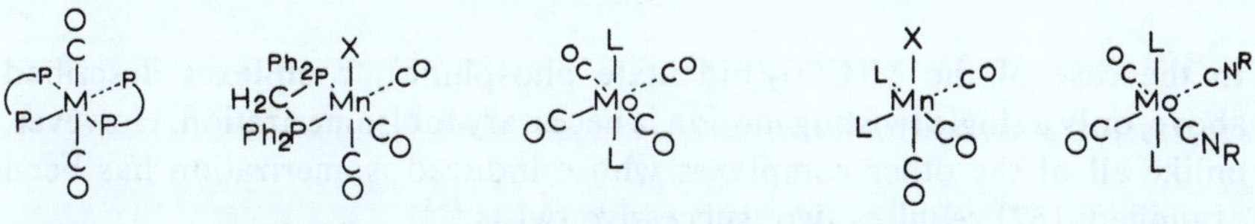

48 **49** **50** $L = PR_3, CR_2$ **51** **52**

In general, the electrochemistry of all of these complexes can be described by the Scheme below (Equation 49).[290]

$$\begin{array}{ccc} cis & \underset{}{\overset{E_1}{\rightleftharpoons}} & cis^+ + e^- \\ k_1 \Big\downarrow\Big\uparrow k_{-1} & & k_2 \Big\downarrow\Big\uparrow k_{-2} \\ trans & \underset{}{\overset{E_2}{\rightleftharpoons}} & trans^+ + e^- \end{array} \tag{49}$$

where E_1 is more positive than $E_2 (E_1 > E_2)$. (Although not defined on the scheme above, "cross-reactions," such as $cis^+ + trans \rightarrow trans^+ + cis$ must also be considered, especially if E_1 and E_2 are quite different.) Two examples of this electrochemical scheme will be considered.

Compounds such as (**48**) can be synthesized with M = Cr, Mo, and W and where the bidentate phosphine is $Ph_2P(CH_2)_nPPh_2$ ($n = 1, 2, 3$).[284-286] According to Scheme 49 above, the *cis* siomer is first oxidized to the cis^+ cation which then rapidly isomerizes to $trans^+$. In general, the rate constants for isomerization of the 17-electron cations are of the order of 20 to 45 s^{-1}, with the trend in lability being Cr < Mo < W. Ligands with a smaller bite angle, $n = 1$, lead to less rapid isomerization than those with more CH_2 groups. The $trans^+$ isomer is then reduced at much less positive potentials ($E_2 - E_1$ = ca. −550 mV) to give the neutral *trans* complex; the latter in turn slowly isomerizes to the neutral *cis* complex with rate constants of the order of 0.1 to 1 s^{-1}.

Rates of electrochemically induced isomerizations are generally found to be solvent independent. This had led to the conclusion that the mechanism for isomerization is a variation of the "Bailar twist" wherein one trigonal face of the molecule is rotated with respect to the other.[293,294]

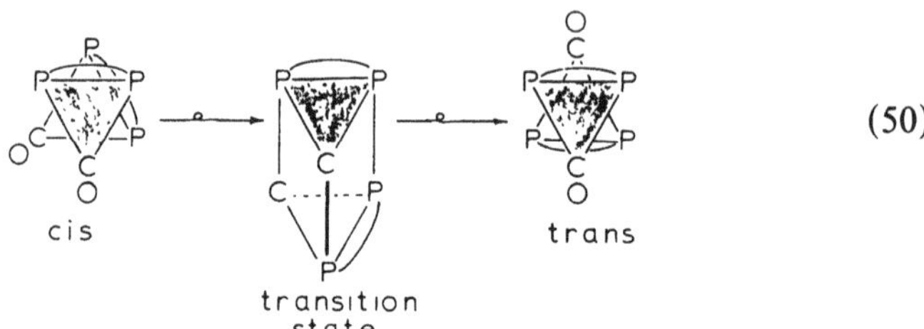

(50)

In the case of the $M(CO)_2$(bidentate phosphine)$_2$ complexes described above, only a single twisting motion is necessary for isomerization. However, unlike all of the other complexes whose induced isomerization has been examined, (**52**) requires two, successive twists.[293]

Trans-carbene complexes such as $M(CO)_4$(carbene)$_2$ (**50**) are quite interesting, because isomerization occurs by an electrochemical process with little or no net current flow.[291,292] The tungsten complex where the carbene was 1,3-dimethyl-4-imidazolin-2-ylidene is typical. The *trans* complex at low temperature gave rise to a small anodic wave (0.14 V) and a larger, reversible wave at +0.337 V. The more positive CV wave was identical to that for the corresponding pure *cis* isomer. Bulk electrolysis of the *trans* isomer at the first wave resulted in very little current flow. This occurred because the $trans^0$ complex was oxidized to $trans^+$ which then rapidly isomerized to the cis^+ complex. Since the cis^+ complex was produced at a potential more cathodic than its reduction potential, it was reduced on formation. Thus, since the rate of the isomerization $trans^+ \rightarrow cis^+$ is large (k_2 in Scheme 49 = ca. 400 s^{-1}), very little net current flow was observed. (Since the difference in E_1 and E_2 (Scheme 49) for these complexes is 0.20 V, it is very likely that the cross reaction $trans + cis^+ \rightarrow cis + trans^+$ is quite important.)

The complexes *cis*- and *trans*-$M(CO)_4(PR_3)_2$ (**50**) are important examples of the pitfalls of electrochemical analysis. Unlike the complexes described above, where the *cis* and *trans* isomers are oxidized at potentials differing by several hundred millivolts, the differences for $M(CO)_4(PR_3)_2$ approximate 50 mV. The result of this is that the CV redox wave of the *cis* isomers, for example, often simply resembles a quasi-reversible wave with a large difference between anodic and cathodic peak potentials.

5.1.2. Isomerization of a Pi Donor Ligand

The stepwise reduction of cyclooctatetraene to the dianion has been thoroughly studied as an example of structural change which occurs as a result of electron transfer. Most interesting is the first reduction step, which is abnormally slow owing to a change from the normal "tub" shape of the olefin to a planar radical anion.[295,296]

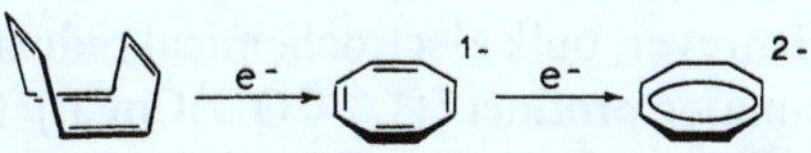

$$\text{COT} \xrightarrow{e^-} [\text{COT}]^{1-} \xrightarrow{e^-} [\text{COT}]^{2-} \quad (51)$$

Cyclooctatetraene (COT) and cyclooctadiene (COD) are four-electron donor, pi-bonded ligands in numerous transition metal complexes. Generally, they are 1,5-bonded in neutral compounds, but Geiger and co-workers have found, in complexes (**53–57**), that electron transfer can lead to changes in their bonding mode from the 1,5 to the 1,3 mode.[295-301]

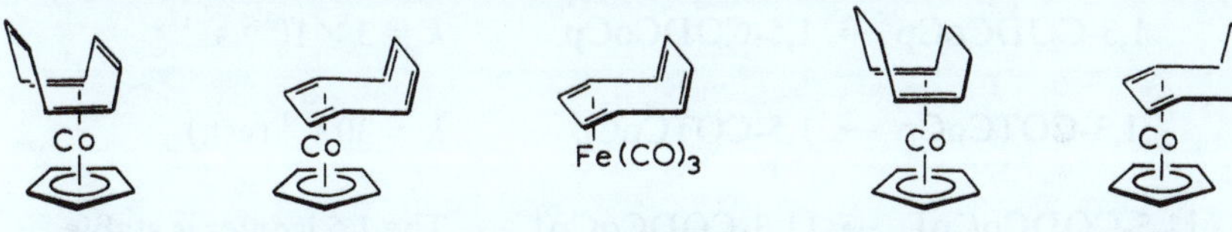

Co Co Fe(CO)$_3$ Co Co

53 **54** **55** **56** **57**

A solution of COTCoCp contains both the 1,5 (**53**) and 1,3 (**54**) isomers, the 1,5 predominating by about 3:1 at room temperature. Reversible one-electron reduction of the 1,3 isomer occurred at −1.8 V, while the 1,5 isomer was irreversibly reduced at −2.0 V. Electrolysis at −2.2 V produced a solution which contained only the anion of the 1,3 isomer, and subsequent oxidation of this anion at potentials more anodic than −1.7 V then led to the neutral complex in the original mixture of isomers. An important question was whether the isomerization occurred simultaneous with or subsequent to electron transfer. Very recent measurements using "fast fourier transform faradaic admittance measurements" show unequivocally that it is indeed an EC process, and that the lifetime of the 1,5 anion is less than a millisecond.[299] Thus, the electrochemical behavior of COTCoCp follows the same general scheme as the octahedral complexes described above.

$$(1{,}3\text{-}\eta^4\text{-COT})\text{CoCp} + e^- \underset{E=-1.8\text{ V}}{\overset{k_s=0.28\text{ cm s}^{-1}}{\rightleftharpoons}} [(1{,}3\text{-}\eta^4\text{-COT})\text{CoCp}]^- \quad (52)$$

$$k_b \; \Big\| \; k_f\text{ (slow)} \qquad\qquad k = 2\times 10^3\text{ s}^{-1}\ \uparrow$$

$$(1{,}5\text{-}\eta^4\text{-COT})\text{CoCp} + e^- \underset{E=-2.0\text{ V}}{\overset{k_s=0.06\text{ cm s}^{-1}}{\rightleftharpoons}} [(1{,}5\text{-}\eta^4\text{-COT})\text{CoCp}]^-$$

As only a relatively minor motion need occur, one can appreciate the notion that isomerization of 1,5-COT to 1,3-COT is readily possible.

However, it is quite surprising that a similar rearrangement occurs in the COD complex (**56**), since actual movement of hydrogen atoms is necessary in this case. A solution of CODCoCp was found to contain only the 1,5 isomer at room temperature, and the isomer was reversibly reduced (in DMF) at −2.45 V. However, bulk electrochemical reduction by one-electron in THF gave as the major product $[(1,3\text{-COD})\text{CoCp}]^-$ ($E_{1/2} = -1.60$ V). In contrast with the COT complexes (see Equation 52), the neutral isomers of CODCoCp were only very slowly interconverted, and the complex could therefore be "trapped" as 1,3-CODCoCp by oxidation of $[1,3\text{-CODCoCp}]^-$ at −1.40 V. Rate constants for interconversion of the neutral and reduced isomers have been determined or estimated, rates for the COT complexes being much higher than the COD analogs.

1,3-CODCoCp → 1,5-CODCoCp		$k = 3 \times 10^{-5}\ s^{-1}$
1,3-COTCoCp → 1,5-COTCoCp		$k = 30\ s^{-1}$ (est.)
$[1,5\text{-CODCoCp}]^-$ → $[1,3\text{-CODCoCp}]^-$		The 1,5 isomer is stable in DMF for *ca.* $\frac{1}{2}$ hour.
$[1,5\text{-COTCoCp}]^-$ → $[1,3\text{-COTCoCp}]^-$		$k > 10^2\ s^{-1}$

To account for these differences, a molecular orbital study was done.[298] This showed that the LUMO of the COT complexes was ligand-centered, being chiefly composed of the four-carbon fragment not bonded to the metal, while the LUMO of the COD complexes are largely metal-based.

In contrast with CODCoCp (**56**), it is very interesting that the isoelectronic and isostructural Ni(II) complex $[1,5\text{-CODNiCp}]^+$ does not convert to the 1,3-bonded isomer on reduction to the neutral complex.[300] Rather, CODNiCp (19-valence electrons) slowly disproportionated to Cp_2Ni (20-valence electrons) and $(COD)_2Ni$ (18-valence electrons).

The iron compound (**55**) contains a 1,3-bonded COT, and the molecule can be reduced in two well-separated, one-electron, chemically reversible steps (−1.24 V and −1.71 V). Unlike the reduction of free COT, both reduction steps are reasonably rapid, because structural rearrangement of the COT apparently does not occur.[296]

Geiger and co-workers have also investigated the unusual class of "triple decker" sandwiches (**58**). Stable complexes of this type usually contain 30 or 34 valence electrons, so that the 36-valence electron dicobalt(I) complex (**58**) was found to be oxidizable in a reversible, two-electron process at about +0.1 V. The reason for the facility of this oxidation is not clear, but an increase in resonance stabilization from flattening the C_8 ring is a possibility. (If one looks on the oxidation as $Co^{+}(C_8H_8)Co^{+} \rightarrow Co^{3+}(C_8H_8^{2-})Co^{3+} + 2e^{-}$, this would lead to a 34-valence electron complex with a planar $C_8H_8^{2-}$ ring.) (**58**) was also reducible in two, one-electron steps, the first one being highly reversible.

58

5.2. Redox-Induced Ligand Substitution

It has been known in organic chemistry for some time that nucleophilic substitution of a halide substituent can be triggered electrochemically, with the reaction following the general scheme[302]

$$\text{Electron transfer} \qquad RX + e^{-} \rightarrow [RX]^{-\cdot} \tag{53}$$

$$\text{Chain initiation} \qquad [RX]^{-\cdot} \rightarrow R^{\cdot} + X^{-}$$

$$R^{\cdot} + Nu^{-} \rightarrow RNu^{-\cdot}$$

$$\text{Chain termination} \qquad RNu^{-\cdot} \rightarrow RNu + e^{-}$$

and/or

$$\text{Chain propagation} \qquad RNu^{-\cdot} + RX \rightarrow RNu + [RX]^{-\cdot}$$

$$\overline{\qquad RX + Nu^{-} \rightarrow RNu + X^{-} \qquad}$$

If the species RX is relatively easily reduced, if the radial anion is long-lived, and if the propagation step is rapid, then a true "electron transfer catalyzed" process will be realized. The number of faradays required to drive the reaction will tend to zero.[303] (Note that the propagation step will be thermodynamically spontaneous, for a reductively driven process, only if the reduction potential of the product is more negative than that of the starting material. The opposite is true for an oxidatively driven process.)

One of the most interesting developments in organometallic chemistry has been the discovery of very efficient chain mechanisms for ligand substitution in metal carbonyls which involve odd-electron, radical intermediates as in the reaction scheme above.[304] Such reactions have been induced electrochemically, both anodically and cathodically, and considerable information is now available.

5.2.1. Substitution of Neutral Sigma Donor Ligands

5.2.1a. Anodically Induced Substitutions. Kochi and his co-workers, in a series of elegant papers, have described oxidatively induced ligand substitutions in molecules, such as (**59**) and (**60**).[206,305–307]

59 **60**

Replacement of the acetonitrile ligands in the starting materials by phosphines or phosphites is a thermodynamically allowed process normally induced thermally or photochemically. However, anodic oxidation using only a trickle of current through the solution induces substitution at a rate much greater than simple thermal substitution. The basic reaction scheme involved is given below (wherein the nonsubstituted ligands are given as Cp and CO).

$$\mathrm{Cp}M(\mathrm{CO})L \xrightarrow{E^0(\text{oxidation})} [\mathrm{Cp}M(\mathrm{CO})L]^+$$

$$[\mathrm{Cp}M(\mathrm{CO})L]^+ + L' \xrightarrow{k_1} [\mathrm{Cp}M(\mathrm{CO})L']^+ + L$$

$$[\mathrm{Cp}M(\mathrm{CO})L']^+ + \mathrm{Cp}M(\mathrm{CO})L \xrightarrow{k_2} \mathrm{Cp}M(\mathrm{CO})L' + [\mathrm{Cp}M(\mathrm{CO})L]^+ \quad (54)$$

$$[\mathrm{Cp}M(\mathrm{CO})L']^+ \xrightarrow{E^0(\text{reduction})} \mathrm{Cp}M(\mathrm{CO})L'$$

The essence of this scheme is that anodic oxidation of a starting material gives a labile, 17-valence electron intermediate which is susceptible to substitution by L'. As an example we shall consider $\mathrm{MeCpMn(CO)_2(NCMe)}$

(**59**), one of the many systems studied by Kochi.[305-307] This compound is oxidized reversibly at +0.19 V in MeCN. However, on adding PPh_3, only an anodic wave for the manganese-containing starting material is observed. In addition, the reversible wave for $MeCpMn(CO)_2(PPh_3)$ is now seen at +0.52 V. The phosphine substitution product is of course produced as the 17-electron cation $(=[CpM(CO)L']^+)$, but, since the reduction potential of this cation is more positive than that for $[CpM(CO)L]^+$, the following reaction occurs.

$$[MeCpMn(CO)_2(PPh_3)]^+ + MeCpMn(CO)_2(NCMe) \rightarrow \qquad (55)$$

$$MeCpMn(CO)_2(PPh_3) + [MeCpMn(CO)_2(NCMe)]^+$$

and leads to chain propagation. Chain termination is induced by reduction of the substitution product at the electrode, a possible reaction since the product is produced at a potential more negative than its reduction potential. In this particular case, however, the chain propagation step was more important. The turnover number or current efficiency (defined as moles of starting material consumed, divided by faradays of charge, passed through the solution) was 1013. (This was one of the highest values observed. More typically, values ranged from 20 or so to several hundred). The values of k_1 and k_2 were determined to be $1.3 \times 10^4\ M^{-1}\ s^{-1}$ and $5 \times 10^5\ M^{-1}\ s^{-1}$, respectively.

A number of unusual features have been observed for these electrocatalytic reactions. First, in ligand substitution reactions of 18-electron molecules, dissociatively activated (D or I_d) processes seem to be the rule. For the 17-electron molecule, however, ligand exchange is apparently associative (A or I_a) with very large rate constants. Second, the rate constant k_1 is very sensitive to the incoming nucleophile; k_1 for PPh_3 is 1000 times greater than that for $P(OPh)_3$. And finally, the lability of the leaving group is greatly magnified in the cation. The k_1 for MeCN substitution by PPh_3 is about 1000 times greater than that for pyridine as the leaving group.

5.2.1b. Cathodically Induced Substitutions. Reduction of an 18-valence electron organometallic compound has also been found to labilize such species to substitution.[303,308-313] For example, reduction of the Mn(I) cation $[Mn(CO)_3(MeCN)_3]^+$ in the presence of two equivalents of a phosphine L ($L = PPh_3$ or PMe_2Ph) gave excellent yields of the hydrido-manganese compound $HMn(CO)_3L_2$.[309]

$$[Mn(CO)_3(MeCN)_3]^+ + 2\,L \xrightarrow{\text{cathodic reduction}} HMn(CO)_3L_2 + MeCN \qquad (56)$$

This result was attributed to a greater lability to substitution for a 19-electron Mn(0) species. That is, there is evidence for the following sequence of reactions.

$$[Mn(CO)_3(MeCN)_3]^+ + e^- \longrightarrow Mn(CO)_3(MeCN)_3 \quad (57)$$

$$Mn(CO)_3(MeCN)_3 + L \xrightarrow{\text{fast}} Mn(CO)_3(MeCN)_2L + MeCN$$

$$Mn(CO)_3(MeCN)_2L + L \xrightarrow{\text{fast}} Mn(CO)_3(MeCN)L_2 + MeCN$$

The final 19-electron product, $Mc(CO)_3(MeCN)L_2$, abstracts an H atom from the solvent (with loss of MeCN) or is anodically oxidized. The overall process cannot be catalytic, however, since the reduction potentials of the products are considerably less negative than that of the starting material; the chain propagation step is not thermodynamically spontaneous.

Other examples of cathodic labilization involve metal clusters.[303,310-313] For example, it has been observed that cathodic reduction of the cluster (**61**) in the presence of PPh_3 led to very rapid substitution of CO for PPh_3 by electron transfer catalysis.

$$\text{61} \xrightarrow{+e^-} [\text{61}]^- \xrightarrow[-CO,\ -e^-]{+L} \text{X-C(R)-Co-Co-L} \quad (58)$$

X = RC, $Co(CO)_3$
Co = $Co(CO)_3$
61

Although less than a 60 percent yield of the desired product could be obtained by refluxing the neutral cluster with PPh_3 in hexane for several hours, almost a 90 percent yield could be obtained in about one minute on cathodically triggering the reaction. (The yield dropped significantly if the reaction was done in a CO atmosphere.) A current efficiency of 2100 was realized.

It is known that reduction of metal clusters places the electron in a delocalized molecular orbital with significant metal contrubution.[25,314,315] It was suggested, therefore, that reduction led to partial metal-metal bond breaking, but the precise mechanism is still a matter of conjecture.

5.2.2. Substitution of Anionic Sigma Donor Ligands: *M-R* and *M-X* Bond Breaking

In the first part of this section, we described the electrochemically induced substitution of one neutral donor ligand by another. We now turn to substitution reactions where the ligand to be replaced is an anionic donor, such as halide, alkoxide, or, most importantly, an organic group such as CH_3^-, $C_6H_5^-$, and so on. Once again, our organization is by periodic group.

In Section 1.4, we described the electrochemistry of "bent metallocenes," complexes such as Cp_2TiCl_2 and $Cp_2Ti(OPh)_2$. The cathodic electrochemistry of such complexes was largely focused on loss of Cl^- or OPh^-. Presumably, the anion was replaced in the titanium coordination sphere by another ligand, but the products were not clear, except perhaps when the reaction was carried out in an atmosphere of CO, and $Cp_2Ti(CO)_2$ was formed.

Group 6B metallocenes have an extensive oxidative electrochemistry and the metal-based products are more clearly defined here. For example, Kochi and co-workers have thoroughly examined some metal hydrides of Group 6B metals and have defined three reaction routes for these complexes following anodic oxidation.[316] For example, $CpMo(CO)_3H$ underwent "oxidative coupling;" the complex gave the well-known dimer $[CpMo(CO)_3]_2$ following one-electron oxidation in MeCN. The postulated reaction sequence is given in Equation 59.

$$CpMo(CO)_3H \xrightarrow{-e^-} [CpMO(CO)_3H]^+ \tag{59}$$

$$[CpMo(CO)_3H]^+ \longrightarrow CpMo(CO)_3 + H^+$$

$$2CpMo(CO)_3 \longrightarrow [CpMo(CO)_3]_2$$

The bent metallocene Cp_2WH_2 was also dimerized on oxidation, but by an "additive dimerization" process. The product was a very interesting ditungsten species with a mono-hydrogen bridge.

$$2cp_2WH_2 \xrightarrow{-2e^-} [Cp_2W(H)\text{-}H\text{-}W(H)Cp_2]^+ + H^+ \tag{60}$$

Finally, anodic oxidation of $Cp_2W(Ph)H$ resulted in "reductive elimination" of the phenyl group, but the tungsten-containing species was not defined.

$$Cp_2W(Ph)N \xrightarrow{-e^-} C_6H_6 + H^+ + ? \tag{61}$$

It was learned, however, that the H atom acquired by the C_6H_5 group was probably not the hydridic hydrogen.

Our research group has also examined the oxidative electrochemistry in MeCN of an extended series of compounds of the type Cp_2MX_2 (M = Mo or W; X = halide or thiolate).[90] The dihalides were oxidized in an electrochemically reversible manner in the range 0.5-0.55 V for M = Mo and 0.3-0.4 V for M = W. When X was Br^- or I^-, the M(V) cations $[Cp_2MX_2]^+$

were labile to substitution, giving the electroactive M(IV) cations $[Cp_2M(NCMe)X]^+$ and Br_2 or I_2. The presence of M(IV) mono-solvento complexes was confirmed by comparison with authentic samples, and electrochemical experiments confirmed the presence of X_2.

$$Cp_2MX_2 \rightleftharpoons [Cp_2MX_2]^+ + e^- \tag{62}$$

$$[Cp_2MX_2]^+ + MeCN \longrightarrow [Cp_2M(NCMe)X]^+ + 1/2X_2$$

$$[Cp_2M(NCMe)X]^+ \rightleftharpoons [Cp_2M(NCMe)X]^{2+} + e^-$$

The mechanism presumed for this reaction was loss of a halogen atom in a reductive elimination to give X_2, a reaction suggested by previous work on the photochemistry of Cp_2MX_2.[317] However, another mechanism, a redox catalyzed process similar to those described above, is also very possible.

Chain Initiation: $Cp_2MX_2 \rightarrow [Cp_2MX_2]^+ + e^-$ (63)

Substitution: $[Cp_2MX_2]^+ + MeCN \rightarrow [Cp_2M(NCMe)X]^{2+} + X^-$

Chain Propagation: $[Cp_2M(NCMe)X]^{2+} + Cp_2MX_2$

$\rightarrow [Cp_2M(NCMe)X]^+ + [Cp_2MX_2]^+$

X^- Oxidation: $X^- \rightarrow 1/2X_2 + e^-$

Since the electrolysis of the Mo dihalides was done at potentials greater than +0.7 V, X^- is produced at a potential more anodic than its oxidation potential, and so it can be oxidized to X_2. This mechanism would therefore lead to a net one-electron oxidation, as observed, and production of the observed products.

The molybdenum and tungsten dithiolates, $Cp_2M(SR)_2$, were also readily oxidized to the M(V) cations. One of the final products was the M(IV) complex $[Cp_2M(NCMe)SR]^+$, as proved by isolation following chemical oxidation. Yet another product isolated was thought to be the thiolate-bridged dimer $[Cp_2M(\mu\text{-}SR)_2MCp_2]^{2+}$, and the by-product RSSR was detected electrochemically. An "oxidatively induced reductive elimination" mechanism similar to that of the dihalides (Equation 64) was proposed.

Oxidation: $Cp_2M(SR)_2 \rightarrow [Cp_2M(SR)_2]^+ + e^-$ (64)

Reductive Elimination of $\cdot SR$ with Substitution by MeCN:

$$[Cp_2M(SR)_2]^+ + MeCN \rightarrow [Cp_2M(NCMe)(SR)]^+ + RSSR$$

Reductive Elimination of $\cdot SR$ with Dimerization of the Metal(IV)-Containing Moiety to give a Thiolate Bridged Dimer:

$$2[Cp_2M(SR)_2]^+ \rightarrow [Cp_2M(SR)]_2^{2+} + RSSR$$

We now recognize, however, that a redox catalyzed mechanism (Equation 63, $X = SR$) is also possible, although it is difficult to see how a thiolate-bridged dimer could arise in this manner.

One of the most important reactions in organometallic systems is the migratory insertion of CO to give metal acyl derivatives (Equation 65).[2,41,318,319] It is generally accepted that the alkyl group migrates to the metal-bound CO, a mechanism supported by theoretical work.[320,321]

$$\underset{\displaystyle R}{\overset{}{M}}\!\leftarrow C\equiv O \xrightarrow{+L} \underset{\displaystyle L\quad R}{M - C=O} \qquad (65)$$

Some years ago, it was observed that some alkylmetal carbonyls can be oxidatively induced to undergo carbonylation,[319] and recent electrochemical work confirms this. Giering and co-workers have found that oxidation CpFe(CO)(L)Me ($L = PPh_3$) in CH_2Cl_2 saturated with CO gave CpFe(L)(CO)(COMe),[322,323] apparently by a radical chain mechanism similar to those studied by Kochi above. That is, oxidation (using only a trickle of current or a trace of chemical oxidant) initiated the chain.

$$CpFe(L)(CO)Me \rightarrow [CpFe(L)(CO)Me]^+ \qquad (66)$$

Methyl migration then occurred with a rate that was estimated to be at least 10^7 faster than carbonylation of the neutral compound.

$$[CpFeL(CO)Me]^+ + CO \rightarrow [CpFe(L)(CO)(COMe)]^+ \qquad (67)$$

Since the reduction potential of $[Fe\text{-}Me]^+$ is +0.380 V, while that of $[Fe\text{-}COMe]^+$ is +0.470 V, the following electron transfer reaction, leading to chain propagation, is thermodynamically spontaneous.

$$[CpFe(L)(COMe)]^+ + CpFe(L)(CO)Me \rightarrow CpFe(L)(COMe) + [CpFe(L)(CO)Me]^+ \qquad (68)$$

The vast difference in rates of carbonylation of the Fe(II) and Fe(III) compounds has been attributed to a substantial reduction in the pi donor ability of the metal center, thereby allowing the bent (or side-on) transition state configuration of the CO to be energetically more accessible.

Anodic oxidation of $[CpFe(CO)(CN)(COMe)]^-$ also apparently involves a methyl migration step.[324] The neutral Fe(III) species was sufficiently stable to observe its esr spectrum, but it eventually decomposed to form acetone in high yield. If the first step in the mechanism is loss of CO, an Fe-Me bond can be formed by methyl migration. Coupling of this Fe^{III}-Me species and the 17-electron Fe^{III}-COMe compound could give rise to the observed ketone. In contrast, oxidized $CpFe(CO)_2(COR)$ was quite unstable, decomposing in the presence of alcohols to give esters in good yields.

Baird and co-workers have carried out extensive studies of the mechanism of oxidative cleavage of Fe-C bonds in the series of compounds CpFeL(CO)R, and some electrochemical results have been reported.[325] Using Bu_4NCl as the supporting electrolyte, irreversible, one-electron oxidation was observed in the range 1-1.3 V (vs. Ag/AgCl), the potential being dependent in the usual manner on the nature of *R* and *L*. (With a noninterfering electrolyte, Bu_4NPF_6, however, a cathodic wave was observed at low temperature and high-scan rate.) Anodic electrolysis of CpFeL(CO)*R* in the presence of Cl^- led to CpFeL(CO)Cl and *R*Cl. If the *R* group was ^{13}C labeled $PhCH_2^*CH_2$, a complete scrambling of methylene groups was observed. In another experiment using *threo*-PhCHDCHD, *threo*-PhCHDCHDCl was isolated with 77 percent configuration retention. Similar results were obtained using $CuCl_2$ as the oxidant.

In view of the foregoing discussion of cyclopentadienyliron carbonyl chemistry, the report of the electrochemistry of $CpFe(CO)_2R$ where $R =$ $-CH_2Fc$, -COFc, and -Fc is quite interesting and should be reconsidered.[326] Two irreversible anodic waves were reported for $CpFe(CO)_2Fc$, and ferrocene was the only product described.

The fundamental process of metal-carbon bond cleavage was also studied in the series of compounds *cis*-$(bipy)_2FeR_2$ (bipy = 2,2′-bipyridyl).[327] The complexes all underwent a one-electron, reversible oxidation at $E^{0\prime}$s of about -1 V and a second, irreversible, one-electron oxidation at about +0.3 V. The neutral Fe(II) species was stable for days in solution. However, the +1, Fe(III)-cation had a halflife of only about 30 minutes for its decomposition by a reaction first order in the cation. The organic products of this decomposition were those expected for a reductive elimination of $R^\bullet$; that is, dialkyls (R_2), alkanes (*R*H), and alkenes (*R*-H) were isolated. When *R* was an ethyl group, for example, thermal decomposition of $[(bipy)_2Fe(Et)_2]^+$ gave butane (45 percent), ethane (48 percent), and ethylene (8 percent). The iron portion of the molecule was left as an insoluble, brick-red precipitate derived from $[(bipy)_2Fe]^+$. Decomposition of the +2, Fe(IV) cation, by contrast, gave only dialkyls (R_2), and it was suggested that this indicates simultaneous ejection of both R groups. The general reaction scheme for these complexes is summarized in Equation 69.

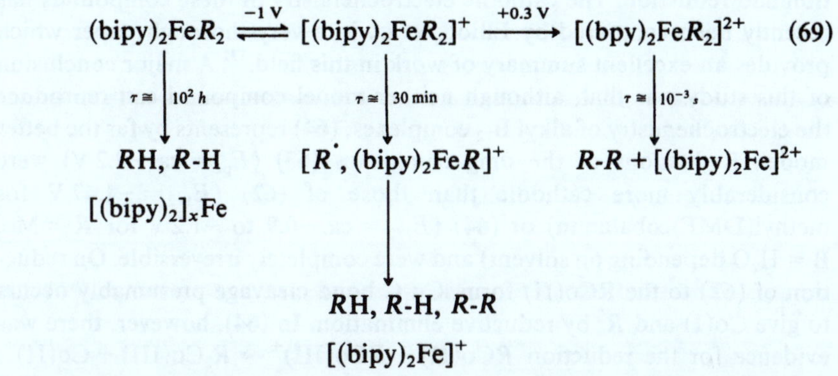

$$(\text{bipy})_2\text{Fe}R_2 \underset{}{\overset{-1\,\text{V}}{\rightleftharpoons}} [(\text{bipy})_2\text{Fe}R_2]^+ \xrightarrow{+0.3\,\text{V}} [(\text{bpy})_2\text{Fe}R_2]^{2+} \quad (69)$$

$$\tau \cong 10^2\,h: \quad (\text{bipy})_2\text{Fe}R_2 \longrightarrow R\text{H}, R\text{-H} \quad [(\text{bipy})_2]_x\text{Fe}$$

$$\tau \cong 30\,\text{min}: \quad [(\text{bipy})_2\text{Fe}R_2]^+ \longrightarrow [R^{\cdot}, (\text{bipy})_2\text{Fe}R]^+ \longrightarrow R\text{H}, R\text{-H}, R\text{-}R \quad [(\text{bipy})_2\text{Fe}]^+$$

$$\tau \cong 10^{-3}\,s: \quad [(\text{bpy})_2\text{Fe}R_2]^{2+} \longrightarrow R\text{-}R + [(\text{bipy})_2\text{Fe}]^{2+}$$

One of the most interesting areas of research in modern inorganic chemistry has been the study of metal-containing compounds of biochemical interest. Organometallic chemists have been particularly intrigued by vitamin B_{12} which contains a stable Co-C bond in some of its derivatives.[328] The electrochemistry of several such compounds (**62**), and of molecules which are potential models for vitamin B_{12} (**63**, **64**), have been extensively studied. The early work on the electrochemistry of these compounds has been reviewed,[329] and the reader is referred to an extensive and recent series of papers by Saveant and co-workers.[330]

Before briefly describing some recent electrochemical work, a word about the nomenclature of this area is useful. Structure (**62**) represents the basic structure of the B_{12} series of compounds, compounds based on the corrin ring system. The basic ligand B is an alpha-5,6-dimethylbenzimidazole nucleotide. When this group is attached, the compounds are referred to as cobalamins. The compound is then named more completely by specifying the type of R group *trans* to B. For example, when R = methyl, the compound would be methylcobalamin. On hydrolysis, the nucleotide side-chain is cleaved to give a series called cobinamides, and B is replaced by water. Model compound (**63**) is based on the dimethylglyoxime ligand, and

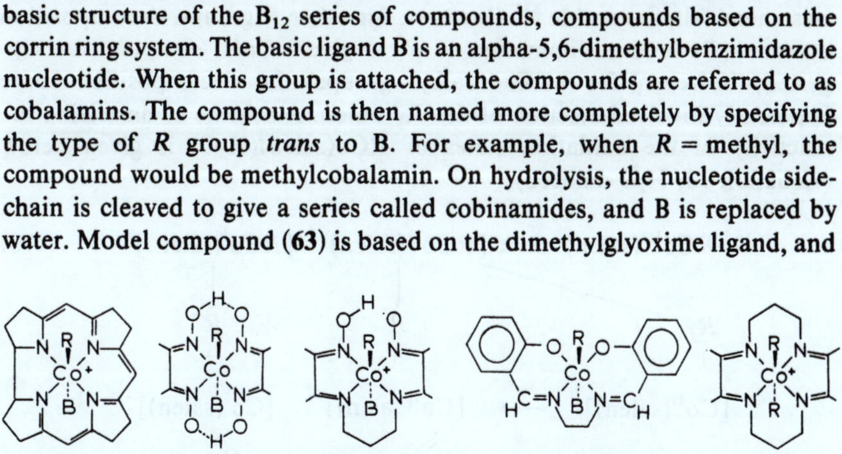

62 **63** **64** **65** **66**

such compounds are often called cobaloximes (abbreviated $Co(dmg)_2$). The model compound (**64**) has been given the abbreviation $Co(C_2(DO)(DOH)_{pn}$.

Cobalamins, cobinamides, and model compounds undergo both oxidation and reduction. The cathodic electrochemistry of these compounds has recently been compared by Elliott, *et al.* in a very thorough paper which provides an excellent summary of work in this field.[331] A major conclusion of this study was that, although neither model compound can reproduce the electrochemistry of alkyl B_{12} complexes, (**64**) represents by far the better model. Reductions of the dmg complexes (**63**) (E_{pc} = ca. −2.2 V) were considerably more cathodic than those of (**62**) ($E_{1/2} = -1.47$ V for methyl(DMF)cobalamin) or (**64**) ($E_{1/2}$ = ca. −0.9 to −1.2 V for R = Me, B = H_2O depending on solvent) and were completely irreversible. On reduction of (**62**) to the RCo(II) form, Co-C bond cleavage presumably occurs to give Co(I) and $R^{\cdot}$ by reductive elimination. In (**64**), however, there was evidence for the reduction $R\text{Co(II)} + R\text{Co(III)}^+ \rightarrow R_2\text{Co(III)} + \text{Co(II)}^+$.

Another thoroughly studied type of organocobalt complex is (**65**) wherein the ligand is the dianion of *N,N'*-ethylenebis(salicylaldimine) (often referred to as salen). The redox chemistry of $R\text{Co}^{III}(\text{salen})H_2O$, and others similar to it, has also been studied and has been compared with $R\text{Co}^{III}[(\text{DO})(\text{DOH})_{pn}]H_2O$ (**65**).[332,333] In general, both types of complexes are reduced in two, one-electron steps to the Co(I) stage, the first stage being generally electrochemically reversible; however, the salen complexes are reduced at much more cathodic potentials than (**64**). For both complexes, there is a linear correlation between $E_{1/2}$ for the Co(III)/Co(II) couple and pK_a of the hydrocarbon ligand, similar to that observed for the reduction of organomercury compounds.[5]

When the axial R group in (**65**) was phenyl, the first reduction step was chemically reversible.[334] However, for other R groups, decomposition occurred. Reductive elimination of $R^{\cdot}$ to give a Co(I) complex was suggested; however, for perfluoroalkyl groups, the decomposition route apparently involved ejection of the R_f^- anion. A scheme summarizing the reactivity of the oxidation states of RCo(salen)(H_2O) is given below (Equation 70; H_2O omitted).

$$
\begin{array}{ccccc}
R\text{Co}^{III}(\text{salen}) & \overset{e^-}{\rightleftharpoons} & [R\text{Co}^{II}(\text{salen})]^- & & [R\text{Co}^{I}(\text{salen})]^{2-} \\
& \swarrow & \downarrow & & \downarrow \\
R_f^- & & R^{\cdot} & & R^- \\
+ & & + & & + \\
[\text{Co}^{II}(\text{salen})] & \xrightarrow{+e^-} & [\text{Co}^{I}(\text{salen})]^- & & [\text{Co}^{I}(\text{salen})]^- \\
& & e^- \searrow\!\!\nwarrow & & \swarrow\!\!\nearrow e^- \\
& & & [\text{Co}^{0}(\text{salen})]^{2-} &
\end{array}
\qquad (70)
$$

A Co(III) compound having two *trans* methyl groups, (**66**), was oxidized irreversibly by one electron at +0.43 V.[335] Only one Co-Me bond was broken in the Co(IV) state, and ethane was formed in high yield; methane, however, was a product in good hydrogen donor solvents, and methyl halide was observed in the presence of a halogen donor. The mechanism proposed to account for these observations, and others, is outlined as follows.

$$Me_2Co^{III} \rightarrow Me_2Co^{IV} + e^- \tag{71}$$

$$Me_2Co^{IV} \rightarrow MeCo^{III} + Me^{\cdot}$$

$$Me^{\cdot} + SH \rightarrow MeH + S^{\cdot}$$

$$2\,Me^{\cdot} \rightarrow Me_2$$

The anodic electrochemistry of the B_{12} model compounds has also been studied owing to the susceptibility of the Co-C bond in the oxidized complex to nucleophilic attack.[336–338] Salen and dmg Co(III) complexes [**65** and **63**] are oxidized in an electrochemically reversible manner by one electron roughly in the range +0.5 to 1 V.[336,337] It is generally thought that the product is best considered a Co(IV) complex, but there is no definitive evidence in this regard. For the dmg complexes, $E_{1/2}$ generally became more positive as R became more electron releasing, a trend which paralleled the rate of chemical oxidation of the complex by $[IrCl_6]^{2-}$.

In aqueous methanol Halpern and co-workers observed that $[RCo^{IV}(dmg)_2H_2O]^+$ decomposed by second order kinetics and gave $RCo^{III}(dmg)_2H_2O$ in a concentration equal to 50 percent of the starting concentration of the Co^{IV} complex.[336] They suggested, therefore, a disproportionation to give a dication which then decomposed.

$$2\,[RCo(dmg)_2H_2O]^+ \rightarrow RCo(dmg)_2H_2O + [RCo(dmg)_2H_2O]^{2+}$$

$$[RCo(dmg)_2H_2O]^{2+} \rightarrow \text{decomposition products} \tag{72}$$

Such a reaction is reminiscent of the disproportionation of $[Cr(CO)_6]^+$.[194]

In the presence of a good nucleophile, however, the Co-R bond is broken and N-R^+ is observed.[337] For example, controlled potential electrolysis of $Bu^nCo(salen)$ in pyridine resulted in the passage of 2 F/mol of complex and the formation of $pyBu^+$ and $[Co(salen)(py)_2]^+$ by the following ECE scheme.

$$BuCo^{III}(salen) \xrightarrow[E_1]{-e^-} [BuCo^{IV}(salen)]^+ \tag{73}$$

$$[BuCo^{IV}(salen)]^+ + 3\,py \longrightarrow Bupy^+ + Co^{II}(salen)(py)_2$$

$$Co^{II}(salen)(py)_2 \xrightarrow[E_2]{-e^-} [Co^{III}(salen)(py)_2]^+$$

$$E_1 > E_2$$

The mechanism of nucleophilic attack on Co-C has been thoroughly examined, and the literature should be consulted for further details.[22]

In the absence of a nucleophile such as X^- or pyridine, the reactivity of the $[RCo^{IV}]^+$ complexes is more complicated, and the scheme in Fig. 6

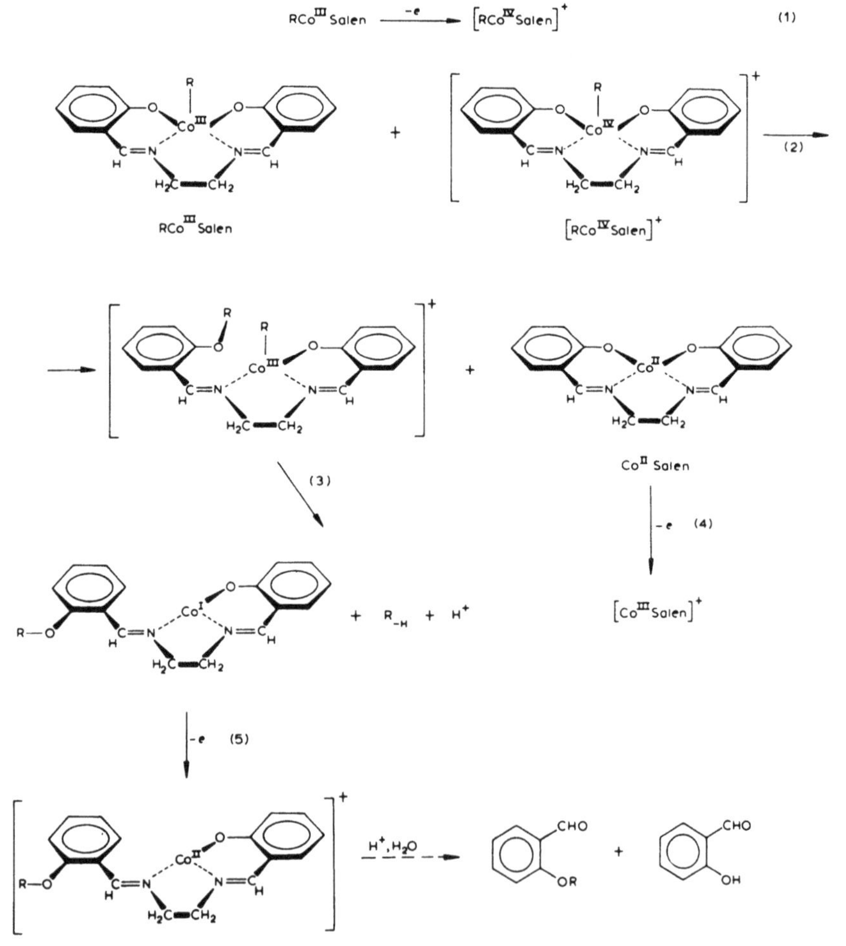

Figure 6. The reaction of anodically oxidized RCo(salen). I. Ya. Levitin, A. L. Sigan, and M. E. Vol'pin, *J. Organometal. Chem.*, **114**, C53 (1976).

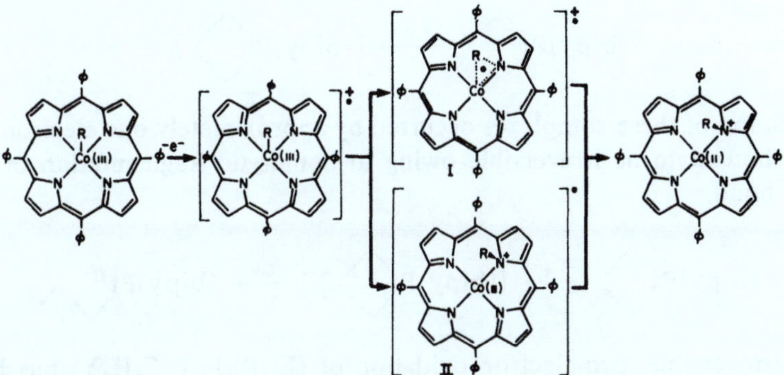

Figure 7. The reaction of an anodically oxidized Co(III)-N alkyl or acyl porphyrin. D. Dolphin, D. J. Halko, and E. Johnson, *Inorg. Chem.*, **20**, 4348 (1981).

was proposed to rationalize the products observed after electrolysis in MeCN.[339] For the R = *n*-butyl complex, for example, coulometry resulted in 1.5 F/mol of starting material, and $[Co^{III}(salen)]^+$, *n*-butene, and *o*-$BuOC_6H_4CHO$ were isolated in equal amounts.

In contrast with the vitamin B_{12} model compounds, the site of one-electron oxidation of the Co(III) prophyrin complex, ethyl(5,10,15,20-tetraphenylporphinato)cobalt(III) is the ligand (Fig. 7).[340] A pi-cation radical is created which is stable at low temperature; however, on warming to room temperature, R migration from cobalt to a ligand nitrogen occurs.

Oxidatively induced reductive eliminations of metal-bonded substituents represent a reaction class broadly important in organometallic chemistry.[22,90] Yet another example of an anodically induced elimination is

$$L_2Ni^{II}(Ar)(R) \xrightarrow{-e^-} [L_2Ni^{III}(Ar)(R)]^+ \rightarrow Ar\text{-}R + [L_2Ni^{I}]^+ \quad (74)$$

where Ar is an aryl group, such as *o*-tolyl, and R = Ar or a methyl group. All such complexes were irreversibly oxidized by one electron and yielded coupling products. When R = Ar = *o*-tolyl, for example, an excellent yield of bitolyl was obtained.[341]

The electrochemistry of platinum metallacyclobutanes provides additional evidence of oxidatively induced elimination, in addition to interesting cathodic behavior.[342] For example, the Pt(IV) platinacyclobutane in Equation 75 was reduced by two electrons with loss of Cl^- to give the Pt(II) metallacycle which was in turn reduced reversibly to the Pt(I) state.

$$(bipy)Pt(Cl)_2\diamond \xrightarrow{+2e^-} (bipy)Pt\diamond + 2\,Cl^-$$

$$(\text{bipy})\text{Pt}\langle\rangle \xrightarrow{-1.67\text{ V}} [(\text{bipy})\text{Pt}\langle\rangle]^-$$

Oxidation of these complexes occurred by approximately one electron and was thought to be irreversible owing to homolytic fragmentation of one Pt-C bond.

$$(\text{bipy})\text{Pt}\langle\rangle \xrightarrow{-e^-} [(\text{bipy})\text{Pt}\langle\rangle]^+ \xrightarrow{\text{fast}} (\text{bipy})\text{Pt}^{\text{II}} \qquad (76)$$

Irreversible, two-electron oxidation of $(Et_3P)_2PtX(C_6H_4Y)$ has been reported, but no product analysis was apparently done.[343]

5.3. Reactions at Coordinated Ligands

In this review, we have already noted a few reactions that occur at coordinated ligands as a result of electron transfer to the metal center (for example, Equations 37–39, 44, 45, and 48). In this section, we describe additional examples of this form of reactivity, as well as reactions wherein the substituent is itself the site of electron transfer.

Transition metal carbene complexes have been widely studied recently owing to their importance in catalyzed reactions, such as olefin metathesis.[2,344] Casey *et al.*, have reported a dependence of $E_{1/2}$ and the $E_{pc} - E_{pa}$ peak separation on the conformation of the aryl ring for cathodic reduction of (**67**) and its analogs.[345] Previous MO calculations had suggested that, while the HOMO energies are little different for an aryl ring co-planar with or perpendicular to the Cr-(carbene C)-O plane, the LUMO energies should differ significantly; the planar conformation was favored by about 2.5 eV. In agreement with this, electrochemical reduction of (**67**) and ring substituted derivatives did indeed show that ortho substituents led to reduction at more negative potentials. It was further suggested that the reason for an increase in ΔE_p as the reduction potential became more negative was a change in geometry on reduction, the structural change giving rise to a slower rate of electron transfer.

$(OC)_5Cr{=}C(C_6H_5)OCH_3$

67

The other electron transfer reactions of metal-carbenes involve ligand protonation following electron transfer. Saveant and co-workers examined the cathodic electrochemistry of several carbene complexes of iron porphyrins, TPPFe:C=Z (TPP = tetraphenylporphyrin; $Z = CPh_2$ and S among others).[346] The general reaction scheme for these complexes is outlined in Equation 77.

$$
\begin{array}{lcl}
Fe^{II}\!:C{=}Z & & \\
\quad +e^- \Big\Updownarrow -e^- & & \\
[Fe^{II}\!:C{=}Z]^- & \underset{-H^+}{\overset{+H^+}{\rightleftharpoons}} & Fe^{II}\text{-}CH{=}Z \\
\quad +e^- \Big\Updownarrow -e^- & & \quad +e^- \Big\Updownarrow -e^- \\
[Fe^{II}\text{-}C{=}Z]^{2-} & \underset{-H^+}{\overset{+H^+}{\rightleftharpoons}} & [Fe^{II}-CH{=}Z]^- \\
 & & \quad +e^- \Big\Updownarrow -e^- \\
 & & [Fe^{II}-CH{=}Z]^{2-}
\end{array} \tag{77}
$$

When the carbene was CCl_2, the transfer of two electrons and a proton led to carbido bridged dimer.

$$TPPFe^{II}\text{-}CCl_2 + 2e^- + H^+ \rightarrow 1/2TPPFe^{II}\!:C\!:Fe^{II}TPP + 1/2CH_2Cl_2 \tag{78}$$

The reactions above are those of metal-carbene complexes. However, a metal-carbene complex was observed as a product from the oxidation of $[RC(=O)M(CO)_5]^-$ (M = Cr(0), W(0)).[347] Anodic electrolysis of $[PhC(=O)Cr(CO)_5]^-$ gave a neutral species whose esr spectrum was very similar to the benzaldehyde radical anion. The oxidized molecule was very unstable, however, decomposing in a first order process to give as the major product the carbene complex $PhC(OH)\!:Cr(CO)_5$. The source of the hydrogen atom was the tetraethylammonium counter ion, and the resulting radical cation in turn gave ethylene. The anodic peak potential for $[PhC(=O)Cr(CO)_5]^-$ was 0.4 V; however, when an acid (CF_3COOH) was added to the solution, the acyl oxygen was protonated to give the carbene complex which underwent a one-electron oxidation at $E_p = 0.5$ V.

Arenechromium tricarbonyls are reduced irreversibly by two electrons (Section 4.2.3). However, when the $Cr(CO)_3$ moiety is pi bonded to an aromatic ketone, a one-electron, reversible reduction is observed, just as is the case for the free ketone.[348,349] For example, in the case of $[MeC(=O)C_6H_5]Cr(CO)_3$, the complex was polarographically reduced by one electron in MeCN at −1.61 V, while acetophenone was reduced at a more cathodic potential, −2.10 V. These results imply that reduction is ligand-based for such molecules, the potential shift being due to the electron withdrawing effect of $Cr(CO)_3$. Later esr studies have showed that the pi electron density in the complexed ketyl radicals is little changed by complexation with $Cr(CO)_3$, but considerable change is indicated in the sigma framework. (In *bis*-benzene chromium complexes, it is the sigma MO of the rings that are involved in metal binding.[167] (See Section 2.)

A thorough study of the cathodic electrochemistry of $(\eta^6\text{-}C_6H_6)Cr(\eta^6\text{-}$benzophenone) has been made, and it was again observed that ligand reduction occurred.[350] In DMF, two closely separated, one-electron waves were observed at −1.91 V and −2.01 V. These latter reductions were presumed to occur at the coordinated benzophenone, which in turn abstracted H^+ from the solvent. The alcoholate can be oxidized back to $[(C_6H_6)Cr(C_6H_5\text{-CO-Ph})]^+$, but it can also slowly decompose to dibenzenechromium and other organic products; benzil was isolated after work-up in oxygen.

$$\mathrm{bzCr(C_6H_5{-}CO{-}Ph)} \xrightarrow{+e^-} [bz\mathrm{Cr(C_6H_5{-}\overset{O^-}{\overset{|}{C}}{-}Ph)}]$$

$$\xrightarrow{+e^-} [bz\mathrm{Cr(C_6H_5{-}\overset{O^-}{\overset{|}{\underset{-}{C}}}{-}Ph)}]$$

$$\xrightarrow{+H^+} [bz\mathrm{Cr(C_6H_5{-}\overset{O^-}{\overset{|}{\underset{\underset{H}{|}}{C}}}{-}Ph}]$$

$$\xrightarrow{+H^+} bz_2\mathrm{Cr^0} + 1/2\mathrm{Ph{-}CO{-}\overset{O^-}{\underset{\underset{H}{|}}{C}}{-}Ph}$$

$$[bz\mathrm{Cr(C_6H_5{-}CO{-}Ph}]^+ \xleftarrow{O_2} [bz\mathrm{Cr(C_6H_5{-}\overset{O^-}{\overset{|}{\underset{\underset{H}{|}}{C}}}{-}Ph}]$$

$$[bz_2\mathrm{Cr}]^+ + 1/2\mathrm{Ph{-}CO{-}CO{-}Ph} \xleftarrow{O_2} bz_2\mathrm{Cr^0} + 1/2\mathrm{Ph{-}CO{-}\overset{O^-}{\underset{\underset{H}{|}}{C}}{-}Ph}$$

$$\mathrm{bzCr(C_6H_5{-}CO{-}Ph)} \underset{e^-}{\overset{-e^-}{\rightleftharpoons}} [bz\mathrm{Cr(C_6H_5{-}CO{-}Ph}]^+ \tag{79}$$

Ferrocenylcarbinols, $FcCR_2OH$, have been studied extensively owing to the fact that they are susceptible to loss of OH and formation of very stable alpha-ferrocenyl carbonium ions.

$$\text{Fc-C}R_2\text{-OH} + H^+ \rightarrow [\text{Fc-C}R_2]^+ + H_2O \tag{80}$$

Such carbonium ions can be polarographically reduced (in the range −0.7 to −1.1 V in water/dioxane) to the neutral radical, and this species then dimerizes.[351] In the case of Fc(Ph)CHOH, the two diastereoisomers of Fc(Ph)CH-CH(Ph)Fc were isolated after electrolysis.

$$[FcCRH]^+ + e^- \rightarrow FcCRH^{\bullet} \rightarrow FcRHC\text{-}CHRFc \qquad (81)$$

In Sections 1 and 3, we noted that some organometallic compounds have been studied in molten salts. The results of an examination of the behavior of ferrocenylmethyl quaternary ammonium derivatives are related to the ferrocenylcarbonium ion work reported above. The compounds $[FcCRHNMe_3]Br$ all underwent reversible oxidation in an $AlCl_3$-1-butylpyridium chloride melt.[352] It was also observed that, if such a solution of $[FcCH_2NMe_3]^+$ was allowed to stand for some time, decomposition occurred to give a blue-green solution. Dissociation to give $[FcCH_2]^+$ and NR_3 was suggested, the blue color possibly arising from a tautomeric form of the carbonium ion $[(Fc^{III})^+CH_2^{\bullet}]$. However, only a trace of dimeric product was observed on product work-up, the major products being ferrocene and methylferrocene.

Chemical and electrochemical reductions of $[CpFe(arene)]^+$ have been found to produce different results. For example, reduction of $[(C_5H_4\text{-}CO\text{-}R)Fe(C_6H_6)]^+$ with KBH_4 transforms the ketone into the alcohol and also reduces the C_6H_6 ligand, changing it from a *hexahapto* to a *pentahapto* ligand.[353] That is, the product is $(C_5H_4\text{-}CHOH\text{-}R)Fe(\eta^5\text{-}C_6H_7)$. In contrast, electrochemical reduction under acidic or basic conditions leads only to reduction of the ketone (Equation 82).

R
C=O
Fe⁺
PH=0, E=−1.0 V
2e⁻ +2H⁺
R
C-H
OH
Fe⁺
(82)

e⁻ + H⁺
pH = 13–14
E = −1.2 V

CROH−CROH
Fe⁺ Fe⁺
NaBH₄
Ph₃C⁺
CROH-CROH
Fe Fe

Finally, in at least two instances electrochemical oxidation has been found to lead to proton loss by a metal-bonded C_5 or C_6 ring and to a concomitant increase in hapticity of the ligand.[354,355] For example, El Murr *et al.* have reported that a cobalt diene complex (**24**) lost H^+ on oxidation to give the cobalticenium salt (Equation 21).[180-182,354] A very similar result was noted for a cyclohexadienyl-molybdenum complex, a reaction explored as a method of substitution of coordinated arene ligands.[355]

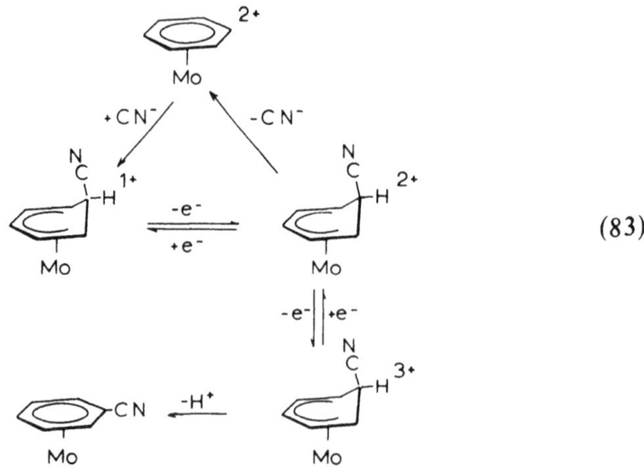

(83)

5.4. The Formation of Metal–Carbon Bonds

It is not our intention to describe fully the electrochemical synthesis of organometallic compounds. A review of the subject which discusses many of the fundamental aspects of the field appeared in 1975.[27] In this section, we wish only to mention a few recent results, especially those involving catalytic processes.

The reaction of cathodically produced, low-valent metal complexes with alkyl halides has been especially thoroughly studied. For example, Savéant and co-workers found that alkyliron porphyrins could be obtained by the reaction of an electrogenerated iron(I) porphyrin with alkyl halides.[356,357] A porphyrin such as TPPFeCl (tetraphenylporphyriniron chloride) can be reduced stepwise from Fe(III) to the Fe(I) state. The reduced complex is then thought to behave as a nucleophile in an S_{N2} type reaction with *RX*. However, in the case of the iron porphyrins, the reduction potential of the Fe(III)-*R* complex thereby produced is less negative than the potential at which TPPFe(I)Cl is produced, and the Fe(III)-*R* complex is reduced to Fe(II)-*R*. (This last point is just opposite of the situation with the corresponding cobalt complexes.)

$$\text{Fe(III)} + e^- \rightarrow \text{Fe(II)} \tag{84}$$

$$\text{Fe(II)} + e^- \rightarrow \text{Fe(I)}$$

$$\text{Fe(I)} + RX \rightarrow \text{Fe(III)-}R + X^-$$

$$\text{Fe(III)-}R + e^- \rightarrow \text{Fe(II)-}R$$

Other porphyrins than TPP were also studied, and, as expected, the reduction potentials became more negative as electron donating substituents were added. Interestingly, the reactivity of the Fe(I) porphyrin paralleled its reduction potential: the more negative the potential, the more reactive the metalloporphyrin toward a given halide. Finally, for a given porphyrin and R in RX, the order of reactivity was $R\text{I} > R\text{Br} > R\text{Cl}$.

The Fe(III)-Fe(II) couple for the alkyliron porphyrin was reversible and was generally found at about −0.9 V. There was some variation in E^0 with the porphyrin ring substituents but little with R. Oxidation of the iron center to Fe(IV) was also observed, but this approached reversibility only at high scan rates.

Compounds such as (**65**) served as good models for vitamin B_{12} owing to the fact that Co-C bond cleavage occurs on electron transfer. This feature has been used to advantage in the catalyzed reduction of alkyl halides.[358,359] The Co(I) complex, $[\text{Co(salen)}]^-$ can be generated by electrolysis from Co^{II}(salen) at potentials less than −1.3 V. In the presence of butyl bromide, for example, an organometallic intermediate is formed.

$$[\text{Co}^{\text{I}}\text{(salen)}]^- + \text{BuBr} \rightarrow [\text{Co}^{\text{III}}\text{(Bu)(Br)(salen)}]^- \tag{85}$$

and this then decomposes to the original Co(II) complex and butene, H_2, and Br^-.

$$[\text{Co}^{\text{III}}\text{(Bu)(Br)(salen)}]^- \rightarrow \text{Co}^{\text{II}}\text{(salen)} + C_4H_8 + 1/2H_2 + Br^- \tag{86}$$

While only a few papers have thus far appeared on the use of electrogenerated cobalt-containing catalysts, there are a number of papers on analogous nickel compounds.[360-369] In all cases organometallic Ni-R intermediates are implicated, and in at least several of these the intermediate complex was observed.[360,361] One case will be considered as an example.

It has been found that cathodic reduction of $[L_2\text{Ni(NCMe)}_4]^{2+}$ ($L = \text{PPh}_3$) in MeCN and in the presence of excess L gives Ni^0L_4.[360] However, addition of an aryl or alkyl halide, PhBr for example, caused the disappearance of the anodic wave of $\text{Ni}L_4$ and the appearance of two, poorly separated waves at about −2 V ($Ag/AgClO_4$). These new waves had a voltammetric profile which corresponded closely to $L_2Ni(Ph)Br$, an oxidative addition product of NiL_4 and PhBr. Electrolysis at the potential of the broad peaks produced biphenyl (with a trace of benzene) in a catalytic manner. The scheme below (Equation 87) was suggested to rationalize most of the observations made in this study. It is based largely on a previous study of the electrochemistry of authentic $L_2Ni(Ph)Br$.[370]

$$[L_2Ni(NCMe)_4]^{2+} + 2\,L + 2e^- \rightarrow NiL_4 + 4\,MeCN \qquad (87)$$

$$NiL_4 + PhBr \rightarrow L_2Ni(Ph)Br + 2\,L$$

$$2\,L_2Ni(Ph)Br + 2e^- \xrightarrow{+L,-Br^-} 2[L_3NiPh]$$

$$2[L_3NiPh] \xrightarrow{+2\,L} 2\,NiL_4 + \text{Ph-Ph}$$

$$2[L_3NiPh] \xrightarrow{+e^-} [L_3Ni(Ph)]^-$$

$$[L_3Ni(Ph)]^- \xrightarrow{+L_2Ni(Ph)Br,\ +3\,L,\ -Br^-} 2\,NiL_4 + \text{Ph-Ph}$$

$$2\,NiL_4 \xrightarrow{PhBr} 2\,L_2Ni(Ph)Br$$

Biphenyl and the catalytically important NiL_4 are produced in two ways: thermal decomposition of the Ni^{I}-Ph complex, and reduction of this complex in the presence of the original oxidative addition product, $L_2Ni(Ph)Br$. The catalytic cycle is maintained by reaction of NiL_4 with additional PhBr.

Although the cycle above was adequate to explain most observations, there were complications: (i) deactivating side reactions were noted, and (ii) alkyl halides produced small amounts of olefin, presumably by a beta-elimination from the intermediate Ni-alkyl organometallic species.

6. CONCLUDING REMARKS

Although our own research has always been in organometallic chemistry, we recognized in the late 1960s that an enormously rich field of research lay in the redox chemistry of organometallic compounds, and Professor Dessy's early survey work gave a glimpse of what was possible. Ten years later, it is now clear that the foundation has been laid: we are aware of the basic aspects of the electrochemical behavior of the various classes of organometallic compounds, of substituent effects, and of many of the major reaction pathways. The way is now clear to make use of this information in devising new synthetic methods and in using redox induced reactions as the basis of catalyzed processes as illustrated in Section 5. It is within this realm that efforts should be concentrated in the next ten years and within which the most significant progress will be made.

ACKNOWLEDGMENTS. There are many people and organizations who have enormously assisted and influenced my work in electrochemistry in many ways. I owe them a great debt of gratitude because, without this support, I could not have endeavored to organize this review. Chief among the people with whom I have worked, and from whom I have learned about electrochemistry, is Professor Richard Reed. His advice, assistance, and moral support over the years are greatly appreciated. There have also been many students who have contributed their talents to our work in electrochemistry: M. H. Garcia, L. Lin, G. Neyhart, J. Peterson, W. van der Sluys, and, especially, W. Vining. Grant support has been obtained from the American Chemical Society-Petroleum Research Fund, the State University of New York Research Foundation, and the Fulbright Commission. Finally, the Chemistry Department of SUNY-Oneonta has very generously supported our work over the past decade.

REFERENCES

1. G. Wilkinson, *J. Organometal. Chem.* **100**, 273 (1975); G. Kauffman, *J. Chem. Educ.* **60**, 185 (1983).
2. J. P. Collman and L. S. Hegedus, *Principles and Applications of Organotransition Metal Chemistry* (University Science Books, Mill Valley, CA, 1980).
3. J. A. Page and G. Wilkinson, *J. Am. Chem. Soc.* **74**, 6149 (1952).
4. R. E. Dessy, W. Kitching, and T. Chivers, *J. Am. Chem. Soc.* **88**, 453 (1966).
5. R. E. Dessy, W. Kitching, T. Psarras, R. Salinger, A. Chen, and T. Chivers, *J. Am. Chem. Soc.* **88**, 460 (1966).
6. R. E. Dessy, T. Chivers, and W. Kitching, *J. Am. Chem. Soc.* **88**, 467 (1966).
7. R. E. Dessy, F. E. Stary, R. B. King, and M. Waldrop, *J. Am. Chem. Soc.* **88**, 471 (1966).
8. R. E. Dessy, R. B. King, and M. Waldrop, *J. Am. Chem. Soc.* **88**, 5112 (1966).
9. R. E. Dessy, P. M. Weissman, and R. L. Pohl, *J. Am. Chem. Soc.* **88**, 5117 (1966).
10. R. E. Dessy, R. L. Pohl, and R. B. King, *J. Am. Chem. Soc.* **88**, 5121 (1966).
11. R. E. Dessy and P. M. Weissman, *J. Am. Chem. Soc.* **88**, 5124 (1966).
12. R. E. Dessy and P. M. Weissman, *J. Am. Chem. Soc.* **88**, 5129 (1966).
13. T. Psarras and R. E. Dessy, *J. Am. Chem. Soc.* **88**, 5132 (1966).
14. R. E. Dessy and R. L. Pohl, *J. Am. Chem. Soc.* **90**, 1995 (1968).
15. R. E. Dessy, R. Kornmann, C. Smith, and R. Haytor, *J. Am. Chem. Soc.* **90**, 2001 (1968).
16. R. E. Dessy and R. L. Pohl, *J. Am. Chem. Soc.* **90**, 2005 (1968).
17. R. E. Dessy and L. Wieczorek, *J. Am. Chem. Soc.* **91**, 4963 (1969).
18. R. E. Dessy, J. C. Charkoudian, T. P. Abeles, and A. L. Rheingold, *J. Am. Chem. Soc.* **92**, 3947 (1970).
19. R. E. Dessy, J. C. Charkoudian, and A. L. Rheingold, *J. Am. Chem. Soc.* **94**, 738 (1972).
20. R. E. Dessy, A. L. Rheingold, and G. D. Howard, *J. Am. Chem. Soc.* **94**, 746 (1972).
21. R. E. Dessy and L. A. Bares, *Acc. Chem. Res.* **5**, 415 (1972).

22. J. K. Kochi, *Organometallic Mechanisms and Catalysis* (Academic Press, New York, 1978).
23. S. P. Gubin, *Pure Appl. Chem.* **23**, 463 (1970).
24. W. E. Geiger, Jr., in *Metal Interactions with Boron Clusters*, edited by R. N. Grimes (Plenum Press, New York, 1982), pp. 239-268.
25. P. Lemoine, *Coord. Chem. Revs.* **47**, 55 (1982).
26. M. M. Bazier, ed., *Organic Electrochemistry* (Dekker, New York, 1973).
27. G. A. Tedoradze, *J. Organometal. Chem.* **88**, 1 (1975).
28. P. M. Treichel and G. E. Dirreen, *J. Organometal. Chem.* **39**, C20 (1972).
29. J. W. Dart, M. K. Lloyd, R. Mason, J. A. McCleverty, and J. Williams, *J. C. S. Dalton* **1973**, 1847.
30. P. M. Treichel and G. J. Essenmacher, *Inorg. Chem.* **15**, 146 (1976).
31. G. J. Essenmacher and P. M. Treichel, *Inorg. Chem.* **16**, 800 (1977).
32. P. M. Treichel and H. J. Mueh, *Inorg. Chem.* **16**, 1167 (1977).
33. J. M. Chatt, C. M. Elson, A. J. L. Pombeiro, R. L. Richards, and G. H. D. Royston, *J. C. S. Dalton* **1978**, 165.
34. W. S. Mialki, R. E. Wild, and R. A. Walton, *Inorg. Chem.* **20**, 1380 (1981).
35. J. Hanzlik, G. Albertin, E. Bordignon, and A. A. Orio, *J. Organometal. Chem.* **224**, 49 (1982).
36. S. Fukuzumi, N. Nishizawa, and T. Tanaka, *Bull. Chem. Soc. Japan* **55**, 2892 (1982).
37. W. S. Mialki, D. E. Wigley, T. E. Wood, and R. A. Walton, *Inorg. Chem.* **21**, 480 (1981).
38. C. Caravana, C. M. Giandomenico, and S. J. Lippard, *Inorg. Chem.* **21**, 1860 (1982).
39. D. A. Bohling, J. F. Evans, and K. R. Mann, *Inorg. Chem.* **21**, 3546 (1982).
40. D. D. Klendworth, W. A. Welters, III, and R. A. Walton, *Organometallics* **1**, 336 (1982).
41. K. F. Purcell and J. C. Kotz, *Inorganic Chemistry* (W. B. Saunders, Philadelphia, 1977).
42. J. W. Lauher and R. Hoffmann, *J. Am. Chem. Soc.* **98**, 1729 (1976).
43. J. D. L. Holloway and W. E. Geiger, Jr., *J. Am. Chem. Soc.* **101**, 2038 (1979).
44. J. D. L. Holloway, W. L. Bowden, and W. E. Geiger, Jr., *J. Am. Chem. Soc.* **99**, 7089 (1977).
45. J. D. L. Holloway, F. C. Senftleber, and W. E. Geiger, Jr., *Anal. Chem.* **50**, 1010 (1978).
46. R. R. Gagne, C. A. Koval, and G. C. Lisensky, *Inorg. Chem.* **19**, 2854 (1980).
47. H. L. Chum, V. R. Koch, L. L. Miller, and R. A. Osteryoung, *J. Am. Chem. Soc.* **97**, 3264 (1975).
48. H. M. Koepp, H. Wendt, and H. Z. Strehlow, *Z. Electrochem.* **64**, 483 (1960).
49. (a) S. P. Gubin, S. A. Smirnova, L. I. Denisovich, and A. A. Lubovich, J. *Organometal. Chem.* **30**, 243 (1971); (b) S. P. Gubin and A. A. Lubovich, *J. Organometal. Chem.* **22**, 183 (1970).
50. T. Kuwana, D. E. Bublitz, and G. Hoh, *J. Am. Chem. Soc.* **82**, 5811 (1960).
51. L. I. Denisovich, N. V. Zakurin, A. A. Bezrukova, and S. P. Gubin, *J. Organometal. Chem.* **81**, 207 (1974).
52. D. N. Hendrickson, Y. S. Sohn, W. H. Morrison, and H. B. Gray, *Inorg. Chem.* **11**, 808 (19972).
53. R. J. Gale and R. Job, *Inorg. Chem.* **20**, 42 (1981).
54. A. F. Diaz, U. T. Meuller-Westerhoff, A. Nazzal, and M. Tanner, *J. Organometal. Chem.* **236**, C45 (1982).
55. E. D. Laganis, R. H. Voegeli, R. T. Swann, R. G. Finke, H. Hopt, and V. Boekelheide, *Organometallics* **1**, 4115 (1982).
56. R. G. Finke, R. H. Voegeli, E. D. Laganis, and V. Boekelheide, *Organometallics* **2**, 347 (1983).
57. W. E. Geiger, Jr., *J. Am. Chem. Soc.* **96**, 2632 (1974).
58. W. E. Geiger, Jr., W. L. Bowden, and N. El Murr, *Inorg. Chem.* **18**, 2358 (1979).
59. S. P. Gubin, S. A. Smirnova, and L. I. Denisovich, *J. Organometal. Chem.* **30**, 257 (1971).

60. N. El Murr, J. E. Sheats, W. E. Geiger, Jr., J. D. L. Holloway, *Inorg. Chem.* **18**, 1443 (1979).
61. J-R. Hamon, D. Astruc, and P. Michaud, *J. Am. Chem. Soc.* **103**, 758 (1981).
62. E. K. Barefield, D. A. Krost, D. S. Edwards, D. G. Van Derveer, R. L. Trytko, and S. P. O'Rear, *J. Am. Chem. Soc.* **103**, 6219 (1981).
63. R. J. Gale and R. Job, *Inorg. Chem.* **20**, 40 (1981).
64. R. J. Wilson, L. F. Warren, Jr., and M. F. Hawthorne, *J. Am. Chem. Soc.* **91**, 758 (1969).
65. R. Prins, A. R. Korswagen, and A. G. T. G. Kortbeek, *J. Organometal. Chem.* **39**, 335 (1972).
66. J. L. Robbins, N. Edelstein, B. Spencer, and J. C. Smart, *J. Am. Chem. Soc.* **104**, 1882 (1982).
67. U. Koelle and F. Khouzami, *Angew. Chem. Intern. Ed. Engl.* **19**, 640 (1980).
68. J. C. Smart and J. L. Robbins, *J. Am. Chem. Soc.* **100**, 3936 (1978).
69. U. Koelle and A. Salzer, *J. Organometal. Chem.* **243**, C27 (1983).
70. Y. Mugnier, C. Moise, J. Tirouflet, and E. Laviron, *J. Organometal. Chem.*, **186**, C49 (1980).
71. H. L. M. van Gaal and J. G. M. van der Linden, *Coord. Chem. Rev.* **47**, 41 (1982).
72. G. Wilkinson and J. M. Birmingham, *J. Am. Chem. Soc.* **76**, 4281 (1954).
73. S. P. Gubin and S. A. Smirnova, *J. Organometal. Chem.* **20**, 229 (1969).
74. K. Andra, *J. Organometal. Chem.* **11**, 567 (1968).
75. S. Valcher and M. Mastragostino, *J. Electroanal. Chem.* **14**, 219 (1967).
76. R. G. Doisneau and J. C. Marchon, *J. Electroanal. Chem.* **30**, 487 (1971).
77. J. E. Bercaw, R. H. Marvich, L. G. Bell, and H. H. Brintzinger, *J. Am. Chem. Soc.* **94**, 1219 (1972).
78. T. Chivers and E. D. Ibrahim, *Can. J. Chem.* **51**, 815 (1973).
79. V. Kadlec, H. Kadlecova, and P. Strouf, *J. Organometal. Chem.* **82**, 113 (1974).
80. E. Laviron, J. Besancon, and F. Huq, *J. Organometal. Chem.* **159**, 279 (1978).
81. N. El Murr, A. Chaloyard, and J. Tirouflet, *J. C. S. Chem. Commun.* 1980, 446.
82. N. El Murr and A. Chaloyard, *J. Organometal. Chem.* **231**, 1 (1982).
83. Y. Mugnier, C. Moise, and E. Laviron, *J. Organometal. Chem.* **204**, 61 (1981).
84. R. S. P. Coutts and P. C. Wailes, *J. Organometal. Chem.* **47**, 375 (1973).
85. M. L. H. Green and C. R. Lucas, *J. C. S. Dalton* **1972**, 1000.
86. M. C. R. Symons and S. P. Mishra, *J. C. S. Dalton* **1981**, 2258.
87. G. P. Pez, *J. Am. Chem. Soc.* **98**, 8072 (1976).
88. A. Chaloyard, A. Dormond, J. Tirouflet, and N. El Murr, *J. C. S. Chem. Commun.* 1980, 214.
89. S. P. Gubin and S. A. Smirnova, *J. Organometal. Chem.* **20**, 241 (1969).
90. J. C. Kotz, W. Vining, W. Coco, R. Rosen, A. R. Dias, and M. H. Garcia, *Organometallics* **2**, 68 (1983).
91. M. F. Lappert and C. L. Raston, *J. C. S. Chem. Commun.*, 1980, 1284.
92. M. F. Lappert, C. J. Pickett, P. I. Riley, and P. I. W. Yarrow, *J. C. S. Dalton* **1981**, 805.
93. A. M. Bond, A. T. Casey, J. R. Thackeray, *Inorg. Chem.* **13**, 84 (1974).
94. Y. Mugnier, C. Moise, and E. Laviron, *Nouv. J. Chim.* **6**, 197 (1982).
95. A. M. Bond, A. T. Casey, and J. R. Thackeray, *Inorg. Chem.* **12**, 887 (1973).
96. A. M. Bond, A. T. Casey, and J. R. Thackeray, *J. C. S. Dalton* **1974**, 773.
97. R. J. Klingler, J. C. Huffman, and J. K. Kochi, *J. Am. Chem. Soc.* **102**, 208 (1980).
98. J. G. Mason and M. Rosenblum, *J. Am. Chem. Soc.* **82**, 4206 (1960).
99. K. Komenda and J. Tirouflet, *Compt. Rend.* **254**, 3093 (1962).
100. G. L. K. Hoh, W. E. McEwen, and J. Kleinberg, *J. Am. Chem. Soc.* **83**, 3949 (1961).
101. W. F. Little, C. N. Reilley, J. D. Johnson, K. N. Lynn, and A. P. Sanders, *J. Am. Chem. Soc.* **86**, 1376 (1964).
102. W. F. Little, C. N. Reilley, J. D. Johnson, and A. P. Sanders, *J. Am. Chem. Soc.* **86**, 1382 (1964).
103. D. W. Hall and C. D. Russell, *J. Am. Chem. Soc.* **89**, 2316 (1957).

104. H. Hennig and O. Gurtler, *J. Organometal. Chem.* **11**, 307 (1968).
105. M. M. Sabbatini and E. Cesarotti, *Inorg. Chim. Acta* **24**, L9 (1977).
106. D. W. Hall, E. A. Hill, and J. H. Richards, *J. Am. Chem. Soc.* **90**, 4972 (1968).
107. (a) J. Kotz and D. G. Pedrotty, *Organometal. Chem. Rev.* **4**, 479 (1969); (b) A. W. Baker and D. E. Bublitz, *Spectrochem. Acta* **22**, 1787 (1966).
108. E. Fujita, B. Gordon, M. Hillman, and A. G. Nagy, *J. Organometal. Chem.* **218**, 105 (1981).
109. T. Ogata, K. Oikawa, T. Fujisawa, S. Motoyama, T. Izumi, A. Kasahara, and N. Tanaka, *Bull. Chem. Soc. Japan* **54**, 3723 (1981).
110. J. E. Gorton, H. L. Lentzner, and W. E. Watts, *Tetrahedron* 1971, 4353.
111. R. J. Gale, K. M. Motyl, and R. Job, *Inorg. Chem.* **22**, 130 (1983).
112. K. H. Pannell, J. B. Cassias, G. M. Crawford, and A. Flores, *Inorg. Chem.* **15**, 2671 (1976).
113. J. Tirouflet, E. Laviron, C. Moise, and Y. Mugnier, *J. Organometal. Chem.* **50**, 241 (1973).
114. S. S. Crawford and H. D. Kaesz, *Inorg. Chem.* **16**, 3193 (1977).
115. J. A. Connor, E. M. Jones, and J. P. Lloyd, *J. Organometal. Chem.* **24**, C20 (1970).
116. J. A. McCleverty, D. G. Orchard, J. A. Connor, E. M. Jones, J. P. Lloyd, and P. D. Rose, *J. Organometal. Chem.* **30**, C75 (1971).
117. M. K. Lloyd, J. A. McCleverty, D. G. Orchard, J. A. Connor, M. B. Hall, I. H. Hillier, E. M. Jones, and G. K. McEwen, *J. C. S. Dalton* **1973**, 1743.
118. J. C. Kotz and E. W. Post, *Inorg. Chem.* **9**, 1661 (1970).
119. J. C. Kotz and W. J. Painter, *J. Organometal. Chem.* **32**, 231 (1971).
120. J. C. Kotz, C. L. Nivert, J. M. Lieber, and R. C. Reed, *J. Organometal. Chem.* **84**, 255 (1975).
121. J. C. Kotz, C. L. Nivert, J. M. Lieber, and R. C. Reed, *J. Organometal. Chem.* **91**, 87 (1975).
122. S. P. Gibin and V. S. Khandkarova, *J. Organometal. Chem.* **22**, 449 (1970).
123. N. El Murr, *J. Organometal. Chem.* **112**, 189 (1976).
124. L. I. Denisovich and S. P. Gubin, *J. Organometal. Chem.* **57**, 87 (1973).
125. L. I. Denisovich and S. P. Gubin, *J. Organometal. Chem.* **57**, 109 (1973).
126. J. Pladziewicz and J. H. Espenson, *J. Am. Chem. Soc.* **95**, 56 (1973).
127. N. El Murr, *J. Organometal. Chem.* **112**, 189 (1976).
128. D. B. Brown, ed., *Mixed Valence Compounds* (D. Reidel Publishing Company, Boston, 1980).
129. M. B. Robin and P. Day, *Adv. Inorg. Chem. Radiochem.* **10**, 247 (1967).
130. N. S. Hush, *Prog. Inorg. Chem.* **6**, 391 (1967).
131. D. O. Cowan, C. LeVanda, J. Park, and F. Kaufman, *Acc. Chem. Res.* **6**, 1 (1973).
132. F. Kaufman and D. O. Cowan, *J. Am. Chem. Soc.* **92**, 6198 (1970).
133. W. H. Morrison, Jr., S. Krogsrud, and D. N. Hendrickson, *Inorg. Chem.* **12**, 1998 (1973).
134. W. H. Morrison, Jr., and D. N. Hendrickson, *J. Chem. Phys.* **59**, 380 (1973).
135. W. H. Morrison, Jr., and D. N. Hendrickson, *Inorg. Chem.* **14**, 2331 (1975).
136. C. LeVanda, K. Bechgaard, and D. O. Cowan, *J. Org. Chem.* **41**, 2700 (1976).
137. M. J. Powers and T. J. Meyer, *J. Am. Chem. Soc.* **100**, 4394 (1978).
138. C. LeVanda, D. O. Cowan, and K. Bechgaard, *J. Am. Chem. Soc.* **97**, 1980 (1975).
139. C. LeVanda, K. Bechgaard, D. O. Cowan, and M. D. Rausch, *J. Am. Chem. Soc.* **99**, 2964 (1977).
140. C. LeVanda, D. O. Cowan, C. Leitch, and K. Bechgaard, *J. Am. Chem. Soc.* **96**, 6788 (1974).
141. J. A. Kramer and D. N. Hendrickson, *Inorg. Chem.* **19**, 3330 (1980).
142. J. B. Flanagan, S. Margel, A. J. Bard, and F. C. Anson, *J. Am. Chem. Soc.* **100**, 4248 (1978).
143. P. Shu, K. Bechgaard, and D. O. Cowan, *J. Org. Chem.* **41**, 1849 (1976).
144. J. A. Kramer, F. H. Herbstein, and D. N. Hendrickson, *J. Am. Chem. Soc.* **102**, 2293 (1980).
145. D. O. Cowan and C. LeVanda, *J. Am. Chem. Soc.* **94**, 9271 (1972).
146. C. LeVanda, K. Bechgaard, D. O. Cowan, U. T. Mueller-Westerhoff, P. Eilbracht, G. A. Candela, and R. L. Collins, *J. Am. Chem. Soc.* **98**, 3181 (1976).

147. G. M. Brown, T. J. Meyer, D. O. Cowan, C. LeVanda, F. Kaufman, P. V. Roling, and M. D. Rausch, *Inorg. Chem.* **14**, 506 (1975).
148. W. M. Morrison, Jr., E. Y. Ho, and D. N. Hendrickson, *Inorg. Chem.* **14**, 500 (1975).
149. Unpublished work, B. P. Sullivan and T. J. Meyer, University of North Carolina, Chapel Hill.
150. D. O. Cowan, P. Shu, F. L. Hedberg, M. Rossi, and T. J. Kistenmacher, *J. Am. Chem. Soc.* **101**, 1304 (1979).
151. R. G. Wollmann and D. N. Hendrickson, *Inorg. Chem.* **16**, 3079 (1977).
152. A. J. Fry and R. L. Krieger, *Tetrahedron Lett.* **52**, 4803 (1976).
153. A. J. Fry, P. S. Jain, and R. L. Krieger, *J. Organometal. Chem.* **214**, 381 (1981).
154. J. C. Smart and B. L. Pinsky, *J. Am. Chem. Soc.* **102**, 1009 (1980).
155. N. Dowling, P. M. Henry, N. A. Lewis, and H. Taube, *Inorg. Chem.* **20**, 2345 (1981).
156. N. Dowling and P. M. Henry, *Inorg. Chem.* **21**, 4088 (1982).
157. J. Kotz, G. Neyhart, W. J. Vining, and M. D. Rausch, *Organometallics* **2**, 79 (1983).
158. S. Colbran, B. H. Robinson, and J. Simpson, *J. C. S. Chem. Commun.* 1982, 1361.
159. H. Zeiss, P. J. Wheatley, and H. J. S. Winkler, *Benzenoid-Metal Complexes* (Ronald Press, New York, 1966).
160. N. Ito, T. Saji, K. Suga, and S. Aoyagui, *J. Organometal. Chem.* **229**, 43 (1982).
161. A. A. Vlcek, *Z. anorg. allgem. Chem.* **304**, 109 (1960).
162. H. Hsiung and G. H. Brown, *J. Electrochem. Soc.* **110**, 1085 (1963).
163. I. A. Korshunov, L. N. Vertyulina, and G. A. Domrachev, *J. Gen. Chem. USSR* **32**, 9 (1962).
164. S. Valcher and G. Casalbore, *K. Electroanal. Chem.* **50**, 359 (1974).
165. S. Valcher, G. Casalbore, and M. Mastragostino, *J. Electroanal. Chem.* **51**, 226 (1974).
166. P. M. Treichel, G. P. Essenmacher, H. F. Efner, and K. J. Klabunde, *Inorg. Chim. Acta* **48**, 41 (1981).
167. S. E. Anderson and R. S. Drago, *Inorg. Chem.* **11**, 1564 (1972).
168. D. M. Braitsch and R. Kumarappan, *J. Organometal. Chem.* **84**, C37 (1975).
169. P. Michaud, J-P. Mariot, F. Varret, and D. Astruc, *J. C. S. Chem. Commun.* **1982**, 1383.
170. E. D. Laganis, R. H. Voegeli, R. T. Swann, R. G. Finke, H. Hopf, and V. Boekelhiede, *Organometallics* **1**, 1415 (1982).
171. R. G. Finke, R. H. Voegeli, E. D. Laganis, and V. Boekelheide, *Organometallics* **2**, 347 (1983).
172. J-R. Hamon, D. Astruc, and P. Michaud, *J. Am. Chem. Soc.* **103**, 758 (1981).
173. A. N. Nesmeyanov, L. I. Denisovich, S. P. Gubin, N. A. Vol'kenau, E. I. Sirotkina, and I. N. Bolesova, *J. Organometal. Chem.* **20**, 169 (1969).
174. N. El Murr, *J. C. S. Chem. Commun.* **1981**, 251.
175. W. H. Morrison, Jr., E. Y. Ho, and D. N. Hendrickson, *Inorg. Chem.* **14**, 500 (1975).
176. H. van Willigen, W. E. Geiger, and M. D. Rausch, *Inorg. Chem.* **16**, 581 (1977).
177. A. Bou, M. A. Pericas, F. Serratosa, J. Claret, J. M. Feliu, and C. Muller, *J. C. S. Chem. Commun.* **1982**, 620.
178. J. Moraczewski and W. E. Geiger, *Organometallics* **1**, 1385 (1982).
179. U. Koelle, *Inorg. Chim. Acta* **47**, 13 (1981).
180. N. El Murr and E. Laviron, *Can. J. Chem.* **54**, 3350 (1976).
181. N. El Murr and E. Laviron, *Can. J. Chem.* **54**, 3357 (1976).
182. N. El Murr, *J. C. S. Chem. Commun.* **1981**, 219.
183. U. Koelle, *J. Organometal. Chem.* **157**, 327 (1978).
184. J. Edwin, M. Bochmann, M. C. Bohm, D. E. Brennan, W. E. Geiger, C. Kruger, J. Pebler, H. Pritzkow, W. Siebert, W. Swiridoff, H. Wadepohl, J. Weiss, and U. Zennack, *J. Am. Chem. Soc.* **105**, 2582 (1983).

185. M. Elian and R. Hoffmann, *Inorg. Chem.* **14**, 1058 (1975); M. Elian, M. M. L. Chen, D. M. P. Mingos, and R. Hoffmann, *Inorg. Chem.* **15**, 1148 (1976).
186. M. King, E. M. Holt, P. Radnia, and J. S. McKennis, *Organometallics* **1**, 1718 (1982).
187. J. A. Butcher, J. Q. Chambers, R. M. Pagni, *J. Am. Chem. Soc.* **100**, 1012 (1978).
188. J. A. Butcher, R. M. Pagni, and J. Q. Chambers, *J. Organometal. Chem.* **199**, 223 (1980).
189. N. El Murr, A. Chaloyard, and W. Klaui, *Inorg. Chem.* **18**, 2629 (1979).
190. B. F. G. Johnson, ed., *Transition Metal Clusters* (John Wiley, New York, 1980).
191. T. J. Meyer, *Prog. Inorg. Chem.* **19**, 1 (1975).
192. (a) C. J. Pickett and D. Pletcher, *J. C. S. Dalton* **1975**, 879; (b) C. J. Pickett and D. Pletcher, *J. C. S. Chem. Commun.* **1974**, 660.
193. C. J. Pickett and D. Pletcher, *J. C. S. Dalton* **1976**, 749.
194. C. J. Pickett and D. Pletcher, *J. C. S. Dalton* **1976**, 636.
195. H. L. Chum, D. Koran, and R. A. Osteryoung, *J. Organometal. Chem.* **140**, 349 (1977).
196. (a) K. F. Purcell and J. C. Kotz, *Inorganic Chemistry* (W. B. Saunders, Philadelphia, 1977), pp. 918-919; (b) W. D. Covey and T. L. Brown, *Inorg. Chem.* **12**, 2820 (1973).
197. D. F. Shriver and A. Alich, *Coord. Chem. Rev.* **8**, 15 (1972).
198. P. Lemoine and M. Gross, *J. Organometal. Chem.* **133**, 193 (1977).
199. M. Diot, J. Bousquet, P. Lemoine,and M. Gross, *J. Organometal. Chem.* **112**, 79 (1976).
200. N. El Murr and A. Chaloyard, *Inorg. Chem.* **21**, 2206 (1982).
201. (a) P. J. Krusic, J. San Filippo, Jr., B. Hutchinson, R. L. Hance, and L. M. Daniels, *J. Am. Chem. Soc.* **103**, 2129 (1981); (b) P. J. Krusic, *J. Am. Chem. Soc.* **103**, 2131 (1981).
202. P. A. Dawson, B. M. Peake, B. H. Robinson, and J. Simpson, *Inorg. Chem.* **19**, 465 (1980).
203. A. M. Bond, P. A. Dawson, B. M. Peake, B. H. Robinson, and J. Simpson, *Inorg. Chem.* **16**, 2199 (1977).
204. D. Montauzon and R. Poilblanc, *J. Organometal. Chem.* **104**, 99 (1976).
205. A. M. Bond, J. W. Bixler, E. Mocellin, S. Datta, E. J. James, and S. S. Wreford, *Inorg. Chem.* **19**, 1760 (1980).
206. J. W. Hershberger, R. J. Klingler, and J. K. Kochi, *J. Am. Chem. Soc.* **104**, 3034, (1982).
207. J. A. McCleverty, D. G. Orchard, J. A. Connor, E. M. Jones, J. P. Lloyd, and P. D. Rose, *J. Organometal. Chem.* **30**, C75 (1971).
208. M. K. Lloyd, J. A. McCleverty, D. G. Orchard, J. A. Connor, M. B. Hall, I. H. Hillier, E. M. Jones, and G. K. McEwen, *J. C. S. Dalton* **1973**, 1743.
209. K. H. Pannell, M. G. D. S. Gonzalez, H. Leano, and R. Iglesias, *Inorg. Chem.* **17**, 1093 (1978).
210. K. H. Pannell and R. Iglesias, *Inorg. Chim. Acta* **33** L161 (1979).
211. P. M. Treichel, G. E. Dirreen, and H. J. Mueh, *J. Organometal. Chem.* **44**, 339 (1972).
212. J. A. Connor, E. M. Jones, G. K. McEwen, M. K. Lloyd, and J. A. McCleverty, *J. C. S. Dalton* **1972**, 1246.
213. P. M. Treichel and J. P. Williams, *J. Organometal. Chem.* **135**, 39 (1977).
214. P. M. Treichel, D. W. Firsich, and G. P. Essenmacher, *Inorg. Chem.* **18**, 2405 (1979).
215. A. C. Sarapu and R. F. Fenske, *Inorg. Chem.* **14**, 247 (1975).
216. B. E. Bursten, *J. Am. Chem. Soc.* **104**, 1299 (1982).
217. C. J. Pickett and D. Pletcher, *J. Organometal. Chem.* **102**, 1975, 327. See also J. Chatt, C. T. Kan, G. J. Leigh, C. J. Pickett, and D. R. Stanley, *J. C. S. Dalton* **1980**, 2032.
218. A. M. Bond, D. J. Darensbourg, E. Mocellin, and B. J. Stewart, *J. Am. Chem. Soc.* **103**, 6827 (1981).
219. A. M. Bond, S. W. Carr, R. Colton, and D. P. Kelly, *Inorg. Chem.* **22**, 989 (1983).
220. J. Chatt, G. J. Leigh, H. Neukomm, C. J. Pickett, and D. R. Stanley, *J. C. S. Dalton* **1980**, 121.
221. A. M. Bond, J. A. Bowden, and R. Colton, *Inorg. Chem.* **13**, 602 (1974).

222. A. M. Bond, R. Colton, and J. J. Jackowski, *Inorg. Chem.* **17**, 105 (1978).
223. N. G. Connelly and K. R. Somers, *J. Organometal. Chem.* **113**, C39 (1976).
224. P. K. Baker, N. G. Connelly, B. M. R. Jones, J. P. Maher, and K. R. Somers, *J. C. S. Dalton* **1980**, 579.
225. S. W. Blanch, A. M. Bond, and R. Colton, *Inorg. Chem.* **20**, 755 (1981).
226. G. Zotti, S. Zecchin, and G. Pilloni, *J. Organometal. Chem.* **181**, 375 (1979).
227. C. A. Streuli, *Anal. Chem.* **32**, 985 (1960).
228. D. de Montauzon and R. Poilblanc, *J. Organometal. Chem.* **93**, 397 (1975).
229. J. Vecernik, J. Masek, and A. A. Vlcek, *J. C. S. Chem. Commun.* **1975**, 736; *Inorg. Chim. Acta* **21**, 271 (1977).
230. G. Zotti, S. Zecchin, and G. Pilloni, *J. Organometal. Chem.* **246**, 61 (1983).
231. G. Pilloni, G. Schiavon, G. Zotti, and S. Zecchin, *J. Organometal. Chem.* **134**, 305 (1977).
232. G. Pilloni, G. Zotti, and M. Martelli, *Inorg. Chem. Acta* **13**, 213 (1975).
233. G. Paliani, S. M. Murgia, and G. Cardaci, *J. Organometal. Chem.* **30**, 221 (1971) and references therein.
234. G. Cardaci, S. M. Murgia, and G. Paliani, *J. Organometal. Chem.* **77**, 253 (1974).
235. N. El Murr, M. Riveccie, and P. Dixneuf, *J. C. S. Chem. Commun.* **1978**, 552.
236. K. Broadley, G. A. Lane, N. G. Connelly, and W. E. Geiger, *J. Am. Chem. Soc.* **105**, 2486 (1983).
237. P. B. Winston, S. J. Neiter Burgmayer and J. L. Templeton, *Organometallics* **2**, 167 (1983).
238. N. El Murr, M. Riveccie, E. Laviron, and G. Deganello, *Tetrahedron Lett.* **1976**, 3339.
239. N. G. Connelly, M. D. Kitchen, R. F. D. Stansfield, S. M. Whiting, and P. Woodward, *J. Organometal. Chem.* **155**, C34 (1978).
240. (a) N. G. Connelly, R. L. Kelly, M. D. Kitchen, R. M. Mills, R. F. D. Stansfield, M. W. Whiteley, S. M. Whiting, and P. Woodward, *J. C. S. Dalton* **1981**, 1317; (b) N. G. Connelly, A. R. Lucy, R. M. Mills, J. B. Sheridan, M. W. Whiteley, and P. Woodward, *J. C. S. Chem. Commun.* **1982**, 1057.
241. N. G. Connelly and M. D. Kitchen, *J. C. S. Dalton* **1976**, 2165.
242. N. G. Connelly and M. D. Kitchen, *J. C. S. Dalton* **1977**, 931.
243. L. I. Denisovich, N. V. Zakurin, S. P. Gubin, and A. A. Ginzburg, *J. Organometal. Chem.* **101**, C43 (1975).
244. P. M. Treichel, K. P. Wagner, H. J. Mueh, *J. Organometal. Chem.* **86**, C13 (1975).
245. P. M. Treichel and D. C. Molzahn, *J. Organometal. Chem.* **179**, 275 (1979).
246. S. N. Milligan, I. Tucker, and R. D. Rieke, *Inorg. Chem.* **22**, 987 (1983).
247. J. A. McCleverty, T. A. James, and E. J. Wharton, *Inorg. Chem.* **8**, 1340 (1969).
248. P. Hydes, J. A. McCleverty, and D. G. Orchard, *J. Chem. Soc. (A)* **1971**, 3660.
249. J. Masek, *Inorg. Chim. Acta Revs.* **3**, 99 (1969).
250. W. Rogers, J. A. Page, and M. C. Baird, *J. Organometal. Chem.* **156**, C37 (1978).
251. W. N. Rogers, J. A. Page, and M. C. Baird, *Inorg. Chem.* **20**, 3521 (1981).
252. P. M. Treichel, D. C. Molzahn, and K. P. Wagner, *J. Organometal. Chem.* **174**, 191 (1979). See also P. M. Treichel and L. D. Rosenhein, *J. Am. Chem. Soc.* **103**, 691 (1981).
253. G. Zotti, R. D. Rieke, and J. S. McKennis, *J. Organometal. Chem.* **228**, 281 (1982).
254. K. Broadley, N. G. Connelly, and W. E. Geiger, *J. C. S. Dalton* **1983**, 121.
255. E. K. Barefield, D. A. Krost, D. S. Edwards, D. G. Van Derveer, R. L. Trytko, and S. P. O'Rear, *J. Am. Chem. Soc.* **103**, 6219 (1981).
256. U. Koelle and H. Werner, *J. Organometal. Chem.* **221**, 367 (1981).
257. J. A. Ferguson and T. J. Meyer, *J. C. S. Chem. Commun.* **1971**, 1544.
258. J. A. Ferguson and T. J. Meyer, *Inorg. Chem.* **10**, 1025 (1971).
259. J. A. Ferguson and T. J. Meyer, *Inorg. Chem.* **11**, 631 (1972).
260. D. E. Sherwood, Jr. and M. B. Hall, *Inorg. Chem.* **17**, 3397 (1978).

261. N. G. Connelly, J. D. Payne, and W. E. Geiger, *J. C. S. Dalton* **1983**, 295.
262. J. A. de Beer, R. J. Haines, R. Greatrex, and J. A. van Wyk, *J. C. S. Dalton* **1973**, 2341.
263. P. D. Frisch, M. K. Lloyd, J. A. McCleverty, and D. Seddon, *J. C. S. Dalton* **1973**, 2268.
264. N. G. Connelly and L. F. Dahl, *J. Am. Chem. Soc.* **92**, 7472 (1970).
265. J. C. T. R. Burckett-St. Laurent, M. R. Caira, R. B. English, R. J. Haines, and L. R. Nassimbeni, *J. S. C. Dalton* **1977**, 1077.
266. M. R. DuBois, R. C. Haltiwanger, D. J. Miller, and G. Glatzmaier, *J. Am. Chem. Soc.* **101**, 5245 (1979).
267. G. J. Kubas, P. J. Vergamini, M. P. Eastman, and K. B. Prater, *J. Organometal. Chem.* **117**, 71 (1976).
268. G. Kubas and P. J. Vergamini, *Inorg. Chem.* **20**, 2667 (1981).
269. R. D. Rieke, J. S. Arney, W. E. Rich, B. R. Willeford, Jr., and B. S. Poliner, *J. Am. Chem. Soc.* **97**, 5951 (1975).
270. S. N. Milligan and R. D. Rieke, *Organometallics* **2**, 171 (1983).
271. R. D. Rieke, I. Tucker, S. N. Milligan, D. R. Wright, B. R. Willeford, L. J. Radonovich, and M. W. Eyring, *Organometallics* **1**, 938 (1982).
272. M. K. Lloyd, J. A. McCleverty, J. A. Connor, and E. M. Jones, *J. C. S. Dalton* **1973**, 1768.
273. N. G. Connelly, Z. Demidowicz, and R. L. Kelly, *J. C. S. Dalton* **1975**, 2335.
274. N. G. Connelly and G. A. Johnson, *J. Organometal. Chem.* **77**, 341 (1974).
275. A. M. Bond, R. Colton, and J. J. Jackowski, *Inorg. Chem.* **18**, 1977 (1979).
276. C. Degrand, A. Radekci-Sudre, and J. Besancon, *Organometallics* **1**, 1311 (1982).
277. V. S. Khandkarova and S. P. Gubin, *J. Organometal. Chem.* **22**, 149 (1970).
278. Valcher and G. Casalbore, *J. Electroanal. Chem.* **50**, 359 (1974).
279. A. Ceccon, C. Corvaja, G. Gaicometti, and A. Venzo, *J. C. S. Perkin 2* **1978**, 283.
280. (a) N. G. Connelly, R. L. Kelly, and M. W. Whiteley, *J. C. S. Dalton* **1981**, 34; (b) N. G. Connelly and R. L. Kelly, *J. Organometal. Chem.* **120**, C16 (1976).
281. N. El Murr, M. Riveccie, and A. Salzer, *Inorg. Chim. Acta* **29**, L213 (1978).
282. M. Bochmann, M. Green, H. P. Kirsch, and F. G. A. Stone, *J. C. S. Dalton* **1977**, 714.
283. E. F. Ashworth, J. C. Green, M. L. H. Green, J. Knight, R. B. A. Pardy, and N. J. Wainwright, *J. C. S. Dalton* **1977**, 1693.
284. F. L. Wimmer, M. R. Snow, and A. M. Bond, *Inorg. Chem.* **13**, 1617 (1974).
285. A. M. Bond, R. Colton, and J. J. Jackowski, *Inorg. Chem.* **14**, 274 (1975).
286. A. M. Bond, B. S. Grabaric, and J. J. Jackowski, *Inorg. Chem.* **17**, 2153 (1978).
287. A. M. Bond, R. Colton, and M. J. McCormick, *Inorg. Chem.* **16**, 155 (1977).
288. A. M. Bond, B. S. Grabaric, and Z. Grabaric, *Inorg. Chem.* **17**, 1013 (1978).
289. A. M. Bond, R. Colton, and M. E. McDonald, *Inorg. Chem.* **17**, 2842 (1978).
290. A. M. Bond, D. J. Darensbourg, E. Mocellin, and B. J. Stewart, *J. Am. Chem. Soc.* **103**, 6827 (1981).
291. R. D. Rieke, H. Kojima, and K. Öfele, *J. Am. Chem. Soc.* **98**, 6735 (1976).
292. R. D. Rieke, H. Kojima, and K. Öfele, *Angew. Chem. Intern. Ed. Engl.*, **19**, 538 (1980).
293. K. A. Connor and R. A. Walton, *Organometallics* **2**, 169 (1983).
294. J. C. Bailar, Jr., *J. Inorg. Nucl. Chem.* **8**, 165 (1958).
295. J. Moraczewski and W. E. Geiger, Jr., *J. Am. Chem. Soc.* **101**, 3407 (1979).
296. B. Tulyathan and W. E. Geiger, Jr., *J. Electroanal. Chem.* **109**, 325 (1980).
297. J. Moraczewski and W. E. Geiger, Jr., *J. Am. Chem. Soc.* **103**, 4779 (1981).
298. T. A. Albright, W. E. Geiger, Jr., J. Morazcewski and B. Tulyathan, *J. Am. Chem. Soc.* **103**, 4787 (1981).
299. M. Grzeszczuk, D. E. Smith, and W. E. Geiger, Jr., *J. Am. Chem. Soc.* **105**, 1772 (1983).
300. G. Lane and W. E. Geiger, Jr., *Organometallics* **1**, 401 (1982).
301. J. Moraczewski and W. E. Geiger, Jr., *J. Am. Chem. Soc.* **100**, 7429 (1978).

302. J. M. Saveant, *Acc. Chem. Res.* **13**, 323 (1980).
303. M. Arewgoda, B. H. Robinson, and J. Simpson, *J. Am. Chem. Soc.* **105**, 1893 (1983).
304. (a) D. R. Kidd and T. L. Brown, *J. Am. Chem. Soc.* **100**, 4095 (1978); (b) H. W. Hoffman and T. L. Brown, *Inorg. Chem.* **17**, 613 (1977); (c) M. Absi-Halabi, J. D. Atwood, N. P. Forbus, and T. L. Brown, *J. Am. Chem. Soc.* **102**, 6248 (1980).
305. J. W. Hershberger and J. K. Kochi, *J. C. S. Chem. Commun.* **1982**, 212.
306. J. W. Hershberger, R. J. Klinger, and J. K. Kochi, *J. Am. Chem. Soc.* **105**, 61 (1983).
307. J. W. Hershberger, C. Amatore, and J. K. Kochi, *J. Organometal. Chem.* **250**, 345 (1983).
308. D. P. Summers, J. C. Luong, and M. S. Wrighton, *J. Am. Chem. Soc.* **103**, 5238 (1981).
309. B. A. Narayanan, C. Amatore, and J. K. Kochi, *J. C. S. Chem. Commun* **1983**, 397.
310. G. J. Bezens, P. H. Rieger, and S. Visco, *J. C. S. Chem. Commun.* **1981**, 265.
311. A. Darchen, C. Mahe, and H. Patin, *J. C. S. Chem. Commun.*, **1982**, 243.
312. C. M. Arewgoda, B. H. Robinson, and J. Simpson, *J. C. S. Chem. Commun.* **1982**, 284.
313. M. Arewgoda, P. H. Rieger, B. H. Robinson, J. Simpson, and S. J. Visco, *J. Am. Chem. Soc.* **104**, 5633 (1982).
314. J. C. Kotz, J. V. Petersen, and R. C. Reed, *J. Organometal. Chem.* **120**, 433 (1976).
315. (a) A. M. Bond, P. A. Dawson, B. M. Peake, P. H. Rieger, B. H. Robinson, and J. Simpson, *Inorg. Chem.* **18**, 1413 (1979); (b) B. M. Peake, P. H. Rieger, B. H. Robinson, and J. Simpson, *Inorg. Chem.* **18**, 1000 (1979); (c) B. M. Peake, P. H. Reiger, B. H. Robinson, and J. Simpson, *J. Am. Chem. Soc.* **102**, 156 (1980).
316. R. J. Klingler, J. C. Huffman, and J. K. Kochi, *J. Am. Chem. Soc.* **102**, 208 (1980).
317. S. M. B. Costa, A. R. Dias, and F. J. S. Pina, *J. Organometal. Chem.* **175**, 193 (1979); **217**, 357 (1981).
318. E. J. Kuhlman and J. J. Alexander, *Coord. Chem. Rev.* **33**, 195 (1980).
319. G. W. Daub, *Prog. Inorg. Chem.* **22**, 409 (1977).
320. H. Berke and R. Hoffmann, *J. Am. Chem. Soc.* **100**, 7224 (1978).
321. J. K. Kochi, *Organometallic Mechanisms and Catalysis* (Academic Press, New York, 1978), p. 560.
322. (a) R. H. Magnuson, R. Meirowitz, S. Zulu, and W. P. Giering, *J. Am. Chem. Soc.* **104**, 5790 (1982); (b) R. H. Magnuson, S. Zulu, W-M. T'sai, and W. P. Giering, *J. Am. Chem. Soc.* **102**, 6887 (1980).
323. R. H. Magnuson, R. Meirowitz, S. J. Zulu, and W. P. Giering, *Organometallics* **2**, 460 (1983).
324. R. J. Klingler and J. K. Kochi, *J. Organometal. Chem.* **202**, 49 (1980).
325. W. N. Rogers, J. A. Page, and M. C. Baird, *J. Organometal. Chem.* **156**, C37 (1978); *Inorg. Chem.* **20**, 3521 (1981).
326. K. H. Pannell, J. B. Cassias, G. M. Crawford, and A. Flores, *Inorg. Chem.* **15**, 2671 (1976).
327. W. Lau, J. C. Huffman, and J. K. Kochi, *Organometallics* **1**, 155 (1982).
328. J. M. Pratt, *Inorganic Chemistry of Vitamin B_{12}* (Academic Press, New York, 1972).
329. D. G. Brown, *Prog. Inorg. Chem.* **18**, 177 (1973).
330. (a) D. Lexa and J-M. Saveant, *J. C. S. Chem. Commun.* **1975**, 872; (b) D. Lexa, J-M. Saveant, and J. Zickler, *J. Am. Chem. Soc.* **99**, 2786 (1977); (c) D. Lexa and J-M. Saveant, *J. Am. Chem. Soc.* **100**, 3220 (1978); (d) N. R. de Tacconi, D. Lexa, and J-M. Saveant, *J. Am. Chem. Soc.* **101**, 467 (1979); (e) D. Lexa, J-M. Saveant, and J. Zickler, *J. Am. Chem. Soc.* **102**, 2654 (1980); (f) D. Lexa, J-M. Saveant, and J. Zickler, *J. Am. Chem. Soc.* **102**, 4851 (1980).
331. C. M. Elliott, E. Hershenhart, R. G. Finke, and B. L. Smith, *J. Am. Chem. Soc.* **103**, 5558 (1981).
332. G. Costa, A. Puxeddu, and E. Reisenhofer, *J. C. S. Dalton* **1975**, 1519.

333. J-P. Costes, G. Cros, M-H. Darbieu, and J-P. Laurent, *Transition Met. Chem.* 7, 219 (1982).
334. D. J. Brockway, B. O. West, and A. M. Bond, *J. C. S. Dalton* **1979**, 1891.
335. (a) W. H. Tamblyn, R. J. Klingler, W. S. Hwang, and J. K. Kochi, *J. Am. Chem. Soc.* **103**, 3161 (1981); (b) R. J. Klingler and J. K. Kochi, *J. Am. Chem. Soc.* **104**, 4186 (1982).
336. J. Halpern, M. S. Chan, J. Hanson, T. S. Roche, and J. A. Topich, *J. Am. Chem. Soc.* **97**, 1606 (1975).
337. I. Levitin, A. L. Sigan, and M. E. Vol'pin, *J. C. S. Chem. Commun.* **1975**, 469.
338. B. Akermark, M. Almemark, and A. Jutand, *Acta Chem. Scand. B* **36**, 451 (1982).
339. I. Ya. Levitin, A. L. Sigan, and M. E. Vol'pin, *J. Organometal. Chem.* **114**, C53 (1976).
340. D. Dolphin, D. J. Halko, and E. Johnson, *Inorg. Chem.* **20**, 4348 (1981).
341. M. Almemark and B. Akermark, *J. C. S. Chem. Commun.* **1978**, 66.
342. R. J. Klingler, J. C. Huffman, and J. K. Kochi, *J. Am. Chem. Soc.* **104**, 2147 (1982).
343. R. Seeber, G. A. Mazzocchin, D. Minniti, R. Romeo, P. Uguagliati, and U. Belluco, *J. Organometal. Chem.* **157**, 69 (1978).
344. M. Gilet, A. Mortreaux, J-C. Folest, and F. Petit, *J. Am. Chem. Soc.* **105**, 3876 (1983).
345. C. P. Casey, L. D. Albin, M. C. Saeman, and D. H. Evans, *J. Organometal. Chem.* **155**, C37 (1978).
346. J-P. Battioni, D. Lexa, D. Mansuy, and J-M. Saveant, *J. Am. Chem. Soc.* **105**, 207 (1983).
347. R. J. Klingler, J. C. Huffman, and J. K. Kochi, *Inorg. Chem.* **20**, 34 (1981).
348. V. S. Khandkarova and S. P. Gubin, *J. Organometal. Chem.* **22**, 149 (1970).
349. A. Ceccon, C. Corvaja, G. Giacometti, and A. Venzo, *J. C. S. Perkin II* **1978**, 283.
350. S. Valcher and G. Casalbore, *J. Electroanal. Chem.* **50**, 359 (1974).
351. J. Tirouflet, E. Laviron, C. Moise, and Y. Mugnier, *J. Organometal. Chem.* **50**, 241 (1973).
352. R. J. Gale, K. M. Motyl, and R. Job, *Inorg. Chem.* **22**, 130 (1983).
353. E. Roman, D. Astruc, and A. Darchen, *J. Organometal. Chem.* **219**, 221 (1981).
354. N. El Murr, Y. Dusausoy, J. E. Sheats, and M. Agnew, *J. C. S. Dalton* **1979**, 901.
355. W. E. Silverthorn, *J. Am. Chem. Soc.* **102**, 842 (1980).
356. D. Lexa, J. Mispelter, and J-M. Saveant, *J. Am. Chem. Soc.* **103**, 6806 (1981).
357. D. Lexa and J-P. Saveant, *J. Am. Chem. Soc.* **104**, 3505 (1982).
358. A. Puxeddu, G. Costa, and N. Marsich, *J. C. S. Dalton* **1980**, 1489.
359. D. Pletcher, *Chem. Soc. Rev.* **1975**, 471.
360. G. Schiavon, G. Bontempelli, and B. Corain, *J. C. S. Dalton* **1981**, 1074.
361. M. Troupel, Y. Rollin, S. Sibille, J. Perichon, and J-F. Fauvarque, *J. Organometal. Chem.* **202**, 435 (1980).
362. K. P. Healy and D. Pletcher, *J. Organometal. Chem.* **161**, 109 (1978).
363. C. Gosden and D. Pletcher, *J. Organometal. Chem.* **186**, 401 (1980).
364. M. Troupel, Y. Rollin, S. Sibille, J-F. Fauvarque, and J. Perichon, *J. Chem. Research* (*S*) **1980**, 24; (*M*), 173.
365. S. Sibille, M. Troupel, J-F. Fauvarque, and J. Perichon, *J. Chem. Research* (*S*) **1980**, 147; (*M*), 2201.
366. M. Trouple, Y. Rollin, S. Sibille, J-F. Fauvarque, and J. Perichon, *J. Chem. Research* (*S*) **1980**, 26.
367. S. Sibille, J-C. Folest, J. Coulombeix, M. Troupel, J-F. Fauvarque, and J. Perichon, *J. Chem. Research* (*S*) **1980**, 268.
368. Y. Rollin, M. Troupel, J. Perichon, and J-F. Fauvarque, *J. Chem. Research* (*S*) **1981**, 322; (*M*), 3801.
369. M. Troupel, Y. Rollin, and J. Perichon, *Nouveau Journal de Chemie* **5**, 621 (1981).
370. G. Schiavon, G. Bontempelli, M. De Nobili, and B. Corain, *Inorg. Chim. Acta* **42**, 211 (1980).

4

Organic Photoelectrochemistry

Marye Anne Fox

1. INTRODUCTION

The term photoelectrochemistry refers to the study of redox reactions which are initiated or assisted by the absorption of a photon at or near the surface of an electrode.[1] Organic photoelectrochemistry then focuses on those transformations which involve carbon-based functional group conversions. As with conventional electrochemical measurements, the emphasis of these investigations can be analytical, mechanistic, or synthetic. Thus, this area of research applies the principles of photoactive electrochemistry as a useful method for studying organic redox reactions. A major objective is to discover processes in which photocatalyzed charge transfer at the electrode surface can govern the rates and/or selectivity of chemical reactions.

1.1. The Photoelectrochemical Cell

In conventional electrochemistry, the most easily characterized electrochemical events are those which occur at the interface of a solid conductor or semiconductor immersed in a liquid electrolyte containing the redox couple of interest. An electrochemical cell is completed when this half-cell is connected with a counterelectrode, usually a conductive solid, in such a way that charge exchange between the two electrolyte solutions can readily occur. A typical cell is shown in Fig. 1. By applying an external

Marye Anne Fox • Department of Chemistry, University of Texas at Austin, Austin, Texas 78712.

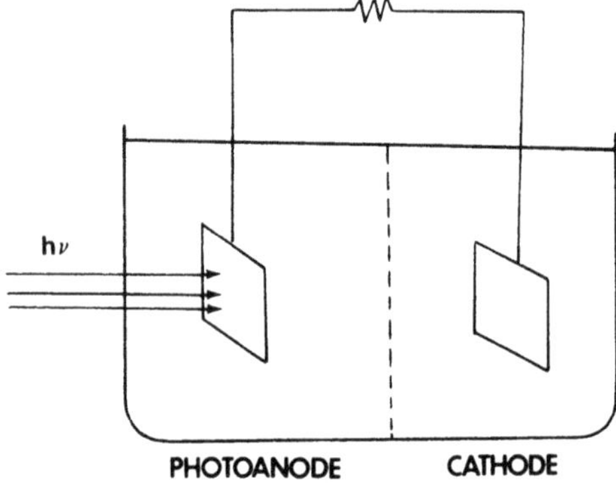

Figure 1. A photoelectrochemical cell.

potential to shift the Fermi level of the working electrode, changes in the equilibrium composition of the anodic and cathodic compartments are induced as electrons move through the external circuit from the anode to the cathode. In lieu of product characterization, the electron exchange can be monitored analytically by observing changes in potential between the two electrodes under constant current conditions or by altering the current density at a constant applied potential.

In the photoelectrochemical experiment, an analogous cell is employed, but the cell equilibrium is perturbed not by current or potential changes from an external power source but rather by photopotentials or photovoltages generated by irradiation of the working electrode. Intimately associated with these photoelectrochemical studies, then, are investigations of the nature of the solid–liquid junction.

At this interface, either the electrode itself or an absorbed sensitizer or substrate can function as the initial light absorber. The goal of organic photoelectrochemistry is a complete description of how the absorption of light perturbs interactions at this interface and ultimately induces observable changes in reagents present in the electrolyte.

1.2. Electron Transfer in the Excited State

It has been recognized for some time among organic photochemists that enhanced redox reactivity, both oxidative and reductive, is observed upon generating an excited state of an organic species. A conceptual depiction of this improved electron transfer ability is shown in Fig. 2. Electronic excitation causes the promotion of an electron to an energetically higher lying orbital, creating a vacancy in a lower-lying orbital. The excited state of an isolated molecule in the gas phase, shown at the right of the scheme, will therefore exhibit both an enhanced electron affinity (since

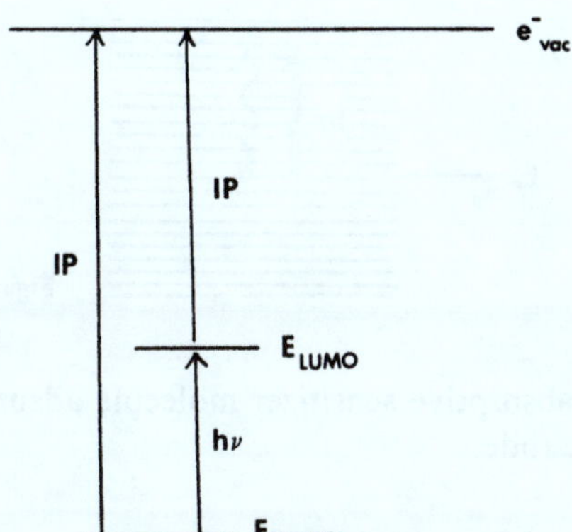

Figure 2. Excited state ionization.

more energy is to be gained by population of the vacancy caused by excitation than by populating the ground state LUMO) and ionization potential (since the higher lying electron is more easily removed from an excited than a ground state) when compared with its ground state. Upon imbedding such a molecule into a solvent or into a polar electrolyte, the highest occupied and lowest unoccupied energy levels will be stabilized by solvation, but not nearly so significantly as will the charged species formed upon ionization or electron attachment. Thus, electron transfers in solvents of reasonable dielectric constant should occur readily upon photoexcitation. In particular, either oxidation or reduction of an excited molecule adsorbed on the electrode surface should be facile.

Light absorption in a semiconductor is quite analogous to excitation of organic species. In a semiconductor, the highest occupied molecular orbital (HOMO) lies very close energetically to a series of other occupied orbitals. The close spacing of these energy levels has prompted chemists to refer to these filled orbitals as the valence band, whose energy can be characterized by estimating the energy of its band edge (E_v). In comparable analogy with lowest unoccupied molecular orbital (LUMO) of a simple organic molecule, the semiconductor contains a series of closely spaced vacant orbitals. These energy levels form a near continuum which is referred to as the conduction band, characterized by a band edge energy (E_c). The gap between the nearly filled valence band and the nearly vacant conduction band is referred to as the energy gap or band gap (E_g).

1.3. The Photoelectrochemical Experiment

Let us consider how a photoelectrochemical oxidation might occur. Two modes of excitation are possible: excitation of a photosensitive electrode in the presence of an optically transparent substrate or excitation of

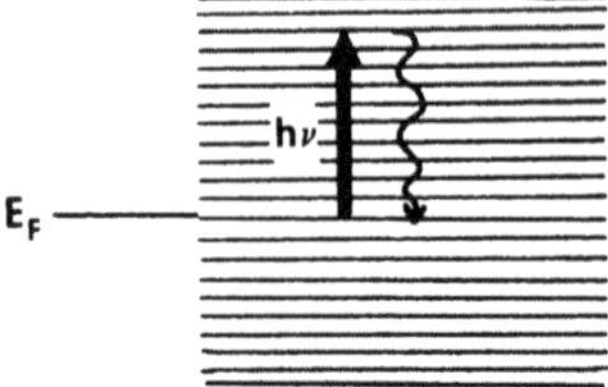

Figure 3. Excitation of a metal electrode.

an absorptive sensitizer molecule adsorbed on the surface of a photoinert electrode.

1.3.1. Irradiated Metal Electrodes

In metals or other conductors, an energetic continuum of states exists, Fig. 3. The Fermi level, the equilibrium potential of the cell, is defined by an equilibration with a redox couple present within the electrolyte. If photons are absorbed by the metal, a rapid electronic transition occurs, but because of the close energetic spacing of the accessible levels, the excited state is degraded extremely rapidly, without effective electron-hole pair separation.

The only photoeffects which can be observed at metal electrodes, therefore, result not from excitation of the electrode but rather by irradiation of absorptive molecules present in the electrolyte. In a typical photogalvanic cell, absorption of light by a molecule in solution initiates an endothermic product-forming sequence in which energy is stored. The observed electrochemical effects are caused by the dark thermal re-equilibration of the system. Since this cell produces electricity as a result of perturbations of the electrolyte composition caused by photoexcitation, only photogalvanic effects are seen with metallic electrodes. Photogalvanic cells are, in general, much less stable and much less efficient energy converters than are the semiconductor liquid junction cells to be discussed below.

1.3.2. Irradiated Semiconductor Electrodes

With a semiconductor electrode, a quantum mechanically defined gap exists between a filled and a vacant series of energy levels, Fig. 4. This gap introduces an energetic discontinuity between the photogenerated electron and the hole caused by its promotion to the higher-lying level. That is, the photoinduced electronic transition from the filled (valence) band to the vacant (conduction) band of a semiconductor generates both a high-potential electron capable of effecting reductions and a highly oxidizing hole, or electron deficient site, capable of initiating oxidations. If these redox sites

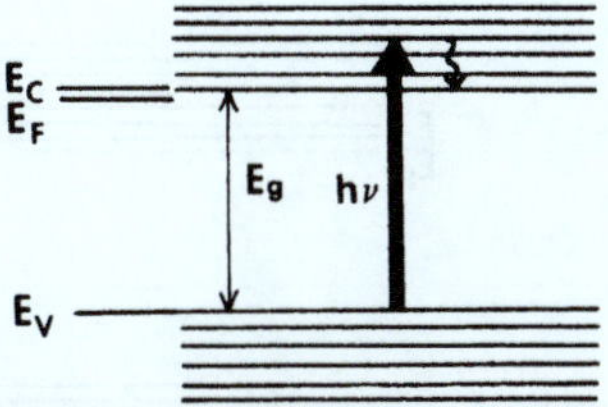

Figure 4. Excitation of a semiconductor electrode.

interact effectively with a reversibly adsorbed redox reagent, organic electrooxidations or electroreductions of appropriate molecules present in the electrolyte can be induced.

The bandgap of a given semiconductor defines a threshold energy for detecting photoeffects. The bandgaps, together with the threshold wavelengths and band edge positions, of several common semiconductors are listed in Table 1.[2,19] As can be readily seen, a number of materials are available for which visible photons can provide bandgap excitation. The smaller is the bandgap of a material, the more efficient in principle will that substance be in utilizing a larger fraction of the energy available from incident solar wavelengths. Unfortunately, it is precisely the low-bandgap materials which are least stable in photoelectrochemical operation, and only the metal oxides (TiO_2, SnO_2, $SrTiO_3$, ZnO, etc.) are sufficiently robust for the long-term irradiations necessary for photocatalysis of chemical redox transformations.

In Fig. 5 is represented the excitation of an *n*-type semiconductor electrode, which has been allowed to equilibrate with a liquid electrolyte which contains the redox couple of interest. Absorption of a photon at the

TABLE 1
Band Positions[a] in Some Common Semiconductors

Semiconductor	E_v (V vs. SCE)	E_c (V vs. SCE)	Eg (V)	λ Eg (nm)
TiO_2	3.0	0.0	3.0	410
SnO_2	3.8	+0.3	3.5	350
ZnO	2.9	−0.1	3.0	390
$SrTiO_3$	3.0	−0.2	3.2	390
CdS	2.2	−0.2	2.4	520
CdSe	1.4	−0.3	1.7	730
CdTe	1.4	0.0	1.4	890
GaAs	0.8	−0.6	1.4	890
GaP	1.4	−0.9	2.3	540
InP	1.2	−0.1	1.3	950

[a] Approximate positions in water at pH 1.

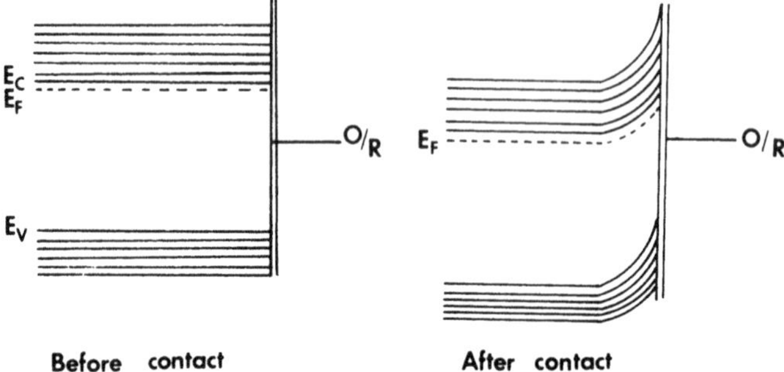

Figure 5. Equilibration of a semiconductor electrode with a dissolved redox couple.

solid-liquid interface initiates an electronic redistribution which ultimately gives rise to a redox sequence in the liquid phase.

The Fermi level is located within the bandgap in intrinsic (undoped) semiconductors. It is the movement of the Fermi level in response to photostimulation which is responsible for photoelectrochemical charge separation. Besides responding to excitation, however, the Fermi level can be shifted by doping. If small quantities of an electron donor are mixed into the semiconductor crystal, partial occupancy of the conduction band results, and the Fermi level increases (Fig. 6). The resulting system is electron-rich (partially *n*egatively charged) and is therefore called an *n*-type material. The proximity of the Fermi level to the conduction band eases the promotion of electrons into the conduction band, thus generating a hole. Oxidations are, therefore, commonly observed on *n*-type semiconductors.

If the intrinsic semiconductor is doped instead with an electron acceptor, the effective Fermi level moves closer to the valence band (Fig. 7), and the material is said to be *p*-type doped in accord with the partial *p*ositive charge. Reductions are often observed upon photoexcitation of *p*-type semiconductors.

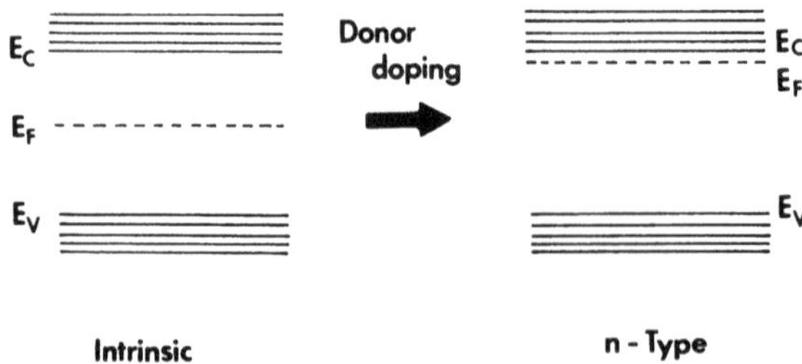

Figure 6. Response of the Fermi level to *n*-type doping.

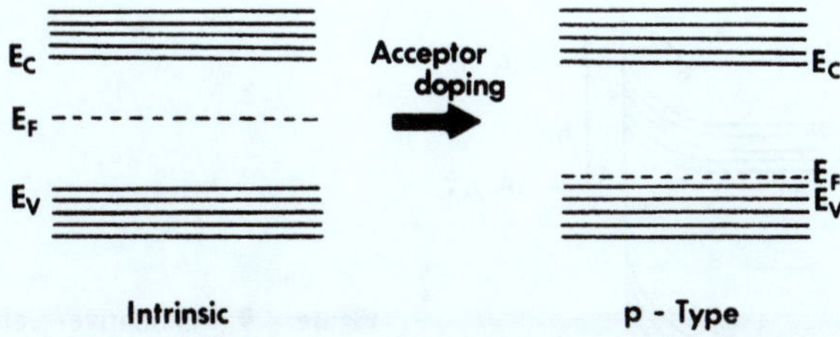

Figure 7. Response of the Fermi level to *p*-type doping.

The Fermi level of a given semiconductor with a fixed dopant can be further manipulated by altering the quantity of the dopant, by applying an external potential, or by immersing the doped semiconductor into a redox-equilibrated electrolyte. Thus, control of the Fermi level by a bias potential allows one additional control of the photoelectrochemical activity of a given electrochemical system.

Irradiation of a semiconductor with light of wavelength sufficiently energetic to induce bandgap transitions creates an electron-hole pair. That is, a vacancy is created in the valence band as the electron moves to the conduction band. If the photogenerated hole lies at a potential positive of the HOMO of the adsorbed molecule, an electrooxidation which was blocked in the equilibrated ground state becomes facile. The ultimate course of this sequence, then, is to transfer an electron from the adsorbed species' HOMO ultimately into the conduction band of the semiconductor, Fig. 8.

1.3.3. Dye Sensitization

In the alternate mode of photoactivation, called dye sensitization (Fig. 9), the absorbed photon causes the promotion of an electron from the dye's highest occupied molecular orbital to its lowest unoccupied molecular orbital. If the HOMO lies within the bandgap of a semiconductor electrode, electron transfer to the conduction band will not be favorable. If the LUMO,

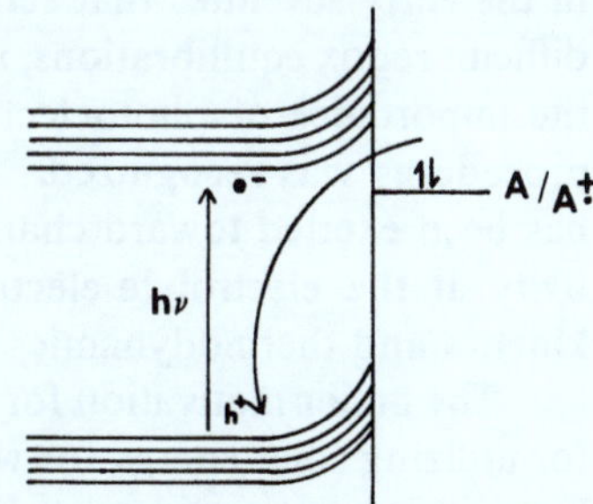

Figure 8. Oxidative electron exchange at an irradiated *n*-type semiconductor electrode/electrolyte system.

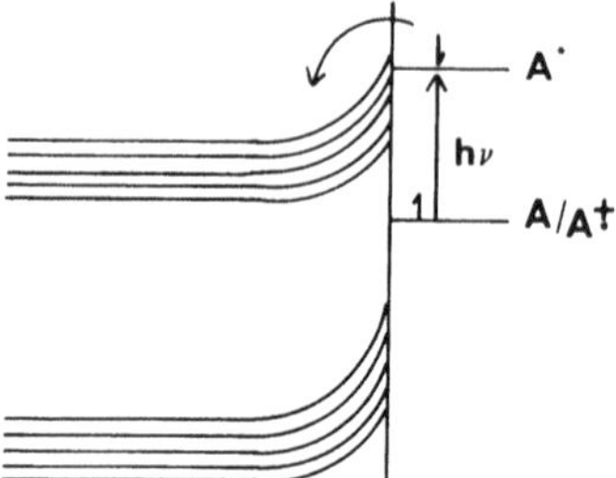

Figure 9. Oxidative electron exchange at a dye-sensitized *n*-type semiconductor electrode.

however, lies at a higher energy, i.e., at a more negative potential than the conduction band edge, electron injection into the semiconductor becomes reasonable. Dye excitation thus renders it highly oxidizable: an electron initially held within its HOMO is transferred to the conduction band of the semiconductor. The adsorbed dye then becomes electron deficient and can interact chemically with nucleophiles, either by bond formation or by further electron transfer. In either case, oxidative product formation ensues.

The simplified picture offered here does not account for dynamic adsorption-readsorption events, nor does it specifically address a number of closely related effects which can profoundly influence the energetics at the semiconductor-electrolyte interface. In particular, a clear depiction of the nature of surface states and their effect on photoelectrochemical efficiencies is still not available.[3] Similarly, the relative importance of hot-carrier processes[4] (where a photogenerated electron or hole reacts before attaining thermal equilibrium with the band edges) and Fermi level pinning[5] (where band bending in a semiconductor contacting a metal seems to be independent of the work function of the metal) is still unresolved.

1.4. Photoelectrochemical Redox Reactivity

The ability of light to bring about enhanced redox reactivity has long been recognized. It is, however, only since Fujishima and Honda's discovery in the early seventies that semiconductor surfaces could be used to effect difficult redox equilibrations, i.e., to generate hydrogen gas from water, that the importance of photoelectrochemical assistance to normal electrolytic procedures was recognized.[6] Since their initial report, a significant effort has been exerted toward characterizing those features which control reactivity at the electrolyte-electrode interface, and which govern both the kinetics and thermodynamics of these redox processes.

The major motivation for these studies has been in devising techniques for utilizing solar energy for water splitting,[7] and indeed great progress has been made both in understanding the fundamental physics of such interfaces and developing novel techniques for the use of these procedures in economically important redox conversions.

Despite this interest in water splitting and in other inorganic redox reactions, for example, chloride oxidation, carbon dioxide reduction, etc., little work has been reported which describes redox reactivity of organic substrates. Since the ability to control redox reactivity will be significantly influenced by the heterogeneous nature of the electrode-electrolyte interface,[8] further studies in this area promise to provide exciting frontiers for both the control of organic redox reactivity and for the discovery of new synthetic routes of potential interest to organic chemists.

2. PRINCIPLES OF PHOTOELECTROCHEMISTRY

Photoelectrochemical events can be initiated either by photoexcitation of the electrode material or by sensitization of the organic substrate at the surface of a poised electrode. In this article, we will discuss both types of photoactivation of redox processes.

2.1. Electron-Hole Pair Formation

Inherent in both techniques for photoactivation is the concept of generation of an electron-hole pair upon absorption of the photon. In photoelectrochemical applications, the excitation is localized at the interface, so the photogenerated carriers need only move small distances before they are converted to chemical redox systems. The promotion of an electron from a filled to a vacant orbital upon absorption of a photon generates both a high-lying electron and a vacancy in a lower-lying orbital. The major obstacle in directing reactivity from such an excited state is the maintenance of the separated electron-hole pair for a time sufficient to allow chemical interception of these redox equivalents and, hence, to initiate the formation of useful products. In the absence of a method for separation of the electron and hole, rapid back electron transfer will ensue with concomitant regeneration of the ground state. Should this reverse electron transfer not be inhibited, little photoactivation will occur.

The major effort, then, in photoelectrochemical schemes involves modification of the solid-liquid junction, the environment in which the electron-hole pair is born, to physically separate the oxidizing and reducing sites and, hence, to inhibit spontaneous-back electron transfer.[8] If an electrical or chemical potential gradient can be established at the interface, the photogenerated charge carriers can be effectively collected, i.e., converted to redox products, and kinetically distinct reactivity from the oxidizing and reducing components of the redox pair could be realized. Since normal excited state lifetimes are very short (ranging typically from nanoseconds to milliseconds in solution), the requirement for modification from usual reactivity may be quite severe. The choice of a particular semiconductor will dictate the energetics of the electron exchange. Whether a given organic reaction will proceed with a given semiconductor will be determined by the wavelength sensitivity threshold ($\lambda >$ band gap, Table 1),[2] the specific valence and conduction band edge positions (determined by flat band measurements[9]), the surface stability (controlled by the chemical reactivity of the semiconductor toward competitive dissolution or corrosion), the thickness of the space-charge layer (controlled by the doping level), and the crystal structure at the interface (single crystal, polycrystalline, or surface treated). Under optimum conditions, the photogenerated e^{-}-h^{+} pair can be converted to oxidizing and reducing equivalent species at the electrode surface. When this electron-hole separation is attained, light energy is stored as chemical potential.

2.2. Band Bending

When a semiconductor is immersed in an electrolyte, charge transfer occurs at the interface because of the difference in the tendency of the two phases to gain or lose electrons. This flow of electrons can be attributed to differences in electrochemical potential of the two phases. When a semiconductor is so immersed, an electrical field forms at the surface with a depth ranging from 2–500 *nm*. The bands of the semiconductor thus bend in response to the electronic demand of the redox couple present in the electrolyte, Fig. 5.

Band bending will occur until the bulk properties of the electrode equilibrate with the redox potential of the electrolyte, which in turn is governed by the Nernst equation. Here, immersion of the semiconductor into the electrolyte induces an electrical field, i.e., creates the space-charge region. This electrical field induces curvature of the bands downward with an electron-rich semiconductor and upward with an electron-poor semiconductor. Under irradiation, the degree of band bending is altered, influencing the efficiency of electron exchange across the interface.

An *n*-type semiconductor, one doped with an electron donor so that extra electrons populate the conduction band, forms its electrical field so that electrons will flow from the interface toward the bulk of the semiconductor. Thus photoexcitation will cause the hole, or oxidizing equivalent, to migrate toward the surface as the photogenerated electron, promoted by light absorption to the conduction band, moves toward the bulk of the electrode. As a result of this electronic motion, *n*-type semiconductors

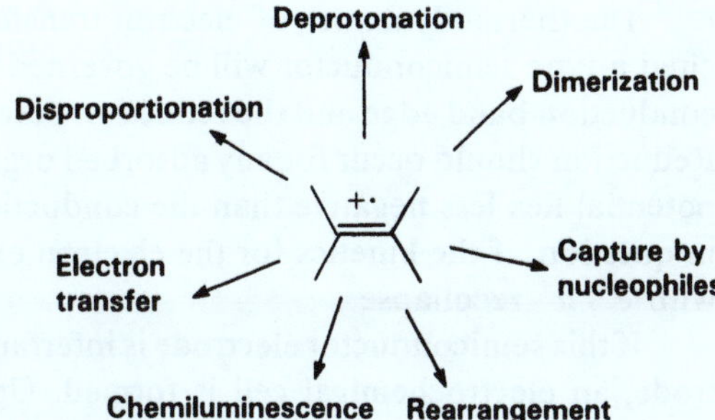

Figure 10. Chemical reactivity of radical cations.

become highly oxidizing and photoelectrochemical oxidations of adsorbed organic substrates become conceivable.

The thermodynamic viability of an electrooxidative transfer will be controlled by the relative potentials of the valence band edge of the excited semiconductor and the oxidation potential of the adsorbed donor. In principle, any organic material having an oxidation potential less positive than the valence band of the semiconductor should act as an electron donor to an excited *n*-type electrode. Electron donation from a neutral organic substrate will generate a radical cation, a reactive species whose chemistry has been extensively discussed.[10] Figure 10 represents some of the numerous chemical routes which are available to radical cations. It is clear that useful chemistry can result if the semiconductor interface can direct reactivity among one of the many assortments of reactions shown as possibilities in Fig. 10.

For a *p*-type semiconductor, one doped with an acceptor so that an electron deficiency exists within the conduction band, the electrical field formed at the electrode-electrolyte interface will be opposite in direction to that of an *n*-type semiconductor, Fig. 11. As a result, electrons will flow from the bulk of the semiconductor to the surface, and reductions will be assisted by photoexcitation of *p*-type semiconductors.

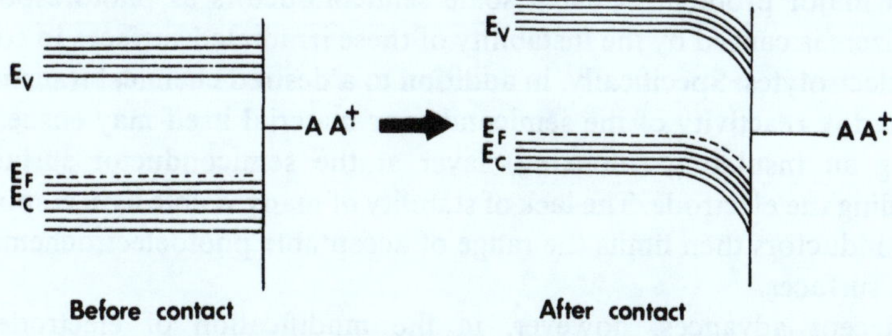

Figure 11. Band bending in a *p*-type semiconductor.

The thermodynamics of electron transfer at the surface of a photoexcited *p*-type semiconductor will be governed by the relative positions of the conduction band edge and the reduction potential of the adsorbed acceptor. Reduction should occur for any adsorbed organic substrate whose reduction potential lies less negative than the conduction band of the semiconductor in question, if the kinetics for the electron exchange allow for competition with e^-/h^+ recollapse.

If this semiconductor electrode is interfaced with a metallic counterelectrode, an electrochemical cell is formed. Upon irradiation of a polarized *n*-type semiconductor, electrons will flow from the irradiated electrode toward the dark metal electrode, inducing a measurable photocurrent. Should the two electrodes be held in separate cells, oxidation will ensue at one surface while reduction occurs at the other. If the oxidation occurs at the irradiated surface, as would be expected with an *n*-type semiconductor, the accompanying reduction will occur in the other half cell at the metal. In parallel fashion, if a photoinduced reduction occurs at the irradiated semiconductor (*p*-type semiconductors), then the accompanying oxidation will ensue at the metallic counterelectrode. Such a cell has specifically accomplished the goal of separation of the electron-hole pairs into two different spatial zones.

This separation is analogous to that which occurs at the *p-n* junction[11] of solid-state photovoltaic cells (for example, at silicon or gallium arsenide) used in responsive solar cells. These find application in products ranging from orbiting satellites to hand-held calculators. In solid-state photovoltaic cells, however, no net chemical transformations occur since the oxidizing and reducing sites are short-circuited. Only if one separates the sites for oxidation and reduction by irreversible redox interaction with a reagent of interest can one hope to find chemical storage by functional group conversion of organic substrates.

2.3. Electrode Stability

A major problem in using some semiconductors as photoresponsive sensitizers is caused by the instability of these irradiated surfaces in contact with electrolytes. Specifically, in addition to a desired chemical transformation, redox reactivity of the semiconductor material itself may ensue, generating an insulating (blocking) layer at the semiconductor surface or corroding the electrode. The lack of stability of many visible light responsive semiconductors then limits the range of acceptable photoelectrochemically active surfaces.

Recent advances, however, in the modification of electrodes by chemical attachment[12] or adsorptive deposition of electroactive catalysts or inert coverages[13] may be successful ultimately in obviating these additional difficulties. By attaching a desired catalytic molecule,[14] by chemically altering the surface composition of the semiconductor, for example, by silicide formation on Si,[15] or by imbedding the semiconductor in a protective polymer layer or film,[16] one can significantly extend the operational life of a given semiconductor, suppress photocorrosion, and improve the rates of a desired oxidation or reduction.

2.4. Powders

A recent extension of these ideas for practical applications of photoelectrochemistry is found in the use of a "short-circuited" photoelectrochemical cell[1a,17] prepared by the deposition of an inert metal (of low-overvoltage characteristics) on a powdered semiconductor. Such an aggregate is represented in Fig. 12. These powders include sites both for photoactivated electron transfer (as on the semiconductor electrodes discussed above) and for the counter redox reaction. Irradiation of these metallized powders still involves electron-hole pair separation and the generation of surface oxidation or reduction sites, but no external current flow accompanies the chemical transformations.

In fact, the native powders themselves can sometimes function as active photocatalysts. A requirement for effective photoelectrochemical conversion on untreated surfaces is that either the oxidation or reduction occurs readily on the dark material, thus scavenging one of the photogenerated charge carriers.

An obvious advantage of such powders is found in their simplicity, their portability, their low cost, and the freedom from the necessity of extensive electrochemical apparati to observe photochemical redox reactions. Nonetheless, irradiated powders do not permit the large spatial separation of the oxidation and reduction sites as do more conventional photoelectrochemical cells. Thus, the oxidized and reduced products are generated in proximate regions, and subsequent dark chemistry, occurring

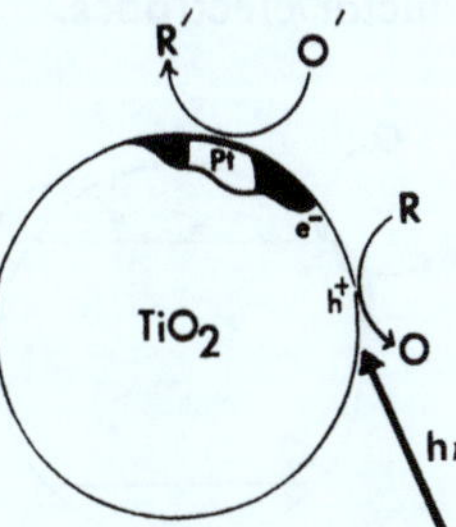

Figure 12. Redox exchange on an irradiated metallized semiconductor powder.

as the oxidized and reduced species migrate toward each other, may lead to undesired complications. Furthermore, the full operation of vectorial electron transfer through the space-charge region may be absent from such irradiated powders, depending on the particle size. It is quite unlikely, for example, that with colloidal suspensions of the photosensitive semiconductors (where optimum surface area per weight is attained) that a substantial space-charge layer can develop.

The distance between redox sites on the particles is still probably much larger than the separation encountered in exciplexes or radical ion pairs formed by redox photocatalysis within solvent cages in homogeneous solution. Under those circumstances, back electron transfer will be very rapid so that deactivation of the photoexcited state will interfere with efficient product formation.

2.5. Microparticulate Systems

The use of very small particulate systems analogous to these short-circuited photoelectrochemical cells can be even further extended. Colloidal forms of the metal oxide semiconductors can often be generated, usually by hydrolysis of an appropriate organometallic precursor.[7a,18] As has been demonstrated by water splitting and by formation of photoinduced oxidation products derived from organic substrates in nonaqueous suspensions, these systems may represent the ultimate miniature photoelectrochemical cell.

Conceptually, the same theoretical picture used for the description of powders can also be used for colloidal semiconductors. As with the powders, these microphotoelectrochemical cells can include metallic catalyst sites analogous to those shown in Fig. 12, absorptive sensitizers (Fig. 13), and/or homogeneous electron transfer relays (Fig. 14). The aggregates obtained in this way thus function in pathways quite similar to those observed with functional micelles and amphiphilic redox relays. It is no surprise, therefore, that many of the same consequences of electron-hole pair separation observed upon irradiation of colloids have also been observed on short-circuited semiconductor particles and single/or polycrystalline semiconductor electrodes.

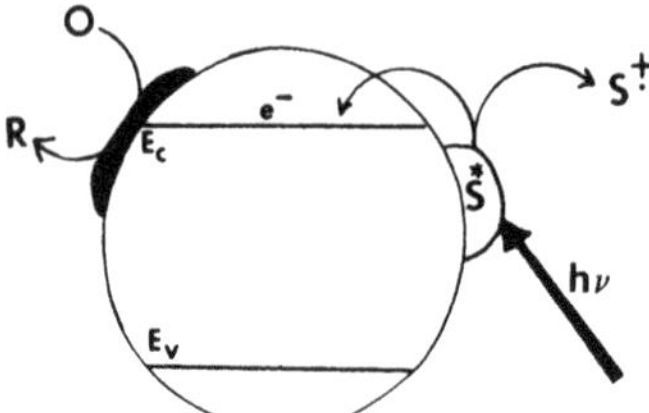

Figure 13. Sensitized redox exchange on a metallized semiconductor colloid.

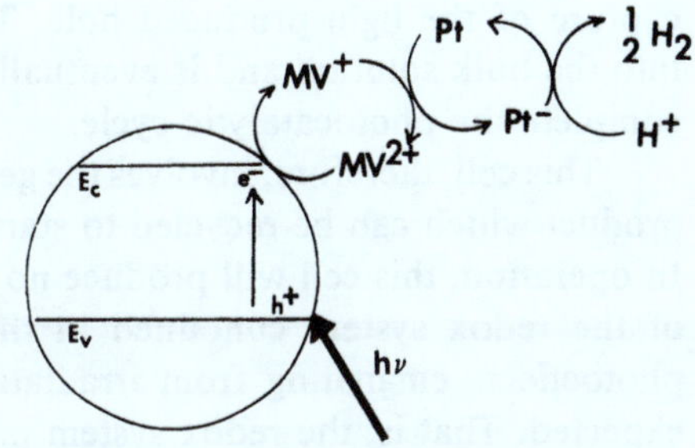

Figure 14. Redox exchange on an irradiated semiconductor in the presence of an electron relay.

3. TYPES OF PHOTOASSISTED REDOX REACTIONS

Redox assistance obtained when electrolytes are placed in contact with an irradiated light-absorbing system can occur in one of several ways.[1a] The light sensitive material may be either the electrode itself, a thin layer of semiconductor on a metal support, or a photosensitive catalyst, for example, an organic dye or transition metal complex adsorbed or covalently attached at the interface. The redox reaction may either be reversible or irreversible, and in the latter case, the initial electron transfer itself may be either endothermic or exothermic in the ground state.

3.1. Photosensitive Semiconductor Electrodes

3.1.1. Photovoltaic Liquid Junctions

A completely chemically reversible (regenerative) cell is called a photovoltaic liquid junction cell. Such a cell, equipped with an *n*-type semiconductor photoelectrode, is represented in Fig. 15. Here bandgap irradiation effects an electron-hole pair separation at the solid-liquid junction. Band bending causes the photogenerated electron to move away from the surface toward the bulk, producing a current flow from the irradiated to the dark electrode. Oxidation of an adsorbed donor accomplishes the chemical

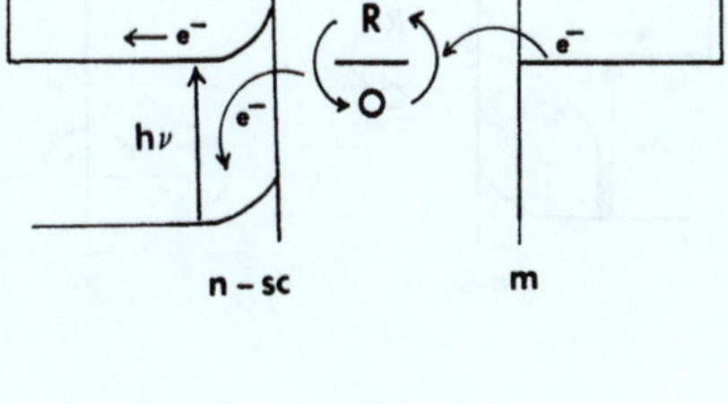

Figure 15. A photovoltaic liquid junction cell.

capture of the light-produced hole. The stable oxidized species migrates into the bulk solution and is eventually reduced at the counterelectrode to complete the photocatalytic cycle.

This cell, therefore, involves the generation of a completely stable redox product which can be recycled to starting material in high-chemical yield. In operation, this cell will produce no change in the chemical composition of the redox system contained in the electrolyte. Only purely physical photoeffects emanating from irradiation of the semiconductor would be expected. That is, the redox system present in the electrolyte functions as a charge transferring relay between the irradiated anode and the dark cathode, the important observable effect involving the conversion of light energy to electricity. The analogy of such cells to the photovoltaic systems discussed above is obvious.

3.1.2. Photoelectrocatalytic Cells

Observable chemistry occurs only if the photoinduced electron exchange is followed by a secondary chemical reaction which renders the redox process irreversible. The chemical transformations initiated by the photoinduced electron transfer can be one of two types, involving either an exothermic but kinetically slow electron exchange ($G < 0$) or a contrathermodynamic ($G > 0$ in the ground state) electron transfer which is driven by photoexcitation and by the occurrence of a rapid secondary chemical step. These two possibilities, called respectively photoelectrocatalytic and photoelectrosynthetic electrochemical cells, are shown as Figs. 16 and 17, respectively.

As with the photovoltaic cell, the reaction sequence in a photoelectrocatalytic cell begins with excitation of the semiconductor, generating an electron-hole pair. The hole induces an oxidation of an adsorbed donor, which is irreversibly converted to a product. The photogenerated electron in the conduction band moves away from the interface and travels through the external circuit to the counterelectrode, where it can effect a reduction of a reducible acceptor. If the reduction potential of the acceptor lies at a

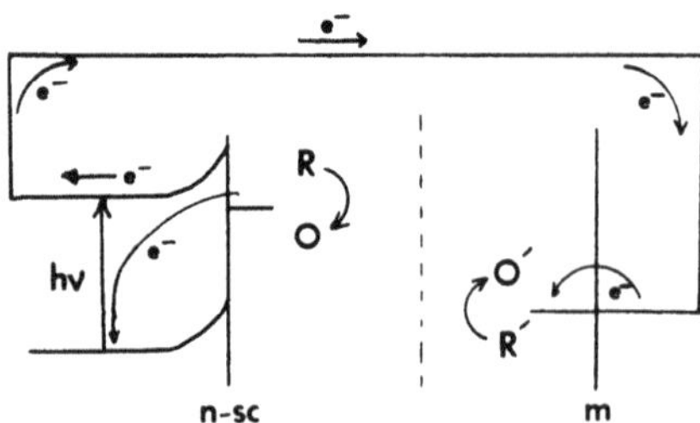

Figure 16. A photoelectrocatalytic electrochemical cell.

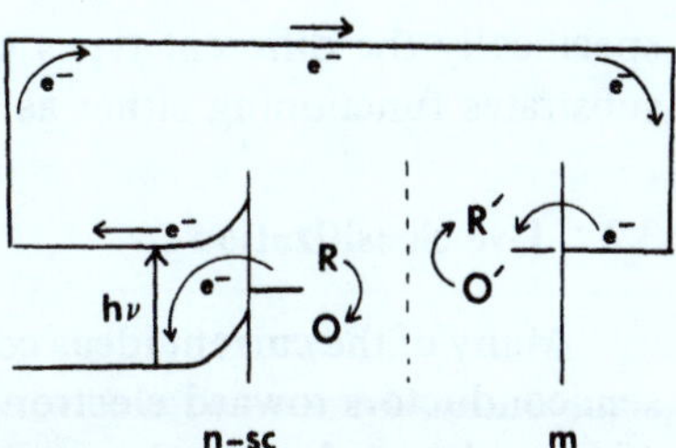

Figure 17. A photoelectrosynthetic electrochemical cell.

lower energy than does the oxidation potential of the donor, the redox exchange is *a priori* exothermic and would occur spontaneously, if a pathway of sufficiently small activation were available. The irradiated semiconductor provides just such a catalytic alternative. In other words, photoexcitation of the semiconductor opens a new route for the overall exothermic transfer of an electron from the HOMO of the donor to the LUMO of the acceptor.

It should be noted that application of the term "photocatalysis" to this cell has been criticized.[19] Although this cell incorporates a solid material which is regenerated after inducing the desired electron exchange (and can therefore be called catalytic), its operation does require a stoichiometric number of photons. Nonetheless, the ability of the cell to function effectively after prolonged use and to make available a single reactive site many times (often for high-effective turnover numbers), in our opinion, justifies the use of this descriptive term.

3.1.3. Photoelectrosynthetic Cells

If the LUMO (reduction potential) of the acceptor lies at a higher energy than does the HOMO (oxidation potential) of the coupled oxidant, the overall electron transfer is contrathermodynamic. The energy required for this electron transfer is available, however, if the photon energy absorbed during the initial excitation of the semiconductor can be funnelled to the task. The endothermicity of permissible reactions is therefore limited to the bandgap energy of the sensitizing semiconductor. Since back-electron transfer between the redox participants is favored thermodynamically, a rapid secondary reaction of either or both of the substrates formed by electron transfer is required, if net observable chemistry is to occur. The operation of such a cell represents an energy storage in which light energy is converted to chemical bonds and the cell is, therefore, termed photoelectrosynthetic.

Both endothermic and exothermic electron transfers have been effectively used for the production of organic photoelectrochemical effects. In the remainder of this article we use these theoretical constructs to treat specifically the different types of photoreactivity emanating from organic substrates functioning either as electron donors or acceptors.

3.2. Dye Sensitization

Many of the current ideas concerning the sensitization of large bandgap semiconductors toward electron-hole pair formation derive from the work of Gerischer and co-workers.[20] In extensive studies of a wide range of dyes held at the surface of either organic or inorganic semiconductors, they have shown that the band theory discussed above for the semiconductor itself can be profitably employed as a model for sensitization, as well.

Either anodic or cathodic photocurrents can be sensitized by excitation of a highly absorptive photocatalyst. The respective electron flow for the excited state sensitization of photocurrents by dyes adsorbed on semiconductors is shown as Figs. 8 and 18. For efficient generation of anodic photocurrent, overlap of the dye LUMO with the semiconductor conduction band must occur, while cathodic photocurrent generation requires energy matching of the dye HOMO with the semiconductor valence band. With highly oxidizable sensitizers, the quantum yield for electron injection can often be near unity.

As the oxidized form of the dye accumulates at the interface of sensitized photoanodic cells, the photocurrents will eventually decrease, unless the stored chemical oxidizing equivalent can be effectively transferred to the electrolyte. This can be readily accomplished if a reactive reducing agent is present. The reduced form of the reagent will then be converted to its oxidized form as the dye at the interface is regenerated, for example, Equation 1.

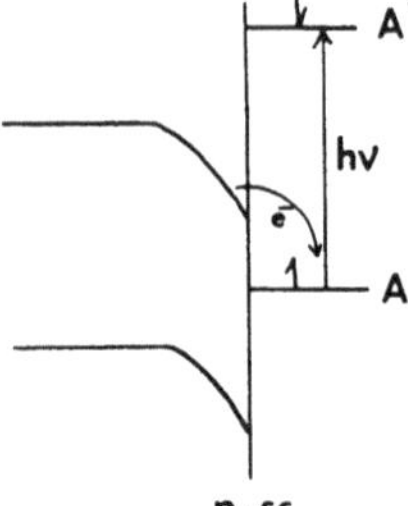

Figure 18. Reductive electron exchange at a dye-sensitized *p*-type semiconductor electrode.

This sequence, called supersensitization, thus effects a net chemical conversion of the dissolved redox couple in parallel to that induced by bandgap irradiation of the semiconductor (cf. Fig. 8).

It is also possible that a single reversible redox couple can function as the oxidant, the reductant, and the photosensitizer. For example, the cyclooctatetraene dianion, a participant in a reversible organic redox couple, can be used as a sensitizer for photogalvanic effects, Equation 2.[21]

$$2\,\mathrm{COT}^{2-} \xrightarrow[-2e^-]{h\nu} 2\,\mathrm{COT}^{\cdot-} \rightleftharpoons \mathrm{COT} + \mathrm{COT}^{2-} \qquad (2)$$

(COT $+2e^-$ returns to the dianion.)

Here, photoinduced electron injection from the excited dianionic dye to the n-TiO_2 electrode initiates the flow of an anodic current. The oxidized product, the monoanion, disproportionates at a rate dependent on the solvent and associated cation to two stable closed shell species, the neutral cyclooctatetraene and the dianionic starting material. The hydrocarbon can migrate to the cathode, where, upon dark reduction, an electrochemical cell is completed. This system can be irradiated, drawing current for the duration, for at least a week without demonstrable chemical degradation.

More detailed study has shown that covalent attachment of the sensitizer can lead to improved photocurrent generation, albeit with some sacrifice in the long-term stability of the system.[22] The improved currents are presumed to arise from a better kinetic control over the relative rates of disproportionation and energy dissipative back-electron transfer.

If the photoelectrochemically oxidized or reduced organic is not stable, net chemistry can also be observed in sensitized photoelectrochemical cells. For example, with a closed shell monanion, a net oxidative coupling, Equation 3,

Photoanode: $\mathrm{Ph_4C_5^-} \xrightarrow[-e^-]{h\nu} \mathrm{Ph_4C_5^{\cdot}} \longrightarrow \mathrm{Ph_8(C_5)_2}$

Cathode: $\mathrm{Ph_4C_5H} \xrightarrow{+e^-} \mathrm{Ph_4C_5H^{\cdot-}} \longrightarrow \mathrm{Ph_4C_5^-} + \tfrac{1}{2}H_2$

Net: $\mathrm{Ph_4C_5H} \longrightarrow \tfrac{1}{2}\,\mathrm{Ph_8(C_5)_2} + \tfrac{1}{2}H_2 \qquad (3)$

could be driven by light absorption either by the anion adsorbed on a single crystal TiO_2 electrode or by the semiconductor itself.[23] The net transformation is endothermic by about 18 kcal/mole, so the transformation shown in Equation 3 represents a semiconductor-mediated energy storage system.

The wavelength response of dye sensitized photoelectrochemical cells closely follows the absorption spectrum of the adsorbed dye. The action spectrum, a measurement of the dependence of the photocurrent on incident wavelength, thus clearly demonstrates that sensitization by adsorbed molecules can expand the region of spectral sensitivity beyond that imposed by the bandgap, i.e., that photocurrents can be induced by sub-bandgap photons.

Unfortunately, back-electron transfer to the oxidized dye sometimes competes very effectively with oxidation of a dissolved substrate and the overall efficiencies of these cells are almost always quite low (although often somewhat higher than those observed for photogalvanic cells).[24] In addition, no dye investigated so far shows sufficient chemical redox reversibility to make practical a long-term photoelectrochemical cell based solely on sensitization. For a cycle to be repeated even a thousand times without loss in efficiency, a rare chemical reversibility of >99.99 percent is required.[25] Relatively few experiments have been described, therefore, in which dye sensitization has been used as a method for organic functional group transformations.

The multitude of available studies of many classes of dyes as potential photosensitizers for electrochemical effects make an extensive discussion of the details of their use impossible here. Among the questions of mechanistic importance to the function of these cells are the identity of the electroactive precursor generated in the photolysis, the nature of dye aggregation at the electrode surface, the reaction sequence in supersensitization, the effect of the electrode surface on the photochemical and photophysical properties of the dye, and the kinetics for the electron exchange. Recent application of photoacoustic techniques,[26] attenuated total reflection,[27] and laser-induced coulostatic measurements[28] promise to bring additional sophistication to this area of investigation.

4. ORGANIC PHOTOELECTROCHEMICAL REACTIONS

Prediction of photoelectrochemically and energetically permissible redox conversions of a given organic substrate requires a knowledge of the reversible oxidation or reduction potentials of that species. Such electrochemical potentials are unfortunately not commonly available, since the radical ions formed by electrooxidation or electroreduction often

TABLE 2
Representative Oxidation Potentials[a] for Some Organic Functional Groups

Compound	Oxid. pot. (V vs. SCE)
O O	1.56
	1.09
	1.21
$\phi - \phi$	1.48
	2.31
CH_3CO_2H	1.6
S	1.51
S	1.84
N	1.82
N	0.76
N H	1.16
$(\phi CH_2)_2NH$	1.08
Et_3N	0.66
$CH_3CH_2C(=O)-N(Me)_2$	1.02

(continued)

TABLE 2—continued

Compound	Oxid. pot. (V vs. SCE)
H_2N— —NH_2	0.18
Ph O Ph	1.29
OH / CH_3	1.21

[a] Measured in acetonitrile.

participate in subsequent rapid chemical reactions. The waves observed in a cyclic voltammetric experiment are, therefore, often irreversible and lack thermodynamic significance. Nonetheless, cyclic voltammetric peak potentials can often provide an approximate energy scaling useful for the relative ordering of the efficiency of electron exchange.

A comparison, then, of the valence band positions of several common semiconductors (Table 1) with the oxidation peak potentials of an array of organic functional groups (Table 2)[29] shows that an electron transfer from most of the organic materials to the excited semiconductor should be thermodynamically permissible, while an analogous comparison of the conduction band edges with typical reduction peak potentials (Table 3)[29] allows for far fewer electroreductions. Consistent with this prediction, far more photoelectrochemical organic oxidations have been characterized than have reductions. Although claims have been made that even the most difficult of inorganic reductions, for example, carbon dioxide[30] and nitrogen,[31] can be observed upon semiconductor photocatalysis, most neutral organic functional groups are reduced at more negative potentials than would commonly be available from simple semiconductor photocatalysis.

The reactions discussed in this section, therefore, emphasize organic oxidations. After a discussion of measurement techniques, these are arranged according to functional group, an ordering not unlike that often encountered in cataloging normal thermal reactivity of these substrates.

TABLE 3
Reduction Potentials of Some Common Organic Functional Groups

Compound	Reduct. pot. (V vs. SCE)
 Br	−2.23
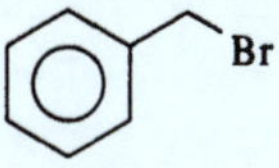 Br	−1.22
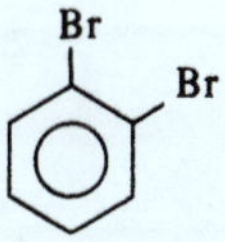 Br, Br	−1.28
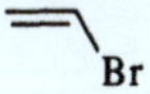 Br	−2.47
 Br	−2.32
Br_3CH	−0.49
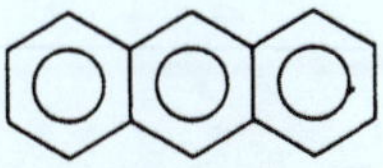	−1.92
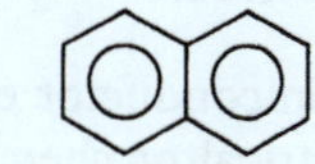	−2.53
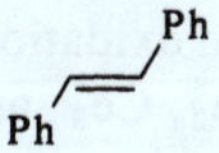 Ph, Ph	−2.19
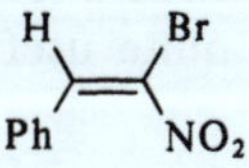 H, Br, Ph, NO_2	−0.4
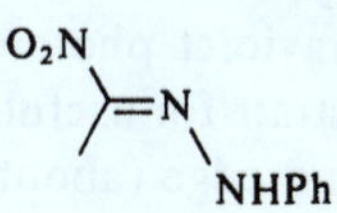 O_2N, N, NHPh (H_2SO_4)	−0.9
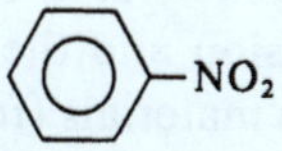 NO_2	−1.1

(continued)

TABLE 3—continued

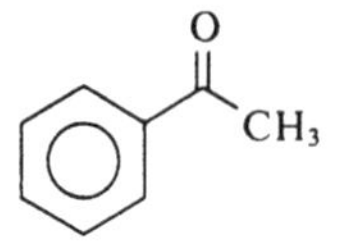 O, CH_3	−1.46
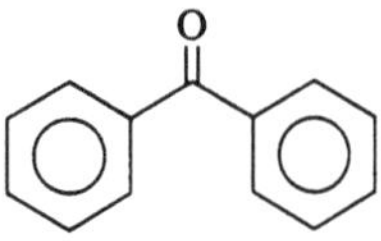 O	−1.21
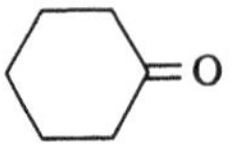 O	−2.45
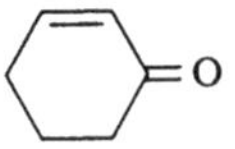 O	−1.55
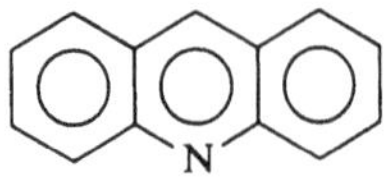 N	− 1.24
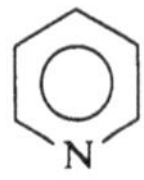 N	−2.07

4.1. Electrochemical Effects at Solid Electrodes

One of the first studies in which semiconductor electrodes were used for the direct conversion of solar to electrical or chemical energy is found in Frank and Bard's report of photoassisted oxidation of several hydroquinones, *p*-aminophenol, iodide, bromide, Fe_2^+, Ce_3^+, and cyanide ion at polycrystalline titanium dioxide.[32] A brief consideration of the details of these photooxidations will be instructive and quite useful for extrapolating to other organic transformations.

Titanium dioxide is a large bandgap material (about 3 V) and, consequently, can be excited only with ultraviolet photons. Nonetheless, its use has been widely investigated as a substrate for useful organic oxidations because of its highly positive valence band edge (about +2.4 V vs. SCE in CH_3CN) and its high reactivity as a photoactivated oxidant. Furthermore, titanium dioxide resists both photocorrosion and the decomposition reactions which plague other smaller bandgap materials (for example, Si, GaAs, GaP, InP, CdS, etc.). These features, together with its low cost and wide availability, make TiO_2 an ideal material for exploratory studies as a photoelectrochemical surface.

4.1.1. Current-Potential Curves

A cyclic voltammetric current-potential curve obtained at the surface of polycrystalline titanium dioxide in contact with an aqueous electrolyte during illumination with long wavelength ultraviolet or visible light is shown in Fig. 19.[32] In the absence of illumination, no anodic current is observed until solvent breakdown occurs (at much higher potentials). Also shown in Fig. 19 are the analogous curves observed at single crystalline TiO_2. With photoassistance, oxygen evolution occurs at potentials just positive of the semiconductor flat-band potential, nearly 0.9 V negative of its thermodynamic potential.

The similarity of the curves obtained at single crystal and polycrystalline electrodes shows that these two crystalline forms can be used reasonably interchangeably. Upon illumination of single crystalline TiO_2, electron-hole pair separation occurs with a quantum yield of approximately 0.7–0.8. A somewhat lower value (~0.6) is observed with polycrystalline materials.[32] Although the exact reproducibility of these current-voltage curves depends somewhat on electrode preparation and on the thickness of the metal oxide layer, similar qualitative characteristics could be easily obtained with

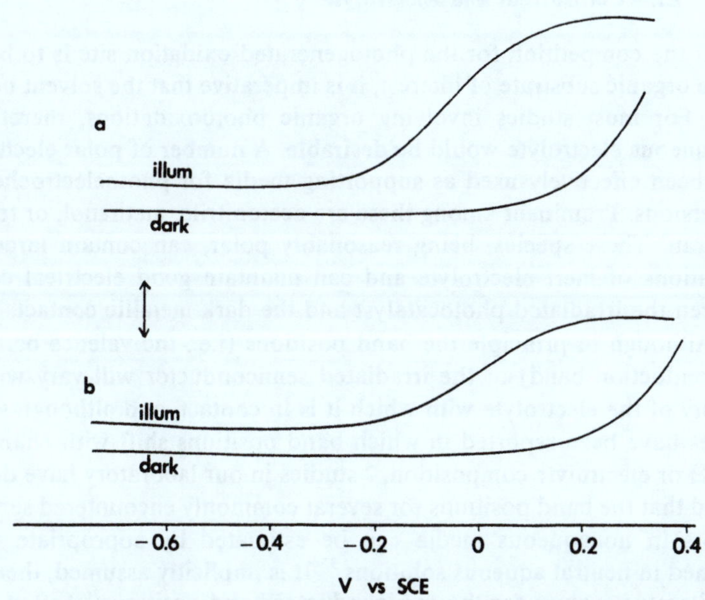

Figure 19. Current-potential curve of aqueous tetrabutylammonium perchlorate at dark and irradiated TiO_2. (a) As a single crystal; (b) in polycrystalline form.

different preparations of the semiconductor. The materials used in these studies were quite robust and gave continued evolution of redox products even after repeated use.

The observed photocurrents can be substantially enhanced in the presence of reducible species. With Cu^{2+} or Fe^{2+}, for example, trapping of the photogenerated electrons on TiO_2 reduces the importance of deleterious e^-/h^+ recombination.[33] In the presence of an oxidizable organic substrate in the electrolyte, either the solvent or the donor can capture the photogenerated hole. With an aqueous electrolyte as the medium for the oxidation, the success with which a given organic donor competes with water for the photogenerated holes is measured by current efficiency, the fraction of the total current which results in net redox reactions of the substrate of interest. In water, current efficiencies significantly less than one are usually obtained. That is, the kinetic competition with water oxidation (to generate hydroxy radicals) is one which most organic substrates lose. In aqueous electrolytes, therefore, most organic products are formed as a consequence of primary generation of OH. In fact, reports from the Bard group on the oxidation of unfunctionalized aromatic hydrocarbons, for example, benzene, strongly suggest that hydroxy radical attack on the aromatic ring initiates the reaction, which ultimately leads to oxidation products.[71]

4.1.2. Effect of Solvent and Electrolyte

If the competition for the photogenerated oxidation site is to be won by the organic substrate of interest, it is imperative that the solvent become inert. For most studies involving organic photooxidations, therefore, a nonaqueous electrolyte would be desirable. A number of polar electrolytes have been effectively used as supporting media for photoelectrochemical conversions. Prominent among these are acetonitrile, methanol, or tetrahydrofuran. These species, being reasonably polar, can contain large concentrations of inert electrolyte and can maintain good electrical contact between the irradiated photocatalyst and the dark metallic contact.

Although in principle the band positions (i.e., the valence band and the conduction band) of the irradiated semiconductor will vary with the identity of the electrolyte with which it is in contact, and although several studies have been reported in which band positions shift with changes in the pH or electrolyte composition,[34] studies in our laboratory have demonstrated that the band positions for several commonly encountered semiconductors in nonaqueous media can be estimated by appropriate values obtained in neutral aqueous solutions.[35] It is implicitly assumed, therefore, that literature values for the semiconductor band positions[36] (often determined in water) can be used as preliminary guides for planning photoelectrochemical synthetic schemes.

4.1.3. Product Analysis

Since many organic compounds can be oxidized at potentials less positive than +2.0 V (Table 2), photoassisted oxidations of many common organic functional groups should occur on irradiated semiconductor electrodes. For example, the onset of oxidation of hydroquinone at titanium dioxide is significantly cathodically shifted by bandgap illumination.[32] The products formed in these photoassisted oxidations are generally the same as those formed upon electrochemical oxidation of the same substrate at an inert metal electrode. For example, halides are oxidized to dihalogens and hydroquinone is oxidized to *p*-benzoquinone on either Pt or irradiated TiO_2, Equation 4.

$$\text{hydroquinone} \xrightarrow[\text{TiO}_2^*]{-2e^-,\ \text{Pt}} p\text{-benzoquinone} \tag{4}$$

The ability to effect these oxidations at significantly less positive potentials than required under normal electrochemical oxidation conditions, however, makes the photoelectrochemical procedure synthetically attractive.

Sometimes complications develop as poised electrodes are exposed to continued irradiation. For example, insoluble polymeric films are formed when TiO_2 is irradiated in contact with aqueous solutions of aniline or N,N-dimethylaniline in a photoelectrochemical cell, even at low concentrations of the amines.[32] An analogous film is formed by either oxidative electrolysis or photoelectrochemical oxidation of pyrrole.[37] With the latter compound, the resulting film is itself conducting and gives rise to a modified surface of catalytically interesting properties. In addition to these primary photooxidation products, additional secondary oxidation products can be sometimes formed upon prolonged irradiation.

4.1.4. Current Doubling

For some two- or multi-equivalent reducing agents, a single photogenerated hole can cause injection of more than one electron into the conduction band. In such cases, examples of a phenomenon called current doubling,[37] a two-step oxidation is required. In the first step, an electron is captured by the photogenerated hole, producing a reactive radical (a radical cation if the reducing agent is a neutral reagent), Equation 5.

$$M + h^{+} \rightarrow M^{\dagger} \tag{5}$$

If this intermediate is oxidized at a potential positive of the conduction band or if a second hole is available in the valence band, a second oxidation can ensue, Equation 6.

$$M^{\dagger} \rightarrow M^{2+} + e^{-} \tag{6}$$

A lack of strict doubling of the observed photocurrent (compared with one-electron oxidant) can be attributed either to overpotential effects or to the competition for holes by lattice ions or dopants.

With a number of organic and inorganic reducing agents, for example, aliphatic alcohols, formates, tartrate, arsenate, borohydride, sulfide, or cyanide, such current doubling phenomena are commonly encountered.[32] With carboxylates, for example, the first oxidation is thought to generate CO_2 and $H^{\bullet}$, with the hydrogen atom acting as the source of the second electron, Equation 7,[38]

$$\begin{aligned} HCO_2^- + h^+ &\rightarrow H^{\bullet} + CO_2 \\ H^{\bullet} &\rightarrow H^+ + e^- \end{aligned} \tag{7}$$

although the electrochemically equivalent sequence (Equation 8) has not been eliminated mechanistically.

$$\begin{aligned} HCO_2^- + h^+ &\rightarrow HCO_2^{\bullet} \\ HCO_2^{\bullet} &\rightarrow H^+ + CO_2 + e^- \end{aligned} \tag{8}$$

The time response of the photocatalytically activated surface is instantaneous. The near constancy of the dark-current value attests to the long-term stability of this metal oxide electrode.[32]

4.2. Indirect Redox Catalysis

In addition to the direct photooxidations described above, there are several alternate routes for photoassisted electrochemical organic synthesis on semiconductor surfaces. These involve an indirect oxidation induced by alteration of the oxidation level of a metal ion to one capable of inducing catalyzed organic redox reactivity or the oxidative generation of a highly reactive inorganic reagent which reacts reversibly with an organic substrate present in the medium. In the former mode, only trace quantities of the electroactive catalyst need be present, but in the latter route, stoichiometric consumption of the photogenerated reagent occurs. Redox processes occurring on the semiconductor surfaces have advantage over comparable homogeneous processes in that low concentrations of the catalyst are required, better control of catalytic efficiency is possible, and cleaner avoidance of secondary effects can be attained.

4.2.1. Metal Ion Catalysis

An example of an indirect metal-mediated oxidation involves the conversion of Ce^{3+} to Ce^{4+}.[32] The latter reagent has been used as an effective oxidant for the conversion of alcohols to ketones.[39] Thus, irradiation of TiO_2 in the presence of cerrous should effect the indirect conversion of dissolved alcohols to the corresponding ketones, Equation 9,

$$\begin{array}{ccc} Ce^{3+} & & Ce^{4+} \\ & TiO_2^* & \\ R_2C{=}O & & R_2CHOH \end{array} \tag{9}$$

even at potentials at which the alcohol itself is electrochemically inert.

A parallel reaction has been reported in the two-electron photooxidation of Tl^{1+} to Tl^{3+}.[40] The latter reagent is effective in inducing a number of significant organic photooxidations: for example, propylene can be selectively converted to a mixture of propylene oxide and acetone with this photoelectrochemical activation, Equation 10.[41]

$$CH_3CH{=}CH_2 \xrightarrow[\text{TlOAc, } O_2]{TiO_2^*} (CH_3)_2C{=}O \;(4) + \text{propylene oxide} \;(1) \tag{10}$$

4.2.2. Generation of Reactive Inorganic Reagents

An example of the second type of indirect photoelectrochemical catalysis involves formation of a reagent which reacts rapidly in a secondary step with a substrate present in the bulk electrolyte. For example, in the photoinduced oxidation of bromide to bromine in the presence of low concentrations of aniline, tribromoaniline can be obtained in excellent chemical yield, Equation 11.[32]

$$2Br^- \xrightarrow{TiO_2^*} Br_2 \xrightarrow{C_6H_5NH_2} 2,4,6\text{-}Br_3C_6H_2NH_2 \qquad (11)$$

This is the product expected from rapid trapping of the photogenerated bromine formed in the photooxidation. An analogous bromination has also been reported for anisole, and this area may be expected to become more important in future studies.

4.3. Direct Oxidative Functional Group Transformations

Many organic materials possess appropriate oxidation potentials to function as electron donors to excited semiconductors. A recent paper[42] entitled "Photocatalytic Hydrogen Production from Water by Decomposition of *p*-Vinylchloride, Protein, Algae, Dead Insects and Excrement" exemplifies the wide range of organic materials which can function as electron donors. The challenge is to use such compounds in a controlled, chemically selective fashion capable of manipulating the relevant functional groups.

4.3.1. Oxidative Cleavages

4.3.1a. Decarboxylations (Photo-Kolbe Reactions). A classic oxidative cleavage involves the splitting off of CO_2 from a carboxylic acid. This reaction, when initiated by electrooxidation to form an alkyl radical which dimerizes, is called the Kolbe electrooxidation. An analogous process can also be induced by long wavelength ultraviolet irradiation of a semiconductor. Bard and co-workers, for example, have studied this photo-Kolbe reaction of acetic acid at the surface of an irradiated n-TiO_2 single crystal or polycrystalline electrode immersed in acetonitrile containing tetrabutylammonium acetate, Equation 12.[43a-e]

$$CH_3CO_2^- \xrightarrow[CH_3CN, O_2]{TiO_2^*} CH_3CH_3 \qquad (12)$$

The true catalytic nature of this conversion could be demonstrated by observing current-potential curves obtained under illumination. The photoassisted oxidation was found to proceed at potentials approximately 2.4 V negative of where acetate oxidation occurs on platinum, and a quantum efficiency of about 65 percent was estimated for this photoconversion. After the reaction, the behavior of the electrode was unaltered.

When the same irradiation was conducted on platinized TiO_2 (anatase) or WO_3 powders in aqueous acetic acid, methane became the major organic product, Equation 13.[43c]

$$CH_3CO_2H \xrightarrow[\text{HOAc, } O_2]{TiO_2^*/Pt} CH_4 \qquad (13)$$

Other saturated carboxylic acids could be similarly photocatalytically decarboxylated, even if the α-carbon of the carboxylic acid were sterically congested, Equation 14.[43d]

$$\text{1-adamantanecarboxylic acid } (CO_2H) \xrightarrow{TiO_2^*} \text{adamantane } (H) + CO_2 \qquad (14)$$

As with the formation of ethane, the overall reaction is exothermic. Biomass has also been suggested as a material susceptible to photo-Kolbe degradation. Lactic acid, for example, can be photodecarboxylated by irradiation of aqueous suspensions of TiO_2, Fe_2O_3, or WO_3.[44]

A more recent detailed study has determined the experimental conditions under which either the radical reduction or radical coupling products could be preferentially formed.[45] A critical factor controlling this ratio may be the overpotential for hydrogen generation on the metal used to partially coat the particle and its degree of presaturation with hydrogen.[46] Rotating ring disc electrode studies have shown that reactivity can also be affected by the kinetic competition by acetate and water for the photogenerated hole on the irradiated powder.[47]

The proposed reaction mechanism for this conversion begins with electron-hole pair separation in the irradiated semiconductor. The organic oxidation is effected by capture of the photogenerated hole at the surface of the semiconductor powder by the adsorbed acid (or carboxylate). Loss of CO_2 by the carboxylate radical leads to the formation of the alkyl radical, which can either dimerize or be reduced to an anion. Protonation of the latter leads to monomeric decarboxylation product.

The critical difference between the irradiated powders and the irradiated electrode surface probably relates to differences on the space-charge layers in the two experiments. In powders, the existence of a significant space-charge region is unlikely, and photoinduced charge separation is thought to occur mainly on the particle surface. Under such circumstances, the newly formed radical remains rather isolated from other radicals generated in the same fashion, persisting on the surface where back-electron transfer ultimately generates the anion. The critical involvement of alkyl radicals could be demonstrated in spin-trapping experiments,[48] but the eventual protonation of an anion was established by deuterium labeling studies.[44]

TiO_2 seems to be the best material for inducing this photo-Kolbe reaction. Schwerzel and co-workers found, for example, that $SrTiO_3$ was only about 20 percent as efficient as rutile in hydrocarbon generation, and that a number of *n*-type semiconductors failed completely in these conversions. They suggested that in many cases decarboxylation may result from direct excitation of the acid or the carboxylate.[49]

4.3.1b. Olefin Oxidations. Olefin oxidations have long been studied electrochemically. That such olefin oxidations can be accomplished on irradiated semiconductor electrodes is obvious from the anodic photocurrent observed when amperometric measurements between an irradiated *n*-type titanium dioxide single crystal and a metallic counterelectrode are monitored.[50] The relevant electrochemical potentials for diphenylethylene oxidation on excited TiO_2 are shown in Fig. 20. As with any organic substrate possessing an oxidation potential less positive than +2.4 V (vs. SCE), diphenylethylene (E_p = +1.6 volts) can serve as an electron donor leading to anodic photocurrent. Single electron transfer from this substrate would

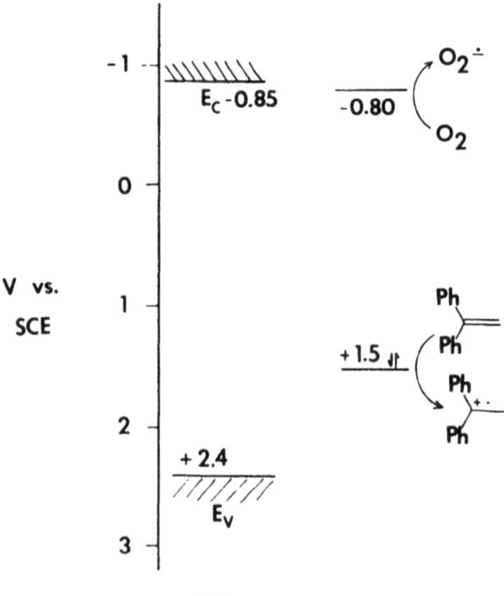

Figure 20. Energy levels for diphenylthylene oxidation on irradiated TiO_2.

generate a radical cation from which subsequent reactivity may be anticipated. Photogenerated electrons promoted to the conduction band upon photoexcitation with long ultraviolet wavelengths are available to effect reduction at the counterelectrode. Exothermic electron transfer from the conduction band to oxygen, generating superoxide, should be favorable, since O_2 reduction occurs at a potential less negative than the conduction band of TiO_2.

If this irradiation is conducted on a semiconductor powder rather than in a photoelectrochemical cell, the olefin radical cation and superoxide would be generated on sites quite near each other on the photocatalyst's surface. It is, therefore, not surprising to find that oxygen incorporation into the hydrocarbon backbone is relatively efficient.

In fact, irradiating suspensions of titanium dioxide powders in nonaqueous solvents, for example, acetonitrile, containing 0.01 *M* concentrations of an oxidizable olefin leads to very efficient oxidative cleavage of the carbon-carbon double bond, forming the carbonyl group. With 1,1-diphenylethylene, Equation 15,[50,51]

$$Ph_2C{=}CH_2 \xrightarrow[CH_3CN,\, O_2]{TiO_2^*} Ph_2C{=}O \quad (\sim 100\%) \tag{15}$$

oxidative cleavage product is isolated in virtually quantitative chemical yield. At partial conversions, small amounts of epoxides and ring-opened compounds are also formed, Equation 16.

$$Ph_2C{=}CH_2 \xrightarrow[DMSO/MeOH,\, O_2]{TiO_2^*} Ph_2C{=}O + \text{(1,1-diphenyloxirane)} + Ph_2C(OH)CH_2OMe \tag{16}$$

The reaction was inhibited by hydroquinone and by 2,6-di-*t*-butyl-4-methylphenol.[51]

Many other olefins can be similarly oxidatively cleaved to carbonyls (Table 4).[50] In general, high chemical yields of products can be isolated. In the one striking exception to this generalization, semiconductor photocatalysis initiates the polymerization of styrene. This observation thus suggests a further important application of these reactions, i.e., as polymerization initiators.

Since the formation of oxidative cleavage products requires the formation of an oxygen-carbon bond at some point along the reaction coordinate, detection or other indirect implication of oxygenated intermediates would

TABLE 4
TiO_2-Photocatalysis of Olefin Oxidative Cleavage[a]

Olefin	Percent conversion	Products (chemical yields %)[b]			
$Ph_2C{=}CH_2$ **1**	85	$Ph_2C{=}O$ (84)	1,1-diphenyloxirane (14)	$Ph_2CH{-}CHO$ (2)	
trans-PhCH=CHPh **2**	85	PhCHO[c] (33)	cis-stilbene oxide (11)	trans-stilbene oxide (42)	cis-PhCH=CHPh (14)
$Ph(CH_3)C{=}CH_2$ **3**	50	$Ph(CH_3)C{=}O$ (100)			
$PhCH{=}CH_2$ **4**	43	PhCHO[d] (17)			
β-pinene **5**	29	myrtenol (CH₂OH) (12)	myrtenic acid (COOH) (12)	nopinone (24)	ketone (24)
cyclohexene **6**	10	2-cyclohexen-1-one (60)	2-cyclohexen-1-ol (40)		

[a] Irradiation in a Rayonet photochemical reactor equipped with RPR-black lights, $\lambda = 350 \pm 30$ nm; room temperature; irradiation time = 6 h.
[b] Yields reported are based on disappeared starting material.
[c] Yield of benzaldehyde is based on the expectation of 2 mol of oxidation product derived from 1 mol of stilbene.
[d] Oligomeric products accounted for the remainder of consumed starting material.

be mechanistically relevant. Independent synthesis of 1,1-diphenyl-dioxetane, followed by treatment with suspended titanium dioxide, (presumably acting as a Lewis acid catalyst) leads to quantitative O-O and C-C cleavage generating benzophenone, exactly the product observed in the photoelectrochemical oxidative cleavage.[50] This dioxetane, therefore, is a permissible intermediate for the oxidative cleavage observed on this photocatalytic surface.

Thus several mechanisms can be envisioned for formation of this intermediate. Specifically, these involve: (1) attack of photogenerated superoxide on the olefinic starting material; (2) back-electron transfer between the photogenerated radical cation and superoxide to form singlet oxygen, which effects a catalytic oxidation of the olefin; or (3) generation of the olefin radical cation in the primary electron transfer, a species which can be captured either by triplet (ground state) oxygen or by photogenerated superoxide to ultimately give the dioxetane.

The lack of reactivity of solubilized superoxide, together with the inability to quench the photocatalyzed reaction with known superoxide quenchers, made the first route unlikely.[50] Contrasting chemical reactivity with photoelectrochemical oxidation and established singlet oxygen pathways weighed against the second pathway.[50,51] In contrast, the involvement of radical cations was implicated by structure-reactivity studies of electron demand,[52] by semi-empirical calculations,[53] and by the similarity of the product distributions obtained by photoelectrochemical routes, and by the usual electrooxidation. Furthermore, flash photolysis experiments conclusively demonstrate the transient formation of both single electron oxidized and reduced species on irradiated TiO_2 colloids.[54] Additional mechanistic details relating to the probable involvement of radical cations have been reviewed elsewhere.[55] Better chemical yields of product were obtained, however, on the irradiated semiconductor powders, than by electrooxidation on solid metal electrodes. It is surely reasonable, therefore, to consider the capture of the photogenerated radical by molecular oxygen or superoxide as important steps in the oxidative cleavage.

Several photocatalytic studies of hydrocarbon oxidations on semiconductor oxides have also implicated lattice oxygen as the source of oxygen atoms in oxidized products,[56] but no comparable studies have yet been completed in solution where exchange with solvent and/or gaseous oxygen may be significant. Since olefin oxidative cleavages can also be observed upon photocatalytic sensitization with CdS,[51] the availability of lattice oxygen is surely not absolute.

The thermodynamics of the initial electron exchange predict that only those olefins whose oxidation potentials lie positive of the valence band of the irradiated semiconductor will efficiently form radical cations upon interaction with the excited semiconductor powder. Since simple olefins display oxidative waves positive of 2.5 V, a different reaction course might be expected. These predictions are borne out in the observed photocatalyzed oxidation of nonarylated olefins.[38] For example, allylic oxidation (rather than C=C cleavage) is observed when cyclohexene is subjected to analogous TiO_2 powder photocatalysis, Equation 17.[50]

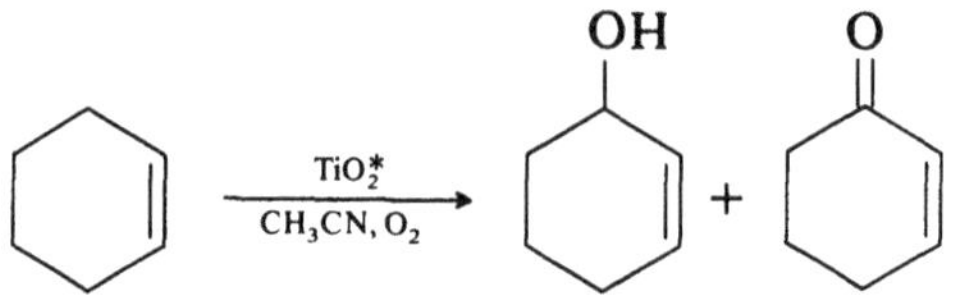

$$\text{cyclohexene} \xrightarrow[CH_3CN,\ O_2]{TiO_2^*} \text{cyclohex-2-en-1-ol} + \text{cyclohex-2-en-1-one} \tag{17}$$

With β-pinene both oxidative cleavage and allylic oxidation are observed, Equation 18.[50]

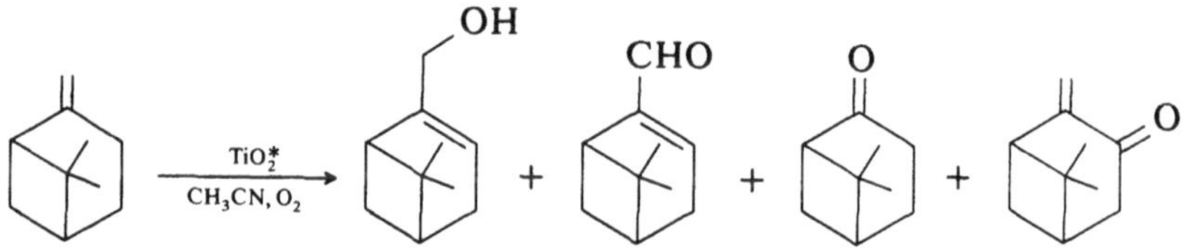

$$\beta\text{-pinene} \xrightarrow[CH_3CN,\ O_2]{TiO_2^*} \text{(CH}_2\text{OH product)} + \text{(CHO product)} + \text{(ketone)} + \text{(enone)} \tag{18}$$

The oxidation of propylene as a gas stream over powdered metal oxides has also been reported, giving CO_2 as well as products of intermediate oxidation level.[57]

It is probable that these reactions proceed not through radical cation formation but rather by capture of the photoelectron by oxygen, perhaps inefficiently, to give superoxide. If this species is protonated, the resulting hydroperoxy radical can initiate a radical chain allylic oxidation.

With other nonarylated olefins, complex mixtures of oxidative cleavage products can be observed. With 3-methyl-1-butene and its isomer 2-methyl-2-butene, the array of products shown in Equations 19 and 20, respectively, were obtained when TiO_2 is irradiated under a gaseous stream containing oxygen and the hydrocarbons.[58]

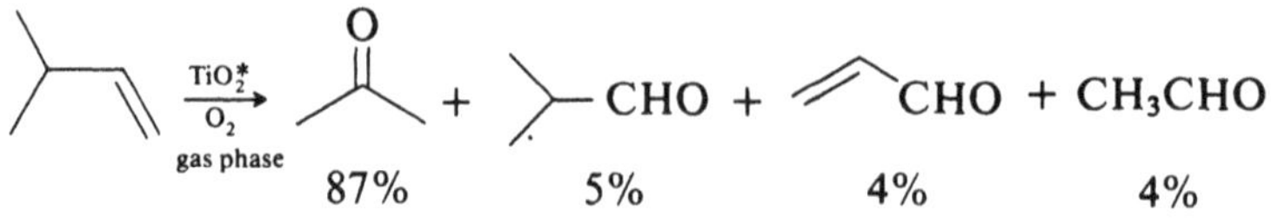

$$\text{3-methyl-1-butene} \xrightarrow[O_2,\ \text{gas phase}]{TiO_2^*} \underset{87\%}{\text{acetone}} + \underset{5\%}{(CH_3)_2CH\text{–CHO}} + \underset{4\%}{CH_2{=}CH\text{–CHO}} + \underset{4\%}{CH_3CHO} \tag{19}$$

$$\text{2-butene} \xrightarrow[\text{gas phase}]{TiO_2^*,\ O_2} \underset{35\%}{\text{2-butanone}} + \underset{11\%}{\text{acetone}} + \underset{3\%}{\text{methylcyclopropane}} + \underset{20\%}{CH_2{=}CHCHO} + \underset{31\%}{CH_3CHO} \qquad (20)$$

No conclusive mechanistic details have yet been established, although the different distributions of products obtained from the two isomers establish that electron transfer equilibration of the organic substrates is not rapid. This result differs from that reported by Kodama and co-workers who found extensive butene isomerization on irradiated ZnO and TiO_2.[59] These authors attributed the observed reactivity on ZnO to O^- centers, and that on TiO_2 to stronger pi-complexation on the surface.

If water is present, olefin reductions can also be observed, for example, Equations 21 and 22.[60]

$$CH_2{=}CH_2 \xrightarrow{TiO_2^*} CH_4 + C_2H_6 + \text{propane}$$

$$\text{propene} \xrightarrow{TiO_2^*} CH_3CH_3 + \text{propane} + CH_4 \qquad (21)$$

$$CH_2{=}CH_2 \xrightarrow{TiO_2^*,\ H_2O} CH_4$$

$$\text{propene} \xrightarrow{TiO_2^*,\ H_2O} CH_4 + CH_3CH_3 \qquad (22)$$

These reactions, reminiscent of Fischer-Tropsch chemistry, again pinpoint the reactivity differences observed when water competes with organic substrates for the electron-hole pair. Analogous reactions are also reported for alkynes.[60,61] This reduction sequence has been suggested as a synthetic route for photodeuteration of alkenes.[62]

The mode of dispersal of the semiconductor is apparently important, for the rate of photooxidative cleavage of 2-methyl-propene on SnO_2 supported on SiO_2 was about one hundred times greater than that observed on crystalline SnO_2, Equation 23,[63]

$$(CH_3)_2C{=}CH_2 \xrightarrow[O_2]{TiO_2^*\ \text{on}\ SnO_2} (CH_3)_2C{=}O \qquad (23)$$

Similarly, TiO_2 grafted on silica gel is much more selective in photoelectrochemical oxidative cleavages than is TiO_2 itself.

4.3.1c. Alkane Oxidations. Most studies of alkane oxidation have been conducted as gas streams over an irradiated TiO_2 powder bed.[64] Under these conditions, complex mixtures of products are usually formed, Equation 24.[58]

$$\text{isobutane} \xrightarrow{TiO_2^*} CO_2 + CH_3CHO + (CH_3)_2CH\text{-}CHO + \text{2-methylbutanal} + (CH_3)_2C{=}O + \ldots$$

15% 16% 5% 5%

$$\xrightarrow[H_2O]{TiO_2^*} \text{HO-alkane} \xrightarrow{-H_2O} \text{alkenes} \xrightarrow{TiO_2^*} (CH_3)_2C{=}O + EtOH$$

$$\xrightarrow{TiO_2^*} CH_3COCH_2CH_3 + CO_2 \quad (24)$$

Such a wide diversity of products is consistent with a photoinitiated radical chain analogous to that suggested for alkene oxidations. That the alkenes formed in this reaction derive from the dehydration of alcohols formed in the primary photooxidation was consistent with studies of the photooxidation of heptane on oxygenated ZnO or TiO_2, where heptanols and heptanones are obtained.[65] In their characterization of the oxidation of isobutane, Pichat and co-workers attributed the formation of acetone and CO_2 to photoactivation of adsorbed oxygen anions, isobutane then reacting in the neutral adsorbed state.[66]

4.3.1d. Arene Oxidations. Even the skeleta of aromatic molecules can be cleaved in these photoelectrochemical oxidations. For example, the photocatalyzed cleavage of substituted naphthalenes proceeds as shown in Equation 25.[67]

$$\text{1-methoxynaphthalene} \xrightarrow[O_2,\ CH_3CN]{TiO_2^*} C_6H_4(CO_2Me)(CO_2H) + \text{phthalic anhydride} \quad (25)$$

The relatively high yields of oxidized product obtained make this route viable as a synthetic alternative for preparation of ortho-substituted benzenes. The relative rates of oxidative cleavage of substituted naphthalenes closely paralleled the oxidation peak potentials of the neutral parents. Dialkylated naphthalenes react similarly, leading to ketones, Equation 26,

$$\text{1,5-dimethylnaphthalene} \xrightarrow[\text{2. } CH_2N_2]{\text{1. } TiO_2^*,\ O_2,\ CH_3CN} \text{methyl 2-acetyl-6-methylbenzoate } (CO_2Me,\ Me) \tag{26}$$

or to noncleaved oxidation products, Equation 27.

$$\text{1,4-dimethylnaphthalene} \xrightarrow[\text{2. } CH_2N_2]{\text{1. } TiO_2^*,\ O_2,\ CH_3CN} \text{1,4-dimethyl-2-hydroxy-3-methoxynaphthalene (Me, OH, OMe, Me)} \tag{27}$$

When an unsymmetrically substituted arene is similarly treated, for example, Equation 28,

$$\text{2,3-dimethylnaphthalene} \xrightarrow[\text{2. } CH_2N_2]{\text{1. } TiO_2^*,\ O_2,\ CH_3CN} C_6H_4(CO_2Me)_2 + Me_2C_6H_2(CO_2Me)_2 \tag{28}$$

cleavage of both rings can be observed with regiochemical preference reflective of the charge density calculated for the radical cation.

Other simple benzene derivatives have similarly been subjected to photocatalytic oxidation. For example, toluene can be oxidized to benzyl alcohol and benzaldehyde, Equation 29,

$$C_6H_5{-}CH_3 \xrightarrow[O_2,\ CH_3CN]{TiO_2^*} C_6H_5{-}CHO \tag{29}$$

upon irradiation of a suspension of powdered anatase in the neat liquid.[68] In the gas phase, the analogous reaction gives aldehydes, Equation 30,[69]

$$R{-}C_6H_4{-}CH_3 \xrightarrow[O_2]{TiO_2^*} R{-}C_6H_4{-}CHO \tag{30}$$

while as an aqueous suspension, both oxidation and coupling products are observed, Equation 31.[70]

$$\text{R-C}_6\text{H}_4\text{-CH}_3 \xrightarrow[\text{H}_2\text{O, O}_2]{\text{TiO}_2^*} \text{HO-C}_6\text{H}_3\text{(R)-CH}_3 + \text{R-C}_6\text{H}_4\text{-CHO} + \text{R-C}_6\text{H}_4\text{-CH}_2\text{CH}_2\text{-C}_6\text{H}_4\text{-R} \quad (31)$$

Even benzene can be oxidized. Upon irradiation of benzene-saturated aqueous suspensions of TiO_2, phenol formation can be observed, Equation 32, although the main product isolated is carbon dioxide obtained by further indiscriminate oxidation.[33,71]

$$\text{C}_6\text{H}_6 \xrightarrow[\text{H}_2\text{O, O}_2]{\text{TiO}_2^*} \text{C}_6\text{H}_5\text{-OH} + \text{CO}_2 \quad (32)$$

Ring hydroxylation can also be observed in the photocatalytic oxidation of benzoic acid.

While the authors propose a mechanism involving radical cation formation in Equation 29, the reactions run in water almost certainly proceed through the hydroxy radical. This species can be formed either via initial solvent oxidation or by dissociation of the adduct formed by trapping the arene radical cation with water.[72] The hydroxy radical generated in this fashion is highly reactive and quite unselective, so the limited control observed in the latter reaction is surely understandable.

In fact, the formation of the hydroxy radical in irradiated aqueous semiconductor suspensions is well-documented. The use of this mechanistic imperative in devising new routes to photo-Fenton reactions is obvious.[73] In particular, exactly analogous products are observed when hydrogen peroxide was added to a dilute aqueous solution of these arenes. Labeling studies showed that the oxygen in the peroxide formed upon irradiation of ZnO under steam is derived not from the lattice oxygen, but rather from oxygen gas.[74]

4.3.1e. Amine Oxidations. Similar considerations apply to the oxidative cleavage of amines. In the photocatalyzed oxidation of toluidines, for example, azo products are formed, Equation 33.[75]

$$CH_3-C_6H_4-NH_2 \xrightarrow[H_2O,\,O_2]{TiO_2^*} CH_3-C_6H_4-N{=}N-C_6H_4-CH_3 \qquad (33)$$

Coupling to form oxidized products is obviously more important than oxidation of the pendant alkyl groups. Such oxidations are reminiscent of the photocatalyzed oxidation of ammonia reported by Pichat and co-workers.[76]

While mechanistic details are still uncertain, preliminary results obtained in our group suggest that amine photoreactivity may indeed be controlled by manipulation of reaction conditions. For example, two pathways, Equation 34,

$$C_6H_5CH_2CH_2CH_2NH_2 \xrightarrow[2.\ H_2O]{1.\ TiO_2^*,\,O_2,\ CH_3CN,\ 0.002\,M} C_6H_5CH_2CH_2CHO$$

$$C_6H_5CH_2CH_2CH_2NH_2 \xrightarrow{1.\ TiO_2^*,\,O_2,\ CH_3CN,\ 0.01\,M} C_6H_5CH_2CH_2CH_2NHCHO \qquad (34)$$

can be observed in the photocatalyzed oxidation of primary or secondary amines, the observed product distribution depending on the initial concentration of the amine and on the nature of the semiconductor photocatalyst.[77] Given the oxidizability of amino-substituted organics and the stability of the aminium radical cation, investigations of the photoelectrochemical oxidation of amines are likely to be quite fruitful.

A related bond cleavage has been described in the CdS-photocatalyzed dealkylation of methylene blue and rhodamine B, Equation 35.[78]

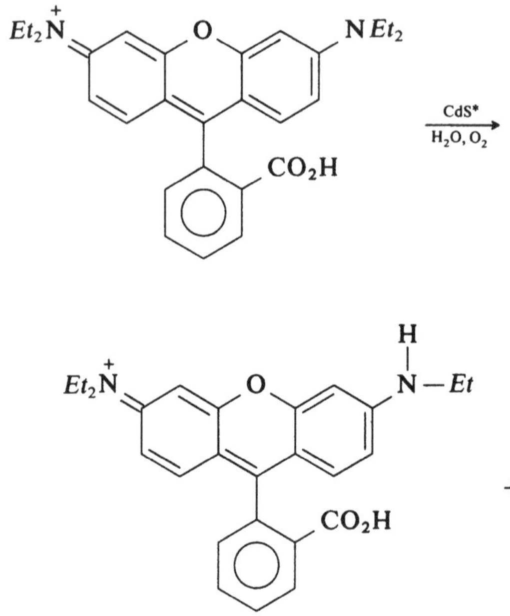

(35)

+ further diethylation products

Again electron transfer sensitization was postulated because of similarities between the CdS-photocatalyzed reaction and that observed with Fe^{+3} sensitization. Singlet oxygen was specifically excluded as an important intermediate.

4.3.1f. Alcohol Oxidations. The conversion of alcohols to aldehydes or ketones, or under more vigorous conditions to carboxylic acids or CO_2, is a common photoelectrochemical reaction.[58,79] For example, acetone and isobutene are formed from isobutene on irradiated TiO_2, but isobutanol leads to formation of isobutyraldehyde, Equation 36.[79a]

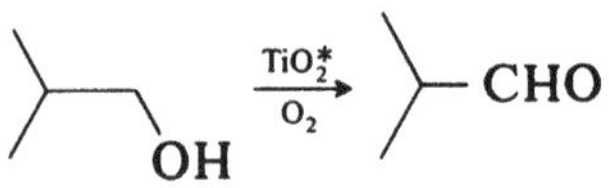

(36)

Incorporation of O_2^{18} into acetone formed from 2-propanol was attributed to exchange with O^- centers and to reactivity from holes photogenerated on diamagnetic ZnO or TiO_2 sites.[80] The efficiency of the reaction depended on both the fractional metal coverage[81] and the identity of the metal deposited on the semiconductor (Pt > Rh > Pd > Ru > Ir > native).[82]

Analogous oxidations have also been reported for polyhydroxylated compounds. For example, carbohydrates[83] can be used as sacrificial electron donors in the photoelectrochemical generation of hydrogen.

The mechanism of these photocatalytic oxidations remains unknown, although the dependence of the rate of oxidation of 2-propanol on the intensity of the irradiation, together with low-quantum yields, emphasizes the importance of e^-/h^+ recombination. That lower activation energies were encountered on pure rutile than on anatase or doped- or coated-TiO_2 reflects the solid-state properties of the catalyst rather than the chemistry of the alcohol, oxidation of which probably involves a radical chain. The photocatalytic nature of these reactions can be easily established, for more than ten turnovers per site can be easily accomplished in the oxidation of isopropyl alcohol.[84]

The reverse reaction, reduction of aldehydes to alcohols, can also be detected under proper conditions. Tinnemans and co-workers observed, for example, that methanol could be formed photoelectrochemically from formaldehyde and water over irradiated $SrTiO_3$ suspensions containing transition metal oxide deposits.[85]

4.3.1g. Other Oxidative Transformations. A variety of additional semiconductor catalyzed functional group oxidations have been reported, although scant mechanistic detail is available for these transformations. Of great potential interest to the organic chemist are the photoinduced oxidation of lactams to imides, Equation 37.[86]

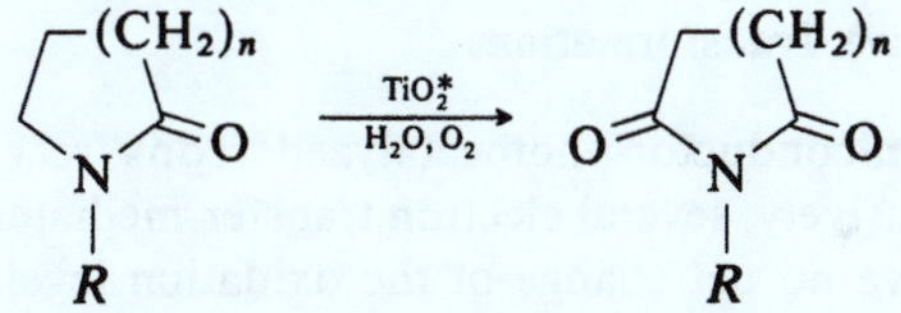

$$(CH_2)_n,\ N\text{-}R \xrightarrow[H_2O,\,O_2]{TiO_2^*} (CH_2)_n,\ N\text{-}R \qquad (37)$$

Five- and six-membered N-acyl amides could be similarly oxidized, Equation 38,

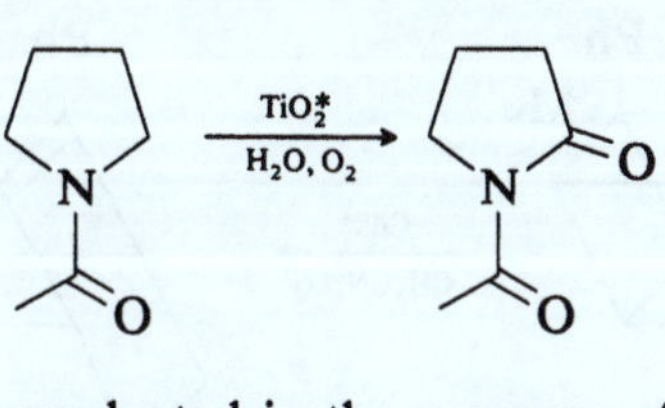

$$\xrightarrow[H_2O,\,O_2]{TiO_2^*} \qquad (38)$$

and if this reaction is conducted in the presence of Cu(II), unsaturation can be induced, Equation 39.

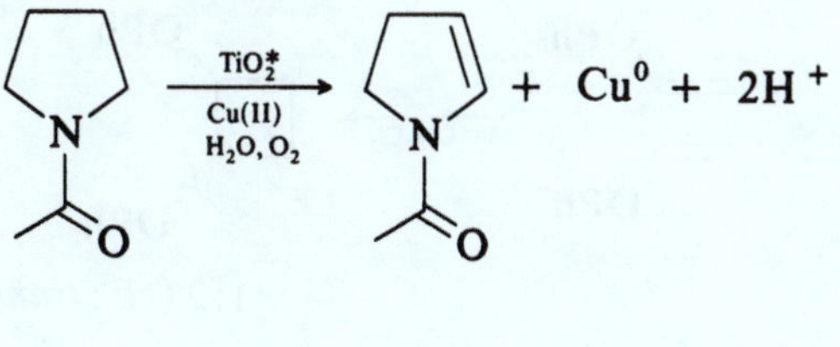

$$\xrightarrow[Cu(II)\;H_2O,\,O_2]{TiO_2^*} \; + \; Cu^0 + 2H^+ \qquad (39)$$

Again, mechanistic details are sparse, but the reactions may reasonably be assumed to begin by solvent oxidation to generate the hydroxy radical whose subsequent dark chemistry leads to the observed products.

In the photocatalyzed reaction of oxalic acid to CO_2, Equation 40,

$$HO_2CCO_2H \xrightarrow[H_2O,\,O_2]{TiO_2^*} CO_2 + H^+ \tag{40}$$

the authors suggest formation of a reactive surface-bound intermediate formed when both holes and electrons are present.[87]

The photoelectrochemical mineralization of chlorocarbons has also been reported, Equation 41.[88]

$$Cl_2C{=}CHCl \xrightarrow[H_2O,\,O_2]{TiO_2^*} Cl_2HC{-}CHO \longrightarrow HCl + CO_2 \tag{41}$$

Although the authors suggest a mechanism in which hydroxy radical directly displaces a vinyllic chloride, a mechanism involving hydroxy radical attack on the double bond to form a radical which initiates a radical chain seems more likely.

4.3.2. Nonoxidative Transformations

Although semiconductor-photocatalyzed redox reactions have been studied most extensively, several electron transfer-mediated pericyclic reactions which involve no net change of the oxidation level of the molecule have also been described. So far, an example of a retro[2 + 2]cycloaddition has appeared, Equation 42[89]

Ph Me Ph Me $\xrightarrow[CH_3CN,\,O_2]{CdS^*}$ Ph Ph Me Me (42)

as has an inefficient [2 + 2]cycloaddition, Equation 43.[90]

OPh OPh $\xrightarrow[CH_2Cl_2]{CdS}$ OPh OPh (43)

1 : 2 (*cis* : *trans*)

The transient involvement of radical cations is assumed because of a parallel sensitization of this same reaction by pyrylium salts[91] and by the known analogy of other retro-[2 + 2]-cycloadditions of cyclobutane radical cations.[92]

A single example of a [4 + 4] photocycloreversion is available, Equation 44.[90]

$$\xrightarrow[CH_2Cl_2]{CdS^*} \quad (44)$$

The authors suggest that these conversions proceed through photocatalytic oxidation to produce a radical cation. Isomerization of this intermediate and recapture of an electron then leads to the observed products. That these conversions involved electron transfer sensitization was established by quenching studies and by the isolation of identical products from alternate, known oxidative routes. Whether a chain reaction is involved in the radical cation isomerization is at present unknown.

A final example of the photosynthetic potential inherent in the irradiation of semiconductor suspensions containing organic molecules is found in Bard's demonstration of amino acid synthesis. Irradiation of platinized TiO_2 suspensions in aqueous ammoniacal methane, Equations 45,[93]

$$NH_3 + 2CH_4 + 2H_2O \xrightarrow[H_2O]{TiO_2^*} H_2NCH_2CO_2H + H_2 + \text{other amino acids} \quad (45)$$

leads to production of glycine, alanine, serine, aspartic acid, and glutamic acid as well as products of intermediate functionality (methanol, ethanol, and methylamine). The potential for using irradiated semiconductors as templates for bringing together simple reactants may be an important area for future investigation.

5. CONCLUSIONS

An appreciation of the importance of photoinduced electron transfer at the surface of irradiated semiconductors as a means for initiating organic redox reactions has only recently evolved. As the above array of permissible reactions suggest, these routes may be very useful pathways for directing specific chemical redox selectivity within multifunctional molecules. Furthermore, the template effect of the photocatalytic surface may allow for more specific control of regio- and stereochemistry than by other homogeneous routes. Determining whether these expectations can be realized represents an exciting area of investigation for the organic electrochemist.

REFERENCES

1. For reviews of the principles of photoelectrochemistry, see (a) A. J. Bard, *J. Photochem.* **10**, 50 (1979); (b) A. J. Bard, *Science* **207**, 139 (1980); (c) A. J. Bard, *J. Phys. Chem.* **86**, 172 (1982); (c) R. Memming, in *Electroanalytical Chemistry*, edited by A. J. Bard (Marcel Dekker, New York, 1979), p. 1; (d) M. S. Wrighton, *Accts. Chem. Res.* **12**, 303 (1979); (e) H. O. Finklea, *J. Chem. Ed.* **60**, 325 (1983); (f) A. J. Nozik, *Faraday Disc.* **70**, 7 (1980); (g) T. Freund and W. P. Gomes, *Catal. Rev.* **3**, 1 (1969); (h) J. A. Turner, *J. Chem. Ed.* **60**, 327 (1983); (i) B. Parkinson, *J. Chem. Ed.* **60**, 327 (1983).
2. A. J. Nozik, *Ann. Rev. Phys. Chem.* **29**, 189 (1978).
3. (a) P. J. Boddy and W. Brattain, *J. Electrochem. Soc.* **109**, 1053 (1962); (b) M. Tomkiewicz, *J. Electrochem. Soc.* **127**, 1518 (1980).
4. D. S. Boudreaux, F. Williams, and A. J. Nozik, *J. Appl. Phys.* **51**, 2158 (1980).
5. A. J. Bard, A. B. Bocarsly, F. R. F. Fan, E. G. Walton, and M. S. Wrighton, *J. Am. Chem. Soc.* **102**, 3671 (1980).
6. A. Fujishima and K. Honda, *Nature* **238**, 37 (1972).
7. (a) C. Leygraf, M. Hendewerk, and G. A. Somorjai, *Proc. Nat. Acad. Sci.* **79**, 5739 (1982) and references therein; (b) K. Kalyanasundaram, E. Borgarello, D. Duonghong, and M. Graetzel, *Angew. Chem. Internat. Ed.* **20**, 987 (1981); (c) S. Sato and J. M. White, *Chem. Phys. Lett.* **72**, 83 (1980); *J. Phys. Chem.* **85**, 336 (1981); (d) A. Heller, *Accts. Chem. Res.* **14**, 154 (1981).
8. A. J. Nozik, *Photoeffects at Semiconductor—Electrolyte Interfaces* (Amer. Chem. Soc. Sympos. Ser. No. 146, American Chemical Society, Washington, 1981).
9. (a) H. Greischer, in *Physical Chemistry: An Advanced Treatise*, edited by H. Eyring, D. Henderson, and W. Jost (Academic Press, New York, 1970), Vol. IXA; (b) E. C. Dutoit, F. Cordon, and W. P. Gomes, *Ber. Bunsenges. Phys. Chem.* **80**, 475 (1976).
10. A. J. Bard, A. Ledwith, and H. J. Shine, *Adv. Phys. Org.* **13**, 156 (1976).
11. D. M. Chapin, C. S. Fuller, and G. L. Pearson, *J. Appl. Phys.* **34**, 129 (1954).
12. (a) R. M. Murray, *Accts. Chem. Res.* **13**, 135 (1980); (b) A. J. Bard, *J. Chem. Ed.* **60**, 302 (1983); (c) M. S. Wrighton, *J. Chem. Ed.* **60**, 335 (1983).
13. (a) T. Kuwana and W. R. Heineman, *Accts. Chem. Res.* **9**, 241 (1976); (b) J. W. Schultze and M. A. Habib, *J. Appl. Electrochem.* **9**, 255 (1979).
14. For example, see S. Chao, J. L. Robbins, and M. S. Wrighton, *J. Am. Chem. Soc.* **105**, 181 (1983).
15. For example, see F. R. F. Fan, R. G. Keil, and A. J. Bard, *J. Am. Chem. Soc.* **105**, 220 (1983).
16. For example, see N. Oyama and F. C. Anson, *J. Electrochem. Soc.* **127**, 247 (1980); T. P. Henning, H. S. White, and A. J. Bard, *J. Am. Chem. Soc.* **104**, 5862 (1982); P. V. Kamat and M. A. Fox, *J. Electroanal. Chem.* **159**, 49 (1983).
17. M. D. Ward, J. R. White, and A. J. Bard, *J. Am. Chem. Soc.* **105**, 27 (1983).

18. (a) M. Graetzel, *Accts. Chem. Res.* **14**, 376 (1981); (b) J. Kuczynski and J. K. Thomas, *Chem. Phys. Lett.* **88**, 445 (1982); (c) J. Moser and M. Graetzel, *Helv. Chim. Acta* **65**, 1435 (1982).
19. L. P. Childs and D. F. Ollis, *J. Catal.* **66**, 383 (1980) and references cited therein.
20. (a) H. Gerischer and F. Willig, *Top. Curr. Chem.* **61**, 33 (1976); (b) M. T. Spitler, *J. Chem. Ed.* **60**, 330 (1983).
21. M. A. Fox and Kabir-ud-Din, *J. Phys. Chem.* **83**, 1800 (1979).
22. J. R. Hohman and M. A. Fox, *J. Am. Chem. Soc.* **104**, 401 (1982).
23. M. A. Fox and R. C. Owen, *J. Am. Chem. Soc.* **102**, 6559 (1980).
24. J. R. Bolton, *Science* **202**, 705 (1978).
25. C. Kutal, R. R. Hautala, and R. B. King, *Sun 2, Proc. Internat. Solar Ener. Soc. Silver Jubilee* **1**, 109 (1979).
26. C. K. N. Patel and A. C. Tam, *Rev. Mod. Phys.* **53**, 517 (1981).
27. H. Gerischer, M. Spitler, and F. Willig, in *Electrode Processes* 1979, edited by S. Bruckenstein (The Electrochemical Society, Princeton, N.J., 1980), p. 115.
28. (a) P. V. Kamat and M. A. Fox, *J. Phys. Chem.* **87**, 59 (1983); (b) J. H. Richardson, S. B. Deutscher, A. S. Maddux, J. E. Harrar, D. C. Johnson, W. L. Schmelzinger and S. P. Perone, *J. Electroanal. Chem.* **109**, 95 (1980).
29. M. M. Baizer, *Organic Electrochemistry* (Marcel Dekker, New York, 1973).
30. (a) T. Inoue, A. Fujishima, S. Konishi, and K. Honda, *Nature* **277**, 637 (1979); (b) M. Halmann, *ibid.* **275**, 115 (1978); (c) CO_3^{2-} reduction: K. Chanrasekaran and J. K. Thomas, *Chem. Phys. Lett.* **99**, 7 (1983).
31. (a) G. N. Schrauzer and T. D. Guth, *J. Am. Chem. Soc.* **99**, 7189 (1977); (b) H. Miyama, N. Fujii, and Y. Nagae, *Chem. Phys. Lett.* **74**, 523 (1980).
32. S. N. Frank and A. J. Bard, *J. Am. Chem. Soc.* **99**, 4667 (1977).
33. M. D. Ward and A. J. Bard, *J. Phys. Chem.* **86**, 3599 (1982).
34. T. Watanabe, A. Fujishima, and K. Honda, *Chem. Lett.* 897 (1974).
35. Kabir-ud-Din, R. C. Owen, and M. A. Fox, *J. Phys. Chem.* **85**, 1679 (1981).
36. For example, see R. Noufi, A. J. Frank, and A. J. Nozik, *J. Am. Chem. Soc.* **103**, 1849 (1981).
37. S. R. Morrison and T. Freund, *J. Chem. Phys.* **47**, 1543 (1967).
38. R. A. L. van den Berghe, W. P. Gomes, and F. Cardon, *Z. Phys. Chem., Neue Folge* **92**, 91 (1974).
39. D. Benson, *Mechanisms of Oxidation by Metal Ions* (Elsevier, New York, 1976).
40. J. A. Switzer, E. L. Moorehead, and D. M. Dalesandro, *J. Electrochem. Soc.* **129**, 2233 (1982).
41. I. Taniguchi, K. Nakashima, H. Yamaguchi, and K. Yasukouchi, *J. Electroanal. Chem.* **134**, 191 (1982).
42. T. Kawai and T. Sakata, *Chem. Lett.* 81 (1981).
43. (a) B. Kraeutler and A. J. Bard, *J. Am. Chem. Soc.* **99**, 7729 (1977); (b) B. Kraeutler and A. J. Bard, *Nouv. J. Chim.* **3**, 31 (1979); (c) B. Kraeutler and A. J. Bard, *J. Am. Chem. Soc.* **100**, 2239 (1978); (d) B. Kraeutler and A. J. Bard, *J. Am. Chem. Soc.* **100**, 5985 (1978); (e) S. Sato, *Chem. Comm.* 26 (1982).
44. M. A. Ratliff and H. L. Chum, *Exten. Abstr. Electrochem. Soc.* **83**, 510 (1983).
45. H. Yoneyama, Y. Takeo, H. Tamura, and A. J. Bard, *J. Phys. Chem.* **87**, 1417 (1983).
46. A. Heller, E. Aharon-Shalom, W. A. Bonner, and B. Miller, *J. Am. Chem. Soc.* **104**, 6942 (1982).
47. K. Hirano and A. J. Bard, *J. Electrochem. Soc.* **127**, 1056 (1980).
48. C. D. Jaeger and A. J. Bard, *J. Phys. Chem.* **83**, 3146 (1979).
49. R. E. Schwerzel, *Ext. Abstr. Electrochem. Soc.* **83**, 513 (1983).
50. M. A. Fox and C. C. Chen, *J. Am. Chem. Soc.* **103**, 6757 (1981); M. A. Fox and C. C. Chen, *J. Photochem.* **17**, 119 (1981).

51. T. Kanno, T. Oguchi, H. Sakuragi, and K. Tokumaru, *Tetrahedron Lett.* **21**, 467 (1980).
52. M. A. Fox and C. C. Chen, *Tetrahedron Lett.* **24**, 547 (1983).
53. C. C. Chen and M. A. Fox, *J. Comput. Chem.* **4**, 488 (1983).
54. (a) M. A. Fox, B. A. Lindig, and C. C. Chen, *J. Am. Chem. Soc.* **104**, 5828 (1982); (b) T. T. Watanabe and K. Honda, *J. Phys. Chem.* **86**, 2617 (1982).
55. M. A. Fox, *Accts. Chem. Res.* **16**, 314 (1983).
56. H. Courbon, M. Formenti, and P. Pichat, *J. Phys. Chem.* **81**, 550 (1977).
57. P. Pichat, J. M. Herrmann, J. Disdier, and M. N. Mozzanega, *J. Phys. Chem.* **83**, 3122 (1979).
58. N. Djeghri and S. J. Teichner, *J. Catal.* **62**, 99 (1980).
59. S. Kodama, M. Yabuta, and Y. Kubokawa, *Chem. Lett.* 1671 (1982).
60. C. Yun, M. Ampo, S. Kodama, and Y. Kubokawa, *Chem. Comm.* 609 (1980).
61. A. H. Boonstra and C. A. H. A. Mutsaers, *J. Phys. Chem.* **79**, 2025 (1975).
62. I. J. S. Lake and C. Kemball, *Trans. Faraday Soc.* **63**, 2535 (1967).
63. L. V. Lyashenko, V. M. Belousov, and F. A. Yampol'skaya, *React. Kinet. Cat. Lett.* **20**, 59 (1982); *Kinet. Katal.* **23**, 662 (1982).
64. M. Formenti and S. J. Teichner, in *Catalysis* (Specialist Periodical Report, The Chemical Society, London, 1979), Vol. 2, p. 87.
65. J. Lacoste, R. Arnaud, and J. Lemaire, *Compt. Rend. Acad. Sci.* **295**, 1087 (1982).
66. J. M. Herrmann, J. Disdier, M. N. Mozzanega, and P. Pichat, *J. Catal.* **60**, 369 (1979).
67. M. A. Fox, C. C. Chen, and J. N. Younathan, *J. Org. Chem.* **49**, 1969 (1984).
68. M. Fujihara, Y. Satoh, and T. Osa, *J. Electroanal. Chem.* **126**, 277 (1981).
69. M. N. Mozzanega, J. M. Herrmann, and P. Pichat, *Tetrahedron Lett.* 2965 (1977).
70. M. Fujihira, Y. Satoh, and T. Osa, *Nature* **293**, 206 (1981).
71. (a) I. Izumi, F. R. F. Fan, and A. J. Bard, *J. Phys. Chem.* **85**, 218 (1981); (b) I. Izumi, W. W. Dunn, K. O. Wilbourn, F. R. F. Fan, and A. J. Bard, *J. Phys. Chem.* **84**, 3207 (1980).
72. M. K. Eberhardt, *J. Am. Chem. Soc.* **103**, 3876 (1981).
73. M. Fujihira, Y. Satoh, and T. Osa, *Bull. Chem. Soc. Japan* **55**, 666 (1982).
74. (a) J. G. Calvert, K. Theurer, G. T. Rankin, and W. M. MacNevin, *J. Am. Chem. Soc.* **76**, 2575 (1954); (b) M. V. Rao, K. Rajeshwan, V. R. PaiVerneken, and J. DuBow, *J. Phys. Chem.* **84**, 1987 (1980).
75. (a) M. A. Hema, V. Ramakrishnan, J. C. Kuriacose, *Ind. J. Chem. B* **16**, 619 (1978); (b) H. Kasturirangan, V. Ramakrishnan, and J. C. Kuriacose, *J. Catal.* **69**, 216 (1981).
76. H. Mozzanega, J. M. Herrmann, and P. Pichat, *J. Phys. Chem.* **83**, 2251 (1979); *Chem. Phys. Lett.* **74**, 523 (1980).
77. M. A. Fox and M. J. Chen, *J. Am. Chem. Soc.* **105**, 4497 (1983).
78. T. Takizawa, T. Watanabe, and K. Honda, *J. Phys. Chem.* **82**, 1391 (1978).
79. For example, see (a) J. Cunningham and P. Meriaudeau, *J. Chem. Soc., Faraday I* **72**, 1499 (1976); (b) R. B. Cundall, R. Rudham, and M. S. Salim, *ibid.*, **72**, 1642 (1976); (c) P. Pichat, M. N. Mozzanega, J. Disdier, and J. M. Herrmann, *Nouv. J. Chim.* **6**, 559 (1982).
80. J. Cunningham, B. K. Hodnett, M. Ilyas, E. M. Leahy, and J. P. Tobin, *J. Chem. Soc., Faraday Trans. I* **78**, 3297 (1982).
81. T. Sakata and T. Kawai, *Chem. Phys. Lett.* **80**, 341 (1981).
82. S. Teratani, J. Nakamichi, K. Toya, and K. Tanaka, *Bull. Chem. Soc. Japan.* **55**, 1688 (1982).
83. T. Kawai and T. Sakata, *Nature* **286**, 474 (1980).
84. (a) P. R. Harvey, R. Rudham, and S. Ward, *J. Chem. Soc., Faraday Trans. I* **79**, 1381 (1983); (b) L. P. Childs and D. F. Ollis, *J. Catal.* **67**, 35 (1981); **66**, 383 (1980); (c) J. Cunningham and B. K. Hodnett, *J. Chem. Soc., Faraday I*' **78**, 3297 (1982); (d) J. Cunningham, B. K. Hodnett, M. Ilyas, E. M. Leahy, and J. P. Tobin, *J. Chem. Soc., Faraday I* **78**, 3297 (1982).

85. A. H. A. Tinnemans, T. P. M. Koster, D. H. M. W. Thewissen, and A. Mackor, *Nouv. J. Chim.* 373 (1982).
86. J. W. Pavlik and S. Tantayanon, *J. Am. Chem. Soc.* **103**, 6755 (1981).
87. J. M. Herrmann, M. N. Mozzanega, and P. Pichat, *J. Photochem.* **22**, 333 (1983).
88. A. L. Pruden and D. F. Ollis, *J. Catal.* **82**, 404, 418 (1983).
89. K. Okada, K. Hisamitsu, and T. Mukai, *Chem. Comm.* 941 (1980).
90. R. A. Barber, P. de Mayo, and K. Okada, *Chem. Comm.* 1073 (1982).
91. K. Okada, K. Hisamitsu, T. Miyashi, and T. Mukai, *Chem. Comm.* 1106 (1982).
92. T. Majima, C. Pac, and H. Sakurai, *J. Am. Chem. Soc.* **102**, 5265 (1980).
93. H. Reiche and A. J. Bard, *J. Am. Chem. Soc.* **101**, 3127 (1979).

5

Structural Effects in Organic Electrochemistry

Wayne E. Britton

1. ELECTROCHEMISTRY OF CYCLOOCTATETRAENE

1.1. Introduction

The intent of this chapter is to examine some of the ways in which the structures of organic compounds can influence their electrochemical properties, and how electrochemistry has been used to unravel structural processes which occur within molecular systems. Three different areas will be examined; however, there are many others which could be included here. The three systems include cyclooctatetraenes, unusual olefins, especially those related to bianthrones, and, finally, vicinal dihalides.

Structural changes can either proceed or follow heterogeneous electron transfer. If they follow, the intermediates often produced by electrochemical oxidation or reduction (typically ions, radicals, or ion radicals) play a direct role in this structural change. The electrode surface can be a factor but often these processes will occur independent of the nature of the surface and, therefore, are not unique heterogeneous electrochemical events. However, electrochemistry is an excellent way to study structural change, because it provides the means both to generate and monitor redox species. In some cases, electrochemistry may be the only reasonable method to obtain information about structural changes associated with electron transfer.

Wayne E. Britton ● Department of Chemistry, University of Texas at Dallas, Richardson, Texas 75083-0688.

1.2. Cyclooctatetraene Structure

Cyclooctatetraenes are a class of $4n$ annulenes with unique properties which are manifested in their electrochemical behavior. The neutral molecule is tub shaped, which allows the sp^2 carbon-to-carbon bond angle to be near the desired 120°—it is actually about 126° in the parent.[1] This tub-shaped molecule undergoes two fundamental structural changes which have been studied theoretically and experimentally. Ring inversion,[2,3] shown in Scheme I, is a process in which the tub turns inside out to form a new tub form via a mechanism thought to involve a planar transition state with D_{4h} symmetry. Bond shifting[2] is the isomerization of the double bonds, also thought to occur through a planar transition state with D_{8h} symmetry. The activation energy for ring inversion is about 13.7 kcal/mol[3] and that for bond shifting is 3.4 kcal/mol higher[4] for the unsubstituted cyclooctatetraene (COT) system.

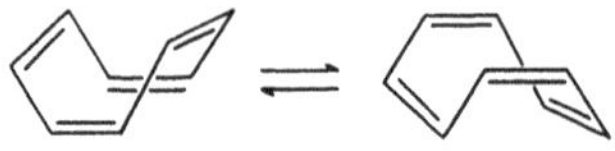

Scheme I

1.3. Structure of Cyclooctatetraene Anion Radical

The main interest in COT electrochemistry is associated with the first reduction, which is thought to form a planar anion radical.[5] Somewhere along the reaction coordinate the molecule flattens, and electron transfer occurs. Just at what point these events take place is of considerable interest. Is the structure of the electron transfer transition state planar, nonplanar, or somewhere in between?

Early electrochemical studies on COT were carried out by Katz, Reinmuth, and Smith,[6] who speculated that the reduction process involved flattening of the COT ring. This was based upon the fact that the first reduction appeared to involve a one two-electron transfer wave in the aqueous dioxane system used. A typical aromatic reduction would exhibit two separate reduction waves (see Fry, chapter 1) with the second electron transfer a half a volt more negative than the first. This is because the second electron transfer must overcome the electron-electron repulsion energy associated with the formation of a dianion. The lack of separation of these waves with COT was implied to be masked by the negative shift in the first wave due to the energy requirement of the ring flattening process.

Strauss, Katz, and Fraenkel[5] carried out COT reductions in ether solvents using alkali metals as the reductants. Their esr results indicate that the electron exchange rate between COT and its anion is slow, while the rate between the anion and dianion is fast. This implies an activation process for the first electron transfer, which they associated with the ring flattening. It also suggests that the mono and dianions have similar structures. A similar conclusion was reached by Carrington and Todd[7] who reexamined the esr spectrum of COT anion radical and several monoalkyl derivatives. It is important to note here that these are solution phase results. The rates of electron transfer reactions at electrodes can be strongly influenced by double-layer effects, which may make their interpretation subject to uncertainties.

The flat D_{8h} structure for the COT anion radical is also supported by the electronic spectrum of the ion,[8] but based upon MINDO-2 calculations, the nonplanar ion is slightly favored over the planar structure.[9] It is also known that flat COT molecules are reduced readily,[10,11] while those which cannot exist in a flat structure are either reduced with great difficulty,[12] or not reduced at all[13] (see Fry, chapter 1). In summary, the experimental data supports a planar structure for the anion radical of COT.

1.4. Electron Transfer Kinetics and the Structure of the Transition State

The earliest electrochemical kinetic study of COT was published by Allendoerfer and Rieger[14] in 1965. An important element of this work is that it was run in DMF and DMSO solvent which reduces the rates of protonations of the COT anions that had occurred in earlier studies. In this media, the reduction follows two one-electron transfer steps. There is some protonation of the COT dianion from residual water, which affords cyclooctatriene. The earlier observations of one two-electron processes can then be explained by facile protonation of the anion radical to give the electrochemically active radical, which accepts a second electron at the same potential as the first, a so-called *ECE* process. It was also pointed out that ion pairing can also cause a change in mechanism to an *ECE* process, but here the chemical reaction symbolized by *C* refers to the disproportionation of two anions to the parent hydrocarbon and the corresponding dianion. Ion pairing, favored both by low-dielectric solvents and alkali metal cations, reduces electron-electron repulsions in the dianion to the point that the free energy for transfer of a second electron is now negative.

Allendoerfer and Rieger[14] extracted the electron transfer kinetics primarily from in-phase ac polarographic currents, measured as a function of temperature. The ac currents afford a parameter which is proportional to the heterogeneous electron transfer rate constant. An Arrhenius plot after correction for the diffusion temperature dependence gives the enthalpy of activation for the electron transfer at the first wave. This value of

7.7 kcal/mole, when combined with the entropy of activation obtained from their estimate of the electron transfer rate constant (0.0085 cm/sec) and application of the absolute rate theory, affords a free energy of activation of 10–11 kcal/mole for the first electron transfer to COT. The similarity of this value with that measured for the flattening of the COT ring system (13.7 kcal)[3] led to the suggestion that the transition state for electron transfer "resembles the planar anion." They were careful to point out that their measurement also includes the solvent reorganization energy for formation of the anion, estimated to be about 2-3 kcal/mole, and this activation energy is therefore not a direct measure of the energy to flatten the COT ring. It was felt that the ring was only partially flattened in the electron transfer transition state.

Allendoerfer and Rieger's results were somewhat qualitative, since the appropriate ac polarographic theory to treat the complex COT reduction had not been fully developed at that time. Huebert and Smith[15] repeated the earlier work and utilized *EEC* theory for fundamental and second harmonic *ac voltammetry* and refined the data of Allendoerfer and Rieger.

The standard rate constant for the second reduction wave in COT was determined from cot ϕ vs. $\omega^{1/2}$ plots. Because the first wave is quasireversible (electron transfer is slow on the timescale of the experiment), it is smaller than the second wave and partly obscured by it, making it difficult to obtain accurate data in the manner described above for the second wave. They instead determined the rate constant for the first reduction from the magnitude of the in-phase total harmonic current, and also from the ratio of peak currents associated with the first and second harmonic ac polarographic waves using quasireversible ac theory. The apparent electron transfer rate constants for the first and second reductions at 25°C are 2×10^{-3} and 1.5×10^{-1} cm/sec, respectively. Even though Allendoerfer and Rieger's results were somewhat qualitative, there is rather good agreement between these two works. Again, the first electron transfer rate is rather slow, with the second one being significantly faster. The electron transfer rate to 1,3,5-cyclooctatriene, which is present in the COT system due to some protonation of the COT dianion, was also measured and found to be 8.3×10^{-2} cm/sec at 25°C. This rate is also significantly greater than that for the first reduction of COT and about the same as that for the second reduction.

With the homogeneous and electron transfer kinetics in hand, Heubert and Smith were able to calculate the reversible potentials for the first and second reduction waves for COT. This required correction to the measured potential of the first wave due to the fact that it is quasireversible, and to the second wave because of a following protonation of the dianion. The $\Delta E_{1/2}$ value is 0.279 volts, which affords a disproportionation equilibrium constant of 2.2×10^{-5} at 25°C in DMF solution. This is about ten orders of magnitude smaller than observed in an ion pairing system.

The electrochemical picture which emerges from these early studies points to an unusually slow electron transfer for the first COT reduction, compared with a typical aromatic hydrocarbon reduction. This appears to be associated with at least partial flattening of the COT ring system in the electron transfer transition state to form a planar anion radical. The second electron transfer is relatively fast, implying that the transition state structure is similar to that of the reactant—the planar COT anion radical.

A paper appeared[16] which was at odds with the general *EEC* scheme cited above for COT reduction. It was observed that only a single one-electron reduction wave was present in THF, with tetramethylammonium ion as the supporting electrolyte, aside from a small wave at very negative potential assigned to the cyclooctatriene reduction. It was concluded that the anion radical formed at the first wave was stable and was not reduced within the available potential range of the solvent system. This conclusion was challenged by several workers[17,20] and shown to be incorrect. Thus, Fry, Hutchins, and Chung[17] showed clearly that even tetraalkylammonium ions, especially small ones, can ion pair with the COT anions and cause waves to shift and tend to merge, even in the rather polar solvent DMSO. Jensen, Ronlan, and Parker[18] also found a dramatic effect on the reduction waves when the size of the tetraalkyammonium cation was changed from tetramethylammonium to tetrabutylammonium, *vide infra*. Smith and Bard[19] found that COT reduction was a one two-electron process in liquid ammonia, and Allendoerfer[20] showed that the anion radical was clearly involved by simultaneous esr and cyclic voltammetric analysis.

1.4.1. Fast Electron Transfer to Cyclooctatetraene

Jensen, Ronlan, and Parker[18] give a slightly different picture of COT electrochemistry. The electrochemical reduction of COT in dry DMF containing tetrabutyl, tetrapropyl, tetraethyl, and tetramethylammonium as the electrolyte cations affords dramatic differences for the electron transfer rates for the first wave as measured by cyclic voltammetry. With the tetramethylammonium cation, the system appears essentially reversible at a scan rate of 131 *mv*/sec at 13.3°C, while with tetrabutylammonium cation, the anodic-to-cathodic peak separation is over 300 mv at 45°C, indicating very slow electron transfer in the latter case. The relatively fast kinetics observed with the tetramethylammonium ion results seems to conflict with earlier kinetic results and casts some doubt on the suggestion that COT partly flattens in the transition state for the first reduction. Furthermore, a similar conclusion was reached by Anderson *et al.*[21] based upon the electrochemical behavior of a heterocyclic analog of COT.

1.4.1a. Electrochemistry of Azocines. The COT heterocyclic analogs alluded to above include 2-methoxyazocine (**1**), and 3,8-dimethyl-2-methoxyazocine (DMMA) (**2**), and other methyl derivatives. Spectroscopic data indicate the azocines are also tub-shaped structures, while their dianions are found to be planar, the primary criterion being a substantial ring current in their nmr spectra. Simple Hückel MO theory[22] predicts a larger delocalization energy for the azocine dianion than the COT dianion.

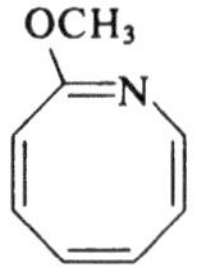

OCH_3 OCH_3

N N

1 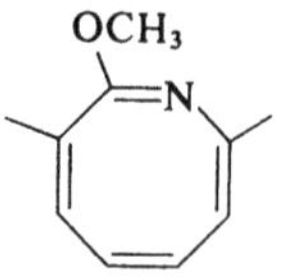 **2**

The electrochemistry of the azocines[21a] shows a one two-electron irreversible (in the electrochemical sense) reduction to afford the stable dianion. The dianion was reported to reoxide to the parent about one volt more anodic than the reduction in either THF or DMF solution. On the basis of this mechanism, the first electron transfer rate was determined for the two azocines above, and the rate constants were found to be about 4×10^{-5} and 1.5×10^{-6}, respectively. These values are considerably smaller than observed for the reduction of COT in DMF. The general behavior is similar to that observed for COT in liquid ammonia,[15] where the process is an irreversible two-electron reduction with a large separation between the anodic and cathodic waves at low temperature.

It was later shown by Jensen, Petterson, Ronlan, and Parker,[21b] however, that the very basic azocine dianions were undergoing a protonation reaction, and this was responsible for the wide peak separation, and not slow electron transfer. They found the reduction of DMMA to be a quasi-reversible ($\Delta E_p = 80\ mv$) two-electron process in dry dimethylformamide containing tetramethylammonium ion. On the basis of the lack of resolution of the wave into two one-electron processes even at temperatures down to −60°C, and simulations of experiments, it was concluded that the second electron transfer was positive of the first. This is consistent with a large delocalization energy for the dianion,[22] which would tend to shift its wave toward (in this case positive of) the first reduction wave. When tetrabutylammonium is used as the counter ion, the oxidation wave of the dianion broadens into what appears to be two waves.

Simulations of different reaction schemes were most consistant with the data for the tetrabutylammonium case when heterogeneous electron transfer to DMMA is slow (k_s for the simulation was 5×10^{-3} cm/s), but formation of the anion radical in solution via disproportionation of the DMMA dianion and DMMA is fast ($k_f = 1.24 \times 10^4 \text{ M}^{-1}\text{ s}^{-1}$, $k_r = 3.64 \times 10^5 \text{ M}^{-1}\text{ s}^{-1}$, also simulation parameters). On this basis, the slow heterogeneous electron transfer to DMMA was ascribed to an electrolyte effect, and not due to a conformational change.

It is important to point out here that fast electron transfer in solution was not observed for COT,[5] according to esr experiments. Also, the barrier for inversion of the azocine system is not known, nor is the structure of the anion radical known. Extrapolation of these results to COT at this time is therefore tenuous.

1.4.2. What Does the Electron Transfer Rate Mean?

What is the structure of the transition state leading to the COT anion radical? The evidence of fast electron transfer found by Parker *et al.*[18] suggests that little structural change occurs prior to the first electron transfer. However, the esr data and several other electrochemical examinations are in apparent conflict with this conclusion. To reconcile this data, it is necessary to delve deeper into the interpretation of electrochemical kinetics.

In order to assess the electron transfer kinetics for COT reduction we will apply the theory of Marcus.[23] According to this theory, the transition state for electron transfer is where the atomic configuration of the reactant and product are identical, with identical solvation and energy, but differing in electronic structure and charge. An infinite number of atomic structures may satisfy these criteria, but only those of lowest energy are important kinetically.

The relationship between the electron transfer rate constant and the free energy of activation is given by:

$$k_f = KZ \exp - (\Delta G^{\dagger}/RT) \tag{1}$$

where K is a transmission coefficient taken to be unity for a nearly adiabatic process, and Z is the collision frequency with the electrode, equal to $(RT/2\pi M)^{1/2}$, with M the molecular weight of the species. The free energy of activation can be broken down into:

$$\Delta G^{\dagger} = (W^* + [\lambda + W - W^* + \exp(E - E^0)]^2)/4\lambda \tag{2}$$

where W^* is the work required to bring the reactant to the pre-electrode site, W is the corresponding term for the product, E is the potential of the electrode, and E^0 the standard couple for the redox system, and $\lambda = \lambda_i + \lambda_0$, the molecular and solvent reorganizational energies, respectively. When $E = E^0$, and the reactant is uncharged, with a high dielectric constant solvent and a large concentration of indifferent electrolyte, this equation reduces to:

$$\Delta G^{\dagger} = (\lambda_i + \lambda_o)/4 \tag{3}$$

Within the limits of the approximations just cited, we can obtain the molecular structural reorganization energy of the transition state relative to the reactant, if we can separate the solvation term. Whether the approximations above are justified is debatable. Basically, what one assumes is that there are no other outside forces acting on the system which will influence the fundamental electron transfer process. Such forces will be discussed further below.

The discussion above suggests that factors other than internal reorganization can contribute to the energy of the activated complex, and the electron transfer rate, as one would expect. These will be divided into two categories and discussed separately. The first group are those factors which influence the rate because of a change in the intrinsic energy of the activated complex, and the second class are those factors associated with the properties of the pre-electrode site which only appear to modify the intrinsic energetics of the activated complex. The second category will be addressed first.

1.4.2a. Properties of the Pre-Electrode Site. Since the solution phase esr rate data[5] are not encumbered by double-layer effects, we might ask how the heterogeneous electrochemical data might be? The fundamental question here is what could make an activated heterogeneous electron transfer process appear kinetically fast when the electrolyte contains tetramethylammonium ion but slow with tetrabutylammonium ion? Up to this point, we have presented electrochemical kinetic data obtained directly from the experimental measurements. Often such kinetic data are massaged in order to correct for so-called double-layer effects. The simplest of these is probably the Frumkin correction.[24] A direct comparison of kinetic data without this correction assumes the reaction occurs outside the double layer of the electrode in the absence of an electric field, which could assist the electron transfer process (or deter it) according to the magnitude of the potential as expressed in the following equation[25]:

$$k_{s\,\mathrm{app}} = k_s \exp[(\alpha n - Z)\phi F/RT] \tag{4}$$

where ϕ is the potential at the pre-electrode site, Z is the charge on the species, and k_s is the rate in the absence of the potential.

It is often assumed, however, that the electron transfer occurs at the outer Helmholtz plane[26]—the plane of closest approach—where the potential may be finite. The actual potential and its gradient depend upon the composition of the solution and the potential on the electrode. With high concentration of the electrolyte, the potential falls rapidly with distance,[27] the opposite being true with low-electrolyte concentration.[27] The size of the electrolyte is also important.[28] Small cations at a negatively polarized electrode lead to a steep field in the double layer, while with large cations the field falls more slowly with distance. The theoretical analysis of the double layer system is a difficult task, because one must make assumptions about molecular size, packing, solvation, and dielectric constant in this region of high-field density,[29] assumptions which are difficult to test experimentally. Furthermore, in many cases it is little more than a guess where the pre-electrode site is located, let alone the conditions which exist there.

Parker and co-workers'[18] basic conclusion about the COT rate data was that a double layer effect appeared to be responsible for the slow electron transfer kinetics when tetrabutylammonium ion served as the electrolyte cation. Since that time, Parker *et al.*[26] have found that the tetramethylammonium ion is adsorbed at the potentials required for reduction of COT. This would mean there is a higher than usual concentration of this ion at the outer Helmholtz plane, and the double layer structure would be perturbed.

It was also found that benzonitrile electron transfer kinetics are influenced by the counter-ion size,[26] but much less dramatically than with COT. Benzonitrile electron transfer rate constants are much greater than those for COT presumably because little structural reorganization is required to reach the transition state. One problem with direct comparison of the electron transfer rates between these two systems is that their reduction potentials differ significantly and, therefore, the double layer structure and the rate correction factor would also be different.[30] But what about other compounds such as tetracene[25,31] which reduce at the same potential as COT with tetrabutylammonium cation, but which has an apparent electron transfer rate 1000 times greater? The answer is that COT is different, and the simplest explanation is that it must be at least partially flat in the transition state.

1.4.2b. Effect of Solvation on the Transition State Energy. Since double-layer corrections are not required for interpretation of the electron transfer kinetics performed homogeneously, it seems compelling that flattening of the ring is partly responsible for the low-electron transfer rate there. How then can the data of Parker *et al.*[22] be rationalized? It can be seen from Equation 3 above, the activation energy for electron transfer is a function of both solvent and internal reorganization. If the internal reorganization remains constant while changing to the small tetramethylammonium ion, then solvation could be decreasing, giving rise to a small overall activation energy and, hence, a large-electron transfer rate constant. Although it was observed that ion pairing was insignificant for these systems as the $E^{0\prime}$ values remained constant with the different ions (Fry *et al.*[21] did observe evidence of ion pairing in DMSO), this does not exclude ion pairing in the transition state. In fact, this suggestion does not seem unreasonable, and such a possibility has been suggested in another context.[29] Ion pairing which would tend to localize charge would be costly in the planar delocalized anion radical. But in a nonplanar transition state with less delocalization, the stabilization gained through ion pairing could well overcome a small loss of delocalization energy. How flat is the transition state? That is difficult to answer, but clearly systems which cannot flatten do not get reduced, meaning either the forward rate constant is very small or the reverse is extremely large.

1.5. The Electrochemical Reduction of Flat Cyclooctatetraenes

A logical question to ask at this point is how would a flat cyclooctatetraene behave electrochemically? Ignoring for the moment possible double-layer perturbations to the electron transfer kinetics, we would expect that the electron transfer rate would be somewhat faster than COT, if flattening was an essential element of the activation process. Indeed, this has been found in two separate cases.[11,12]

Bard *et al.*[11] studied the electron transfer kinetics to four COT derivatives. sym-Dibenzo-1,5-cyclooctadiene-3,7-diyne (DBCOD), (**3**), with the two triple bonds in the COT ring has been shown by X-ray analysis to be a planar structure in the crystal, while sym-dibenzocyclooctatetraene, (**4**),

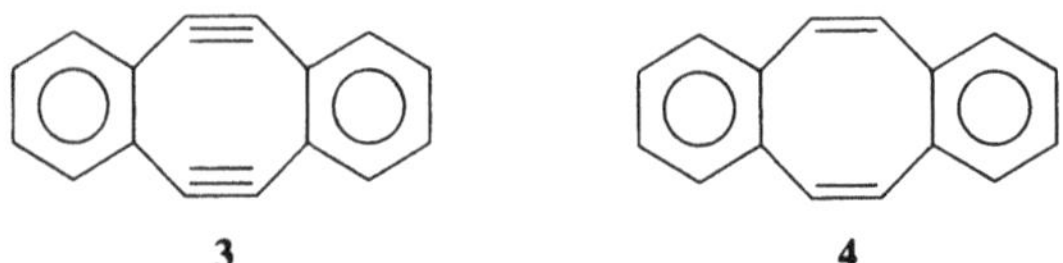

3 **4**

is a tub-shaped molecule similar to cyclooctatetraene itself. The apparent electron transfer rate constants for these compounds were found to be 0.05 and 0.004 cm/sec, respectively. The flat compound is reduced more rapidly than the tub molecule, which would be expected if the molecule was partially flattened in the transition state. Furthermore, the reduction potential of DBCOD is essentially the same as that for COT, which means that simple double-layer corrections would not change this outcome. Although the electrolyte concentration was quite high in the last case, the kinetics are rather insensitive to electrolyte concentration for the tetrabutylammonium cation which was used here.

From the measured rate data, Bard and co-workers calculated the difference for the free energy of activation for DBCOT and DBCOD, assuming work and solvation differences would cancel. The free energy of activation difference attributed to structural reorganization was found to be about 6 kcal/mole. This number is similar to that obtained by Allendoefer and Rieger[14] (10 kcal) which was calculated from the temperature dependence of the rate. While the agreement is not excellent, there are certainly differences expected here because of the different systems used and the different approaches and assumptions involved. It is significant that both numbers are positive and not inconsequential.

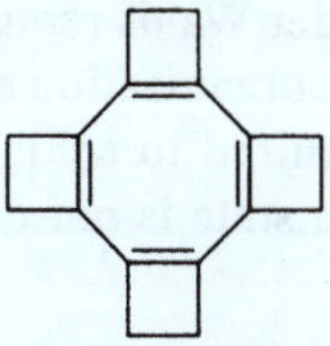

5 (16F)

Another flat COT system was studied by Britton, Ferraris, and Soulen.[12] Perfluorotetracyclobuta-1,3,5,7-cyclooctatetraene (FCOT), (**5**), was shown by X-ray analysis to be a flat molecule in the crystal. The apparent electron transfer rate constant for the first reduction of FCOT is about 35 times greater than for COT, 0.07 cm/sec, and remarkably close to the 0.05 cm/sec value found by Bard *et al.* for DBCOD. Substitution of the rate for FCOT and COT into expression 3 which assumes the work terms can be neglected, affords a difference of 10 Kcal/mole for $\lambda_i + \lambda_o$. If one assumes the O terms are equal for COT and FCOT, then the 10 kcal/mole represents the molecular reorganization energy required to reach the COT transition state. It seems likely that systems which do undergo extensive structural change, change their molecular volume as well and, therefore, require additional solvent restructuring. Thus, the 10 kcal calculated above is probably an upper limit.

Here again, we have neglected double-layer corrections. Yet, it is satisfying that FCOT and DBCOD have similar rates for reduction, since the compounds are of similar structure, while COT has a significantly lower rate, because its structure is somewhat different than the former compounds and must undergo some ring flattening to reach the anion radical product. If double-layer corrections could be applied to FCOT and BDCOD, the "corrected" rates would very likely differ significantly as the reduction potentials for these compounds are separated by about 2.4 volts!

1.6. Conclusions

In summary, the COT system appears to undergo a structural reorganization prior to electron transfer. The reorganization energy is manifested in a slow-electron transfer, as few species have sufficient energy to reach the activated state. While the actual structure of the transition state is not known, it is approximately 6–10 kcal/mole above that for the neutral tub-shaped species based upon experimental estimates. Of course the actual structure depends upon the shape of the potential surface. This structure appears to be a compromise between stabilization gained due to resonance, which is maximized when the system is completely flat, and the energy cost associated with bending forces and van der Waals repulsion in the flat species.[32] Since all of the measurements give reorganization energies which are considerably lower in energy than that required to achieve a flat D_{4h} or D_{8h} structure, it is certain that the transition state is not completely flat either.

2. ELECTROCHEMISTRY OF UNUSUAL DOUBLE BONDS

2.1. Introduction

Electrochemical reduction or oxidation of carbon-to-carbon double bonds represents a class of reactions in which structural reorganization in the molecule can take place following electron transfer. The redox reaction lowers the double bond order allowing the olefin to twist. The intermediate ion radical can then be converted back to a neutral form via electron transfer at the electrode or with a redox compound present in the solution, possibly the parent molecule. (The particular pathway chosen will depend upon the

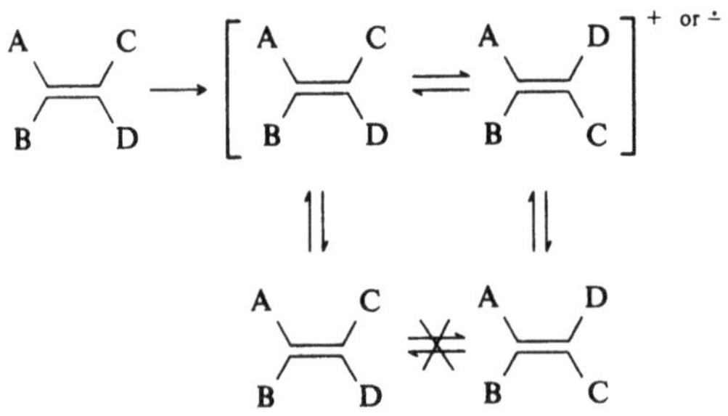

Scheme II

kinetic and thermodynamic parameters of the system.) The neutral compound produced may end up having a different structure than the starting compound. This isomerization often affords the more stable olefin. The generalized picture is shown in Scheme II.

2.2. Activated Olefins

A number of interesting examples of double-bond isomerization stimulated by electrochemical reduction have been reported by Bard and coworkers.[33] They[33e,f] studied the diethyl fumerate/diethyl maleate systems in dimethylformamide solution. Diethyl fumerate undergoes a one-electron reversible reduction at −1.41 V vs. SCE to afford the anion radical. However, the reduction of the corresponding *cis* isomer, diethylmaleate, is an irreversible process which occurs at $Ep = -1.61$ V, and affords an anodic peak at the same position as that for oxidation of the diethylmaleate anion radical. This indicates the DEM anion radical isomerizes and upon oxidation gives DEF.

$$\text{DEM} \xrightarrow{e} [\text{DEM} \rightleftharpoons \text{DEF}]^{\cdot-} \rightarrow \text{DEF} + e \quad (5)$$

The rate constant for the isomerization of DEM was found to be about 10 sec^{-1}.[33e] The driving force for the isomerization is apparently due to coulombic repulsion of the two carbonyl oxygens in the *cis* form for the anion, a factor which raises the energy of the *cis* anion compared with the *trans* form. Dietz and Peover[34] found that *cis* stilbene behaves similarly, although the isomerization is slower and is presumably sterically driven.

When a sample of DEM containing a small amount of DEF was electrolyzed at the diffusion plateau for DEF but positive that for DEM, the concentration of DEF increased.[33e] This was attributed to the net conversion of DEM to DEF catalyzed by $DEF^{\cdot-}$. Similar electrochemical behavior was observed for the *cis* and *trans* forms of dibenzoyl ethlylene.

2.3. Tetraphenylethylene (TPE)

The tetraphenylethylene system is also very interesting and has been recently studied in detail by Grzeszczuk and Smith.[35] Using mainly Faradaic admittance data, they were able to extract all the essential rate and thermodynamic parameters for the reduction processes in this rather complicated system. The major complication is due to the fact that the first and second reduction potentials are virtually identical, $E^{0\prime} = -1.24$ V and −1.28 V vs. SCE at 278 K, respectively. Hence, to evaluate the rate and equilibrium constants, one must therefore take into account the homogeneous processes, such as disproportionation. In spite of this, all parameters were evaluated, principally through application of working curves for the plausible mechanisms.

The disproportionation rate constant for TPE anion radical was found to be $2 \times 10^7\ M^{-1}\ s^{-1}$ in DMF at 278 K, and the first and second heterogeneous standard rate constants are 0.10 and 0.008 cm/sec, respectively. The second rate constant is very low and considerably smaller than that for the first reduction.

The low second-electron transfer rate constant suggests that considerable structural reorganization is necessary to reach the electron transfer transition state, while the first reduction being fast suggests little structural reorganization is involved. As usual, this assumes there is nothing special about the energy of activation for solvation, since the structural and solvent reorganization terms cannot be separated—see Section 1.4.2b. Since this is the usual practice when comparing systems of similar structure (in this case we are suggesting that solvation changes in tetraphenylethylene are similar to those of other aromatic compounds), we will speculate on the molecular reorganization which may be operating. If energies of activation for solvation change as a consequence of a molecular reorganization in the transition state, these could in principle raise or lower the energy of the transition state by comparison with a structurally rigid model. In either case, a reduction in the electron transfer rate would be a manifestation of the molecular reorganization, while the actual magnitude of the effect would be divided between the solvation and structural reorganization terms.

How might we account for the electron transfer kinetics in molecular terms? Since the first electron transfer is reasonably fast, probably little structural change is taking place. Thus, the anion radical remains close to a planar structure, with perhaps some small twisting of the four aromatic rings to avoid steric interactions. This geometry would tend to preserve the partial double-bond order for the ethylene portion of the structure and facilitate delocalization of the electron over the entire molecular system. As we shall see below when we discuss bianthrones, this is an important consideration. Upon addition of the second electron, it is advantageous to have the two electrons far removed from one another to reduce electron-electron repulsion. This could be achieved by allowing the two halves of the molecule to twist about the ethylene, so that they are perpendicular with respect to one another. Such a structure with one charge localized in each half of the molecule, and hence noninteracting, may account for the fact that the first and second electron transfers occur at the same potential. Typically, the separation between the first and second wave in aromatic systems is around 0.5 V (see Fry, chapter 1). A twisted structure for the TPE dianion has been suggested in ion pairing systems.[36] Such a structural reorganization may also account for the slow electron transfer at the second wave. Double layer effects upon the first and second electron transfer should be similar since their $E^{0\prime}$ values are virtually identical and they each change charge by one unit.

2.4. Bianthrones and Related Compounds

Recent investigations,[37–40] principally by Evans and co-workers,[37] have revealed the nature of structural changes occurring in bianthrones and related olefins upon electron transfer. Bianthrones are unusual olefins, and a brief discussion of their properties will be given.

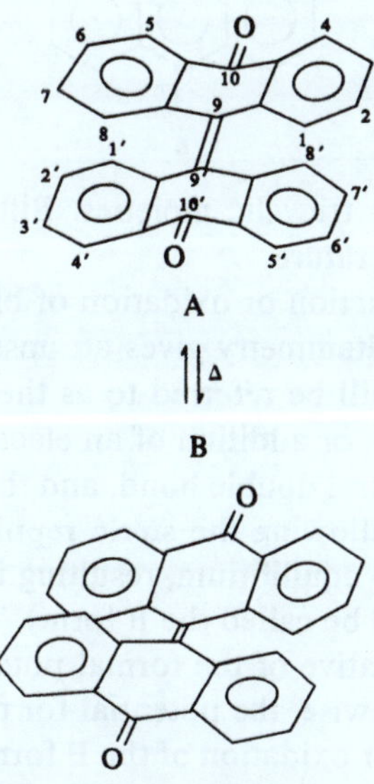

Scheme III

Bianthrone itself has two stable forms which thermally interconvert—see Scheme III. The more stable form on the top is a doubly folded structure in which each anthracenyl is bent away from the other and folded toward a plane which passes through the 9,10,9′,10′ positions. This relieves steric crowding of the 1,8′ and 1′,8 hydrogens, allowing maximum overlap of the double bond between the 9,9′ carbons. Upon heating, the ring systems twist and become planar, relieving the steric congestion but at a cost of overlap of the central double bond. It is estimated that the angle of twist about the 9,9′ carbons is 57 degrees[41] for bianthrone. The interconversion of these two structures is responsible for the thermochromatic properties of bianthrone and related compounds. If substituents other than hydrogen are bonded at the 1,1′,8, or 8′ positions, obviously the equilibrium position will be shifted, which eliminates the thermochromatic effects when thermal decomposition preceeds this transition.[41,42] For example, 1,1′-dimethylbianthrone is not thermochromic,[37g] because in this case the twisted form is thermodynamically unstable.

Lucigenin (**6**) (10,10′-dimethyl-9,9′-biacridinium ion)[37d,39b] is a related system in which the twisted form is stable, but which does not have a double

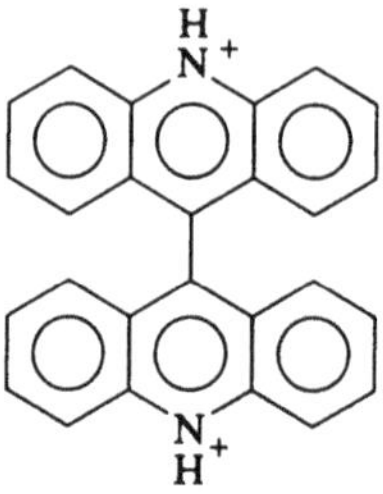

6

bond connecting the two tricyclic moieties. Bifluorenyl is also a twisted structure[43] at room temperature.

Electrochemical reduction or oxidation of bianthrone and substituted bianthrones via cyclic voltammetry gives an unstable ion radical with the bifolded structure (this will be referred to as the A form). Removal of an electron from the HOMO, or addition of an electron to the LUMO, lowers the bond order in the central double bond, and the energy balance between the two forms is upset, allowing the steric repulsions at the 1,1′ and 8,8′ positions to dominate the equilibrium, resulting in the rapid conversion to the twisted form (this will be called the B form). The potential[37d] necessary for reduction of A is negative of the formal potential for reduction of the B form of the anion. Likewise, the potential for oxidation of the A form is more positive than that for oxidation of the B form[37d] of either the monocation or the neutral species. Thus, as soon as A is reduced or oxidized, it rapidly isomerizes to the B form, and a second electron is transferred immediately forming the diion with the B structure. Upon reversing the voltage scan direction, the B diion is converted to the corresponding monoion, and then in a second electron transfer, to the neutral B form. Thus, the unstable B form of the neutral bianthrone can be generated and studied electrochemically. This is summarized in Scheme IV.

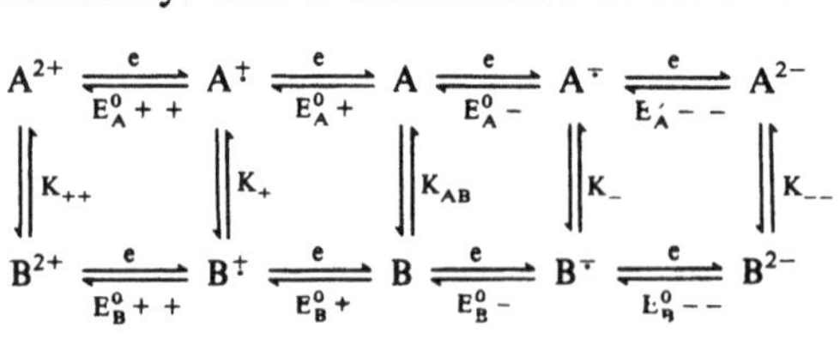

$$
\begin{array}{ccccccccc}
A^{2+} & \underset{E^0_{A++}}{\overset{e}{\rightleftharpoons}} & A^{+\cdot} & \underset{E^0_{A+}}{\overset{e}{\rightleftharpoons}} & A & \underset{E^0_{A-}}{\overset{e}{\rightleftharpoons}} & A^{-\cdot} & \underset{E^0_{A--}}{\overset{e}{\rightleftharpoons}} & A^{2-} \\
\Updownarrow K_{++} & & \Updownarrow K_{+} & & \Updownarrow K_{AB} & & \Updownarrow K_{-} & & \Updownarrow K_{--} \\
B^{2+} & \underset{E^0_{B++}}{\overset{e}{\rightleftharpoons}} & B^{+\cdot} & \underset{E^0_{B+}}{\overset{e}{\rightleftharpoons}} & B & \underset{E^0_{B-}}{\overset{e}{\rightleftharpoons}} & B^{-\cdot} & \underset{E^0_{B--}}{\overset{e}{\rightleftharpoons}} & B^{2-}
\end{array}
$$

Scheme IV

It is possible to directly measure the formal potentials for the two oxidations and reductions of B for bianthrone and several derivatives by cyclic voltammetry. The reversible potential for reduction of 1,1′-dimethylbianthrone in the A form has been measured at −56°C via cyclic voltammetry.[37d] The data which describe the kinetics and the thermodynamics for these systems have been compiled using variable temperature voltammetry, spectroelectrochemistry, pulse radiolysis, and other techniques. Digital simulations of the various processes have been used to extract many of the thermodynamic and kinetic parameters and show the consistency between postulated mechanism and all of the data.

2.4.1. The A/B Equilibrium Constant in 1,1′-Dimethylbianthrone

Several aspects of Scheme IV and the data associated with the species here are quite interesting and deserve comment. For the systems studied, all electron transfers to or from the B forms appear fast, indicating that large structural changes are not occurring prior to electron transfer. The equilibrium constants, K_{AB}, have been measured spectrophotometrically, with the exception of that for 1,1′-dimethylbianthrone. The very small value for the latter compound was determined indirectly[37g] using both voltammetry data and potentiometry, as spectrophotometric measurement does not have the required sensitivity.

A given amount of the 1,1-dimethylbianthrone was coulometrically reduced to the B^{2-} form, and the equilibrium potential measured. The difference in free energy between B^{2-} and B is determined directly by cyclic voltammetry. Combination of these values in a thermodynamic cycle affords the Gibbs free energy difference between A and B, as shown in Equations 6–9.

$$A + 2e \rightleftharpoons B^{2-} \qquad \Delta G_7 \tag{6}$$

$$B^{2-} \rightleftharpoons B^{-} + e \qquad \Delta G_8 \tag{7}$$

$$B^{-} \rightleftharpoons B + e \qquad \Delta G_9 \tag{8}$$

$$A \rightleftharpoons B \qquad \Delta G = \Delta G_7 + \Delta G_8 + \Delta G_9 \tag{9}$$

The small equilibrium constant obtained ($K = 8 \times 10^{-6}$) indicates a greater energy for the twisted B form here relative to that form for bianthrone itself ($K = 2.2 \times 10^{-3}$), yet the B structure would appear to have the less severe steric restrictions. This must mean that the loss of *p* orbital overlap at the 9,9′ positions in the B structure for 1,1′-dimethylbianthrone is even more costly than the methyl hydrogen interactions of the folded structure.

Interestingly, the rate constant for conversion of the B to A forms for bianthrone and 1,1′-dimethylbianthrone are essentially the same.[37e] If the transition state is characterized by methyl/hydrogen eclipsing-like interactions, this should raise the transition state energy for 1,1′-dimethylbianthrone over that for bianthrone, where hydrogens would eclipse. Either the transition state has some other structure, or there is an equally large gain in energy for the A form.

2.4.2. Activation Energy for B to A Conversion in Lucigenin

Ahlberg, Hammerich, and Parker[39b] have measured the activation energy for the B to A conversion in lucigenin using double-potential step chronoamperometry, and find the value to be 16.4 kcal/mol. They point out that this is significantly smaller than the estimated activation energy for the racemization of optically active 2,2′-*bis*-methoxycarbonyl-9,9′-bianthryl, (7), of 42 kcal/mole,[44] yet one might envision these processes to be similar.

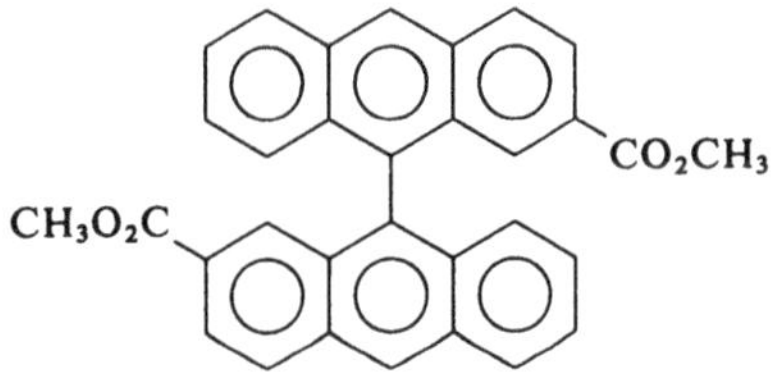

7

This disparity led to the suggestion that the lucigenin configurational change does not involve a simple rotation about the central bond, but is a more complicated process. They discuss the possibility of the B form folding in order to allow the two hydrogens to pass one another during the change from the B to A form. Others[41] have evaluated this process in more detail.

2.4.3. Comparison of A/B Equilibria for the Neutral and Charged Forms

It is also of interest to compare the free energies for the B to A conversions with the same process for the corresponding singly charged species.[37d] This could provide information about bond order changes occurring with the 9,9′ double bond upon oxidation or reduction. The comparison requires the equilibrium constants (K_{AB}) for the neutral systems, which can be measured spectrophotometrically or electrochemically, as described above. The reversible potentials for the reduction and oxidation of the A structures are also needed, but these are irreversible. They must be computed from the appropriate peak potentials by taking into account the effect of the followup isomerization which shifts the peak potential positive. With

EC theory, the rate constants, the transfer coefficients, the peak potentials, and the isomerization rate constants, the formal potentials can be calculated.

Evans and Busch[37d] find that for bianthrone, 3,3′-dimethyl-, 1,1′-dimethyl-, and 3,3′-dimethoxybianthrone, the equilibrium constants for A/B isomerizations are about 150 to 4200 times greater for the anions (K_-, Scheme IV) than the cations (K_+, Scheme IV) at room temperature. (Note in both cases, the B form is strongly favored for the ions.) Qualitatively, it might be expected that the opposite would be true, since simple anions prefer to be pyramidal, thus favoring the bent A form, and cations are planar, thus favoring the flat but twisted B form. Obviously, these are not simple ions, the real situation being much more complex. What this data suggests is that the cations have more double-bond order than the anions, shifting the equilibrium toward the A form for these systems.

However, the opposite shift in equilibrium is found for 10,10′-dimethyl-9,9′-biacridylidene and dixanthylene.[37d] Here again, the B structure is greatly preferred for either ion; however, the equilibrium constant for the cation is greater than that for the anion by more than 10^5 in both cases. Apparently, the cation tends to force the rings into planarity, raising the energy of the A structure, while the anion forms are relatively happy being folded in the A conformation.

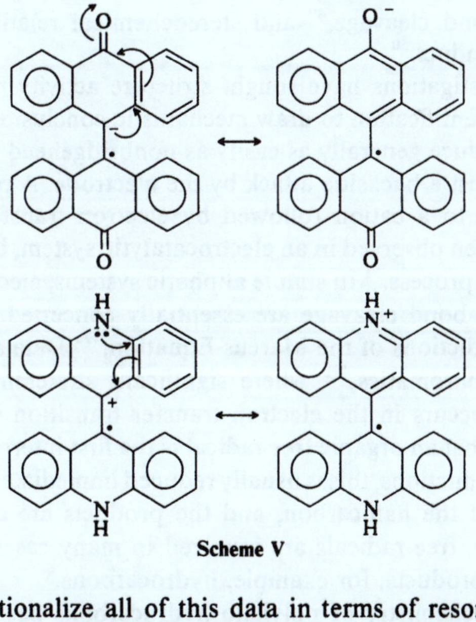

Scheme V

One can rationalize all of this data in terms of resonance structures for the various ions. Scheme V illustrates this, using bianthrone and 10,10′-dimethyl-9,9′-biacridylidene for the two different cases. Here, we compare the ability of the two systems to delocalize their respective charges, thus lowering double-bond order at the 9,9′ positions, which at the same time creates the rigid conjugated system shown in the upper three rings to the right for both systems. This delocalization would be favored by the B form when the rings can achieve the requisite planar structure. If we now change the charges on the ions in Scheme V, we find that the charges are not readily accommodated at the 10 positions as above. This makes the double-bond order more important for these ions, and thus the relative equilibrium tilts back the opposite way.

3. THE ELECTROCHEMISTRY OF VICINAL DIHALIDES

3.1. Introduction

The electrochemistry of alkyl and aryl halides has been one of the more intensively investigated areas over the years.[45] Mechanistic questions have been concerned with the number of electrons involved in the rate-determining step for a particular substrate,[46] the orientation of the carbon halogen bond with respect to the electrode,[47] the SN1/SN2 character of the reaction,[47a,48] the existence of anion radical intermediates vs. concerted electron transfer and bond cleavage,[49] and stereochemical relationships between reactant and product.[50]

These investigations have sought structure activity relationships and used product identification to draw mechanistic conclusions. Thus, bridgehead halides reduce generally as easily as nonbridgehead halides,[47a] which may argue against a backside attack by the electrode. A predissociation of an alkyl halide to a cation followed by electron transfer (an SN1-type reaction) has been observed in an electrocatalytic system, but not in a direct electrochemical process.[51] In simple aliphatic systems, electron transfer and carbon halogen bond cleavage are essentially concerted.[49c,d] This is consistent with predictions of the Marcus Equation.[49a] Data support a similar picture for haloaromatics,[49b] where significant stretching of the carbon halogen bond occurs in the electron transfer transition state. While it is usually agreed that an organic free radical is the first intermediate in carbon halogen bond reductions, this is usually reduced immediately at the potential for reduction of the halocarbon, and the products are derived from the anion. However, free radicals are involved in many cases,[52] and can give carbanion-like products, for example, hydrocarbons.

The electrochemistry of polyhalo hydrocarbons has raised questions about the possible interaction between the two or more halogens,[53,54] and the nature of their potential interaction in the electron transfer transition state. The reduction potential of vicinal dihalides is known to depend upon the orientation of the two halogens with respect to one another and has been suggested to involve concerted elimination of halides in certain systems. Recent elegant investigations have clarified the stereochemical picture of these latter reactions and demonstrate the power of electrochemistry to extract conformational data which would be difficult or impossible to obtain in other ways. We will now focus our attention on these vicinal systems.

3.2. Dependence of Reduction Potential upon Dihedral Angle

In the early 1960s, Zavada, Krupicka, and Sicher[54] found that the potential for reduction of a series of conformationally immobile vicinal dihalides was strongly dependent upon the dihedral angle between the two halogens. When the angle was 180 degrees (anticonformation for the halogens) between the two halogens, they were reduced at a remarkably low potential; however the opposite was true when the dihedral was 90 degrees, this system being hardest to reduce. When the angle was 0 degrees, reduction was also quite facile, although less so than for the 180 degree angle situation. See compounds **8** and **9** for examples. It has been suggested that the transantiplanar conformation followed a concerted reduction to give the olefin, while the molecules in which the two carbon-bromine bonds were not in the same plane followed a stepwise path, involving a carbanion intermediate.[55]

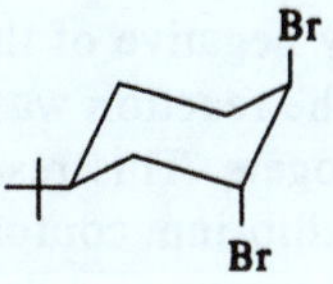

Br Br Br Br

8 (−0.86 v) 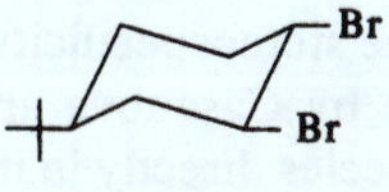 **9** (−1.67 v)

3.3. Stereochemistry of Reduction of Vicinal Dibromides

Casanova and Rogers[56] examined the electroreduction of seven vicinal dibromides and isolated the products of reduction. They were interested in the stereochemistry of the reduction and sought evidence for an intermediate carbanion. No evidence for an anion intermediate was found, and the reductions proved to be stereospecific. For example, *dl* and *meso*-2,3-dibromobutane give quantitatively *cis* and *trans*-2-butene, respectively, even in the presence of $10M$ methanol added to trap an anion which might be involved. Isomerically pure diequatorial 1,2-dibromo-4-*tert*-butylcyclohexane gives 97 percent and 96 percent 4-*tert*-butylcyclohexene in dimethylformamide with and without added 10*M* methanol, respectively. Clearly, if an intermediate carbanion is involved here, it is not long lived. But such is the case for the reductive cyclization of 1,3-dihalides where intermediate carbanions were clearly involved, yet could not be effectively trapped with water.[53f]

Brown, Middleton, and Threlfall[57] have examined *meso* and *dl*-3,4-dibromohexane and 2,5-dimethyl-3,4-dibromohexane in liquid ammonia and dimethylformamide using several different cathodes. They find that antielimination is generally preferred, except in cases where it is not an energetically viable conformation. For example, the half-wave potential for reduction of *meso*-3,4-dibromohexane is about 200 *m*V easier than that for the corresponding *dl* compound, while the difference is about 1 volt for the stereoisomeric 2,5-dimethyl-3,4-dibromohexanes. This can be understood by considering the conformations for the isomers. A simplified molecular mechanics approach was used to estimate the relative energies for the various conformations. *meso*-2,5-Dimethyl-3,4-dibromohexane leads to *trans*-2,5-dimethylhex-3-ene through a relatively low-energy transition state in which the bromines are anti. However, the *dl* isomer affords the same olefin, but it must do so through the relatively low energy (conformational energy) transition state in which the bromines are gauche, requiring a greater voltage. The conformation in which the bromines are anti is high in energy, and presumably it does not have an opportunity to be kinetically viable on the experimental time scale.

As the potential during the reduction of the stereoisomeric 3,4-dibromohexanes was made progressively negative of their respective half-wave potentials, the stereospecificity of the reaction was lost, in contrast to observations made by Casanova and Rogers. This result was ascribed to reduction of the species directly in its equilibrium conformation. Thus *meso* gives a 50:50 mixture of *cis* and *trans* 3-hexene, while *dl* affords a 30:70 distribution, respectively. Both stereoisomeric 2,5-dimethyl-3,4-dibromohexanes give nearly quantitative yields of the *trans* olefin independent of the applied potential, indicating one conformation exists predominately in solution.

These workers ascribed the differences in energy for reduction of the different stereoisomers to the energies of the antibonding orbitals involved in the reductions. They inferred that this energy would vary with the dihedral angle associated with the two carbon bromine bonds. This contrasts with the conclusions of Evans and co-workers, who show that the voltage difference for reduction of different conformations results from kinetics of electron transfer—*vide infra.*

3.4. Electrochemical Measurement of Conformational Equilibria in Vicinal Dibromides

Evans and co-workers[58] have utilized the fact that the different conformations of vicinal dihalides have different reduction potentials to study the conformational equilibrium of 1,2-dibromocyclohexane. This idea has been applied to the electrochemical study of conformational equilibria in tetraalkylhydrazines by Evans and Nelsen and co-workers,[59] which has been reviewed.[60] The principle approach is to apply temperature-dependent variable scan rate cyclic voltammetry and use digital simulation fitting of the data to extract thermodynamic and kinetic parameters.

In the study of 1,2-dibromocyclohexane, Klein and Evans[58b] find that as the temperature of the system is lowered and the cyclic voltammetry scan rate is increased, the single irreversible reduction wave diminishes in size with a second peak appearing at a more negative potential. At −90°C at a scan rate of 1 V/*s*, the two peaks become invariant with scan rate, which means that the equilibration is frozen out under these conditions. Assuming that these peaks represent the two possible conformations for the halogens (diaxial and diequitorial), the equilibrium constant and rate constants for the interconversions were extracted using digital simulation. Based upon the equilibrium constant of 0.55 measured at −90°C, the rate for the diequitorial to diaxial conversions was calculated to be 29 sec^{-1} at −60°C. The free energy of activation for the interconversion obtained from data at −60, −70, and −80°C for the equitorial-to-axial process is about 11 kcal/mol, with the enthalpy and entropy of activation being 9.9 kcal/mol and −5.1 eu., respectively.

O'Connell and Evans[58a] extended the vicinal halide studies to include *trans*-1,2-dibromocyclohexane, 1,1-dimethyl-*trans*-3,4-dibromocyclohexane, *meso*- and *dl*-1,2-dibromo-1,2-diphenylethane, and 2,3-dibromo-2,3-dimethylbutane, and others. Low-temperature electrochemistry is generally similar to that described for the cyclohexane system above; however, equilibrium constants and rates are of course different. One exception was that *trans*-3,4-dibromo-*cis*-1-methylcyclohexane gives only one wave, even at temperatures up to 50°C, apparently because of the unfavorable 1,3-diaxial interaction between the methyl and a bromine preventing the adoption of that conformation. It was also concluded that the barrier for 2,3-dibromobutane was less than 7 kcal/mol as only one peak was observed down to −135°C. This certainly seems reasonable. Finally, *meso* and *dl*-3,4-dibromo-2,5-dimethyl hexane each exhibit only one wave up to 80°C, due apparently to isopropyl/isopropyl and isopropyl/bromine interactions.

3.5. The Reason for the Dependence of Reduction Potential upon Dihedral Angle

The question of why the two conformations for these various vicinal halides reduce at significantly different potentials (as much as 0.7 volts for the case of diaxial- and diequatorial-1,2-dibromocyclohexane) was raised.[56a] Since the free energy difference between the two conformations is around 1 kcal for most compounds, and since in some cases the same olefin is produced from either conformation quantitatively, the difference in potential for the reduction does not have a thermodynamic origin. It, therefore, must be due to either the electron transfer rate constant and/or the transfer coefficient. The fact that the peak potentials are totally irreversible, allows calculation of the relative rates for the two reductions from the following relation:

$$k_a/k_e = (\alpha_a D_a/\alpha_e D_e)^{1/2}\exp[-\alpha_e nF/RT(E_{pe} - E_{pa})] \qquad (10)$$

Substituting the measured values for the transfer coefficients (α_a and α_e); the peak potentials for the two conformers (E_{pa} and E_{pe}); and with the reasonable assumption that the diffusion coefficients (D_a and D_e) are equal for the two conformers give a reduction rate ratio of 1.6×10^5 for 1,2-dibromocyclohexane (diaxial : diequitorial).

We see that the remarkable difference in reduction potential for the two different conformers in this system is due to a large difference in the electron transfer rates.[57a] Since the "ground state" energies are approximately equal, then the transition state energies differ significantly, assuming there is not a significant double-layer effect on the two different conformers, which would have different dipole moments.[28] If this latter assumption is correct, this result is in complete agreement with the suggestion that the co-planar system is reduced in a concerted process in which an unstable intermediate is not involved. When co-planarity is not achieved, the mechanism follows a different path involving a possible intermediate. The intermediate would likely be the bromo carbanion, whose charge is not effectively delocalized because of the orthogonality of the anion orbital with respect to the carbon halogen bond.

One last interesting observation should be mentioned. This occurred in connection with a study by Nelson *et al.*[61] with two dihalonorbornenes. It was observed that compound **10** was reduced at a very positive potential (−0.9 V vs. SCE), analogous to reactions of antico-planar dihalides we have discussed above. However, compound **11** was not reduced until beyond −1.5 V.

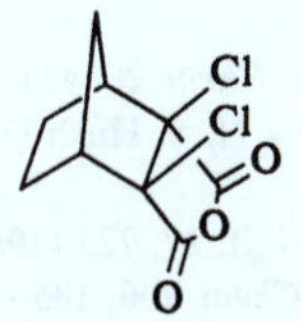

10 (−0.9 v) 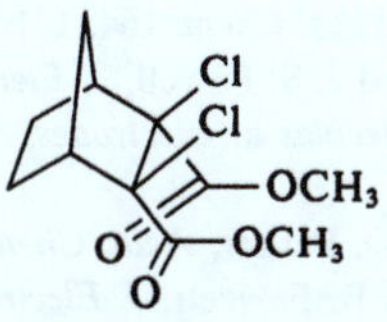 11 (< −1.5 v)

ACKNOWLEDGMENTS. I express my sincere thanks to Dennis Evans for many helpful comments.

REFERENCES

1. G. I. Fray and R. G. Saxton, *The Chemistry of Cyclo-octatetraene and Its Derivatives* (Cambridge University Press, New York, 1978).
2. L. A. Paquette, *Tetrahedron* **31**, 2855 (1975); (b) L. A. Paquette, *Pure Appl. Chem.* **54**, 54 (1982).
3. F. A. L. Anet, *J. Am. Chem. Soc.* **84**, 671 (1962).
4. (a) F. A. L. Anet, A. J. R. Bourn, and Y. S. Lin, *J. Am. Chem. Soc.* **86**, 3576 (1964); (b) J. M. Gardlik, L. A. Paquette, and R. Gleiter, *J. Am. Chem. Soc.* **101**, 1620 (1979), and references therein.
5. H. L. Strauss, T. J. Katz, and G. K. Fraenkel, *J. Am. Chem. Soc.* **85**, 2360 (1963).
6. T. J. Katz, W. H. Reinmuth, and D. E. Smith, *J. Am. Chem. Soc.* **84**, 802 (1962).
7. A. Carrington and P. F. Todd, *Mol. Phys.* **7**, 533 (1964).
8. P. I. Kimmel and H. L. Strauss, *J. Phys. Chem.* **72**, 2813 (1968).
9. M. J. S. Dewar, A. Harget, and E. Haselbach, *J. Am. Chem. Soc.* **91**, 7531 (1969).
10. H. Kojima, A. J. Bard, H. N. C. Wong, and F. Sondheimer, *J. Am. Chem. Soc.* **98**, 5560 (1976).
11. W. E. Britton, J. P. Ferraris, and R. L. Soulen, *J. Am. Chem. Soc.* **104**, 5322 (1982).
12. (a) R. D. Rieke and R. A. Copenhafer, *J. Electroanal. Chem.* **56**, 409 (1974); (b) D. L. Taggart, W. Peppercorn, and L. B. Anderson, *J. Phys. Chem.* **88**, 2875 (1984); (c) L. A. Paquette, S. V. Ley, R. H. Meisinger, R. K. Russell, and M. Oku, *J. Am. Chem. Soc.* **96**, 5806 (1974).
13. J. E. Garbe and V. Boekelhiede, *J. Am. Chem. Soc.* **105**, 7384 (1983).
14. R. D. Allendoerfer and P. H. Rieger, *J. Am. Chem. Soc.* **87**, 2336 (1965).
15. B. J. Huebert and D. E. Smith, *J. Electroanal. Chem.* **31**, 333 (1971).
16. D. R. Thielen and L. B. Anderson, *J. Am. Chem. Soc.* **94**, 2521 (1972).
17. A. J. Fry, C. S. Hutchins, and L. L. Chung, *J. Am. Chem. Soc.* **97**, 591 (1975).
18. B. S. Jensen, A. Ronlan, and V. D. Parker, *Acta. Chem. Scand.* B, **29**, 394 (1975).
19. W. H. Smith and A. J. Bard, *J. Electroanal. Chem.* **76**, 19 (1977).
20. R. D. Allendoefer, *J. Am. Chem. Soc.* **97**, 218 (1975).
21. (a) L. B. Anderson, J. F. Hansen, T. Kakihana, and L. A. Paquette, *J. Am. Chem. Soc.* **93**, 161 (1971); (b) B. Jensen, T. Pettersson, A. Ronlan, and V. D. Parker, *Acta. Chem. Scand.* B, **30**, 773–776 (1976).
22. J. F. Hansen, T. Kakihana, and L. A. Paquette, *J. Am. Chem. Soc.* **93**, 168 (1971).
23. R. A. Marcus, *Can. J. Chem.* **37**, 155 (1959).

24. A. N. Frumkin, *Z. Phys. Chem.* **164**, 121 (1933).
25. (a) M. E. Peover and J. S. Powell, *J. Electroanal. Chem.* **20**, 427 (1969); (b) M. E. Peover in *Reactions of Molecules at Electrodes*, edited by N. S. Hush (Wiley, New York, 1971), p. 265.
26. E. Ahlberg and V. D. Parker, *Acta. Chem. Scand.* B, **37**, 723 (1983).
27. A. Baranski and W. R. Fawcett, *J. Electroanal. Chem.* **100**, 185 (1979).
28. W. R. Fawcett, *J. Electroanal. Chem.* **22**, 19 (1969).
29. W. R. Fawcett and S. Levine, *J. Electroanal. Chem.* **65**, 505 (1975).
30. A. J. Bard and L. R. Faulkner, *Electrochemical Methods* (John Wiley and Sons, New York, 1980), p. 540.
31. J. M. Hale, in *Reactions of Molecules at Electrodes*, edited by N. S. Hush (Wiley, New York, 1971), p. 229-257.
32. N. L. Allinger, J. T. Sprague, and C. J. Finder, *Tetrahedron* **29**, 2519 (1973).
33. (a) L. R. Yeh and A. J. Bard, *J. Electrochem. Soc.* **124**, 189 (1977); (b) J. Phelps and A. J. Bard, *J. Electroanal. Chem.* **68**, 313 (1976); (c) L. R. Yeh and A. J. Bard, *J. Electroanal. Chem.* **81**, 333 (1977); (d) E. Laviron and Y. Mugnier, *J. Electroanal. Chem.* **93**, 69 (1978); (e) A. J. Bard, V. J. Puglisi, J. V. Kenkel and A. Lomax, *Discuss. Faraday Soc.* **56**, 353 (1973); (f) V. J. Puglisi and A. J. Bard, *J. Electrochem. Soc.* **120**, 748 (1973).
34. R. Dietz and M. E. Peover, *Disc. Faraday Soc.* **45**, 154 (1968).
35. M. Srzeszczuk and D. E. Smith, *J. Electroanal. Chem.* **162**, 189 (1984).
36. M. Szwarc, *Ions and Ion Pairs in Organic Reactions* (Wiley Interscience, New York, 1972), Vol. 1, p. 112, 113.
37. (a) B. A. Olsen and D. H. Evans, *J. Am. Chem. Soc.* **103, 839** (1981); (b) P. Neta and D. H. Evans, *ibid* **103**, 7041 (1981); (c) D. H. Evans and X. Naixian, *J. Electroanal. Chem.* **133**, 367 (1982); (d) D. H. Evans and R. W. Busch, *J. Am. Chem. Soc.* **104**, 5057 (1982); (e) B. A. Olsen, D. H. Evans and I. Agranat, *J. Electroanal. Chem.* **136**, 139 (1982); (f) D. H. Evans and N. Xie, *J. Am. Chem. Soc.* **105**, 315 (1983); (g) D. H. Evans and A. Fitch, *J. Am. Chem. Soc.* **106**, 3039 (1984); (h) T. Matsue, D. H. Evans, and I. Agranat, *J. Electroanal. Chem.* **163**, 137 (1984).
38. (a) Z. R. Grabowski and M. S. Balasiewicz, *Trans. Faraday Soc.* **64**, 3346 (1968); (b) R. B. Czochralska, A. Vincenz-Chodowska, and M. S. Balasiewicz, *Diss. Faraday Soc.* **45**, 145 (1968); (c) M. E. Peover, *Discuss. Faraday Soc.* **45**, 177 (1969).
39. (a) O. Hammerich and V. D. Parker, *Acta. Chem. Scand.* B **35**, 395 (1981); (b) E. Ahlberg, O. Hammerich, and V. D. Parker, *J. Am. Chem. Soc.* **103**, 844 (1981).
40. (a) J. Heinze, *Angew. Chem. Intern. Ed. Engl.* **20**, 202 (1981); (b) Z. R. Grabowski and M. S. Balasiewicz, *Trans. Faraday Soc.* **64**, 552 (1968).
41. R. Korenstein, K. A. Muszkat, and S. Sharafy-Ozeri, *J. Am. Chem. Soc.* **95**, 6177 (1973).
42. (a) E. Harnik, *J. Chem. Phys.* **24**, 297 (1956); (b) J. H. Day, *Chem. Rev.* **63**, 65 (1963).
43. N. A. Bailey and S. E. Hull, *Acta Crystallogr.* B **34**, 550 (1978).
44. C. Koukotas and L. H. Schwartz, *Chem. Commun.* **1400** (1969).
45. (a) A. J. Fry, *Synthetic Organic Electrochemistry* (Harper and Row, New York, 1972); (b) N. L. Weinberg and H. R. Weinberg; (c) M. Rifi and F. H. Covitz, *Introduction to Organic Electrochemistry* (Marcel Dekker, New York, 1974); (d) M. M. Baizer, *Organic Electrochemistry: An Introduction and a Guide* (Marcel Dekker, New York, 1973); (e) J. Casanova and L. Eberson, *The Chemistry of the Carbon-Halogen Bond Reduction* (Wiley, New York, 1973); (f) C. K. Mann and K. K. Barnes, *Electrochemical Reactions in Non-Aqueous Systems* (Marcel Dekker, 1970); (g) P. J. Elving, *Rec. Chem. Prog.* **14**, 99 (1953).
46. A. Streitwieser and C. Perrin, *J. Am. Chem. Soc.* **86**, 4938 (1964); (b) J. W. Sease, F. G. Burton, and S. L. Nicol, *J. Am. Chem. Soc.* **90**, 2595 (1968); (c) F. L. Lambert, *J. Org. Chem.* **31**, 4184 (1966).

47. J. Pearson, *Trans. Faraday Soc.* **44**, 683 (1948); (b) J. W. Sease, P. Chang, and J. Groth, *J. Am. Chem. Soc.* **86**, 3154 (1969); (c) F. Lambert and K. Kobagaski, *J. Am. Chem. Soc.* **82**, 5324 (1965); (d) R. S. Abeywickrema and E. W. Della, *Aust. J. Chem.* **34**, 2331 (1981).
48. F. L. Lambert, A. H. Albert, and J. P. Hardy, *J. Am. Chem. Soc.* **86**, 3155 (1964).
49. (a) L. Eberson, *Acta Chem. Scand.* B **36**, 533 (1982); (b) C. P. Andrieux, C. Blocman, J. M. Dumas-Bouchiat, and J. M. Saveant, *J. Am. Chem. Soc.* **101**, 3431 (1979); (c) P. Neta and D. Behar, *J. Am. Chem. Soc.* **103**, 103 (1981); (d) E. Canadell, P. Karafiloglon, and L. Salem, *J. Am. Chem. Soc.* **102**, 855 (1980).
50. (a) R. Annino, R. E. Erickson, J. Michalovic, and B. McKay, *J. Amer. Chem. Soc.* **88**, 4424 (1966); (b) A. J. Fry and M. Mitnick, *J. Am. Chem. Soc.* **91**, 6208 (1969); (c) A. J. Fry and R. W. Moore, *J. Org. Chem.* **33**, 1283 (1968; (d) A. J. Fry and R. G. Reed, *J. Am. Chem. Soc.* **93**, 553 (1971); (e) A. J. Fry and R. G. Reed, *J. Am. Chem. Soc.* **94**, 8475 (1972); (f) R. E. Erickson, R. Annino, M. D. Scanion, and G. Zon, *J. Am. Chem. Soc.* **91**, 1767 (1969).
51. C. P. Andrieux, A. Merz, J. M. Saveant, and R. Tomahogh, *J. Am. Chem. Soc.* **106**, 1957 (1984).
52. (a) F. M'Halla, J. Pinson, and J. M. Saveant, *J. Am. Chem. Soc.* **102**, 4120 (1980), and references therein; (b) B. C. Willett, W. M. Moore, A. Salajegheh, and D. G. Peters, *J. Am. Chem. Soc.* **101**, 1162 (1979); (c) B. C. Willett and D. G. Peters, *J. Electroanal. Chem.* **123**, 291 (1981); (d) M. S. Mbarak and D. G. Peters, *J. Org. Chem.* **47**, 3397 (1982); (e) R. Shao, J. A. Cleary, D. M. La Perriere, and D. G. Peters, *J. Org. Chem.* **48**, 3289 (1983).
53. (a) S. Wawzonek and J. H. Wagenknecht, *J. Electrochem. Soc.* **110**, 420 (1963); (b) M. R. Rifi, *Collect. Czech. Chem. Commun.* **36**, 932 (1971); (c) M. R. Rifi, *Tetrahedron Lett.* **1043**, (1969); (d) M. R. Rifi, *J. Amer. Chem. Soc.* **89**, 4442 (1967); (e) A. J. Fry and W. E. Britton, *Tetrahedron Lett.* **4363** (1971); (f) A. J. Fry and W. E. Britton, *J. Org. Chem.* **38**, 4016 (1973); (g) W. F. Carroll, Jr. and D. B. Peters, *J. Am. Chem. Soc.* **102**, 4127 (1980).
54. J. Zavada, J. Krupicka, and J. Sicher, *Coll. Czech. Chem. Commun.* **28**, 1664 (1963).
55. A. J. Fry, *Fortsch. Chem. Forsch.* **34**, 1 (1972).
56. (a) J. Casanova and H. R. Rogers, *J. Org. Chem.* **39**, 2408 (1974); (b) J. Casanova and H. R. Rogers, *J. Amer. Chem. Soc.* **96**, 1942 (1974).
57. O. R. Brown and P. H. Middleton, and T. L. Threlfall, *J. Chem. Soc. Perkin Trans. II* **955** (1984).
58. (a) K. M. O'Connell and D. H. Evans, *J. Am. Chem. Soc.* **105**, 1473 (1983); (b) A. J. Klein and D. H. Evans, *J. Amer. Chem. Soc.* **101**, 757 (1979).
59. S. F. Nelsen, L. Echegoyen, E. L. Clennan, D. H. Evans, and D. A. Corrigan, *J. Am. Chem. Soc.* **99**, 1130 (1977).
60. (a) S. F. Nelsen, *Acc. Chem. Res.* **14**, 131 (1981); (b) D. H. Evans and S. F. Nelsen, in *Characterization of Solutes in Nonaqueous Solvents*, edited by G. Mamantov, (Plenum Press, New York, 1977), pp. 131–154.
61. S. F. Nelsen, E. F. Traiecedo, and E. D. Seppanen, *J. Am. Chem. Soc.* **93**, 2913 (1971).

6

Modified Electrodes

Masamichi Fujihira

1. INTRODUCTION

In electrochemistry, an electrode or an electrode system is defined in a broad sense as a system in which an ionically conductive phase, i.e., an electrolyte, is in contact with an electronically conductive phase, which can be a metal or a semiconductor.[1] Of course the metal, graphite, or semiconductor, dipped in the electrolyte, itself is called an electrode in a narrower sense. These systems are illustrated pictorially in Fig. 1.

Investigation of chemically modified electrodes was started as a means of fabrication of a functional electrode by immobilizing deliberately a chemical on an electrode surface. The electrode thus modified displays the chemical, electrochemical, optical, etc., properties of the immobilized molecule.[2] In 1975, Murray *et al.*[3] and Miller *et al.*[4] reported independently their first paper on the modified electrode, in which a molecule with a specific function was bound covalently on the electrode surface. This opened a new field of chemically modified electrodes. Lane and Hubbard[5,6] utilized strongly adsorbed molecules at electrode surfaces to study the basics of electrochemical reactions. Since 1975,[7] the modification of electrode surfaces has been studied very extensively and chemicals with various properties have been immobilized. These properties include fast outer-sphere electron transfer, chirality, electron transfer mediator catalysis, adsorption of trace molecules or ions from solutions for analysis, dye sensitization for a semiconductor photoelectrode, and inhibition of corrosion, etc.[2]

Masamichi Fujihira • Department of Chemical Engineering, Tokyo Institute of Technology, Tokyo 152, Japan.

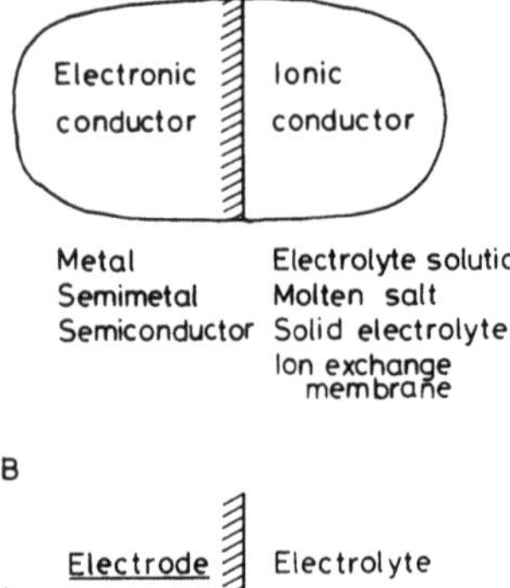

Figure 1. Definition of an electrode. A: an electrode (system) in a broad sense in which an ionically conductive phase (electrolyte) is in contact with an electronically conductive phase. B: an electrode in a narrow sense which is a part of electrode system and is an electronically conductive phase itself.

In the initial period of this research, functional molecules have been bound covalently[8-17] or adsorbed strongly[18,19] on the substrate electrode material in the form of monomolecular layers. However, it was soon realized through the pursuit of better efficiencies that modifications with a monolayer are not sufficient. Due to this situation, the modification with polymer layers[20-23] to increase the concentration of active sites has become popular. The polymer coats may function as a part of the electrode itself in the narrow sense,[24] as for example electronically conductive materials,[25-28] or they may absorb the ionically conductive electrolyte containing high concentrations of redox species.[29,30] In the former case, the electrochemical charge transfer reaction proceeds at the polymer-solution interface, while in the latter, the change in the applied potential results in a change in the potential difference at the electrode–polymer interface and, therefore, the electrochemical charge transfer reaction occurs at this interface. The electron transfer reaction between the redox species in the polymer layer and species in solution then proceeds either at the polymer–solution interface or inside of the polymer layer depending upon their reaction and mass transfer rates. The interconnection[30] of these rates will be described later in more detail. The coated layers in the latter case can be regarded as a thin part of the solution phase.

It is worthwhile at this point to consider briefly the interfacial region of an electrode in the broad sense in connection with the above discussion on the polymer-coated electrodes. In Fig. 2, a side view of an electrode system is schematically illustrated. In contact with an electrode phase (region I), there is a transfer layer (region II) in which the charge transfer process proceeds under the influence of the electrical field across the interface. Next to this layer, a so-called diffuse double layer (region III) exists, where

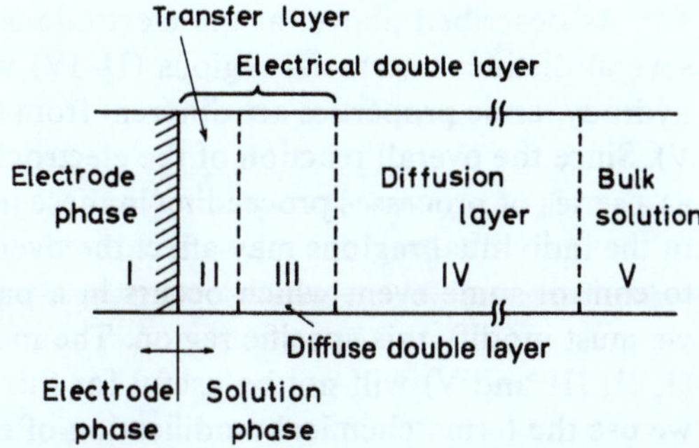

Figure 2. A side view of an electrode (system) in the broad sense. Region I, an electrode phase; II, a transfer layer; III, a diffuse double layer; IV, a diffusion layer; V, a bulk solution. The region, II + III, is called an electrical double layer.

charge neutrality does not hold. As a whole, regions II and III are often called an electrical double layer. When electric currents are flowing, outside of the electrical double layer there exists a diffusion layer (region IV). In the diffusion layer electrical neutrality is satisfied, but the composition of the constituents differs from that of the bulk solution (region V). The mass transfer rate of the electroactive species depends on their concentration gradients in region IV. If there is no convection in a solution, the thickness of the diffusion layer will expand infinitely into the solution. Convection limits the thickness of layer IV, but the thickness is usually far beyond that of the electrical double layer. When the charge transfer reaction is preceded or followed by chemical reactions in solution and the overall electrode process is kinetically controlled, we can assume the existence of a reaction layer,[31] as shown in Fig. 3. The thickness of the reaction layer, μ, is related to the mean lifetime, τ, and the diffusion coefficient, D, of the reactive species in solution by the equation

$$\mu = (D\tau)^{1/2} \tag{1}$$

Although the reaction layer is not illustrated in the side view of Fig. 2, the thickness of the layer can be much thinner than the diffusion layer in the ordinary convectional conditions, if the chemical reaction is fast.

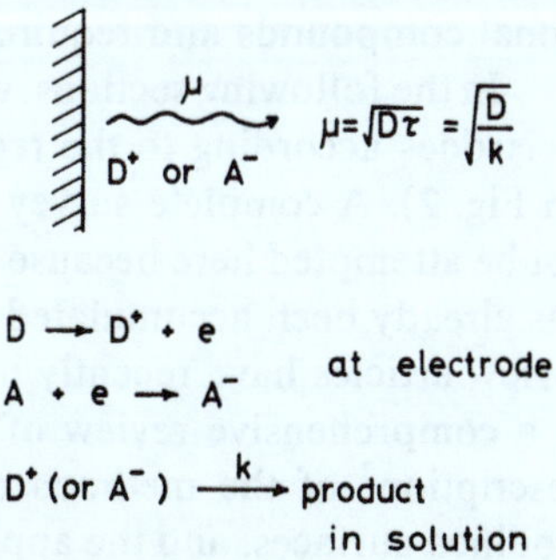

μ

D^+ or A^-

$\mu = \sqrt{D\tau} = \sqrt{\frac{D}{k}}$

$D \rightarrow D^+ + e$

$A + e \rightarrow A^-$ at electrode

$D^+ \text{ (or } A^-) \xrightarrow{k}$ product in solution

Figure 3. The thickness of the reaction layer, μ.

As described above, at the electrode-electrolyte boundary there exist several distinct interfacial regions (II–IV) whose chemical, electrical, and hydrodynamic properties are different from those of the bulk phases (I and V). Since the overall reaction of the electrochemical event can be described as a series of processes proceeding in these interfacial regions,[32] any change in the individual regions may affect the overall electrode process. If we try to control some event which occurs in a particular region, say region IV, we must modify this specific region. The modification of the other regions (I, II, III and V) will not be useful for this purpose. Consequently, when we use the term "chemical modification of electrodes," we are speaking in the broad sense. Monolayer modification is limited to the control of processes which proceed in regions II and III; the phenomena related to the diffusion layer and the reaction layer cannot be controlled effectively by the modification of the surface monolayer.

The electrode modified with a redox polymer can promote catalysis by incorporating a high concentration of mediators within the reaction layer. It is easy to understand the demand placed upon this modification. The redox mediator must be firmly incorporated in the film, but the polymer layer should not be so rigid that the mass transfer through the film becomes the rate-determining step of the overall process. At present, it is a fairly difficult task to fix the redox species in the polymer matrix and simultaneously have facile diffusion of the redox charge with high permeability of the substrate into the film. The catalytic reactivity of such mediators is not very high despite their high concentration in the film. Application of selective permeability in the polymer film[33] is another example of control by the modification of the diffusion layer. This cannot be attained by the monolayer molecular design. On the other hand, species selective electrolysis by regulation of the activation overpotential can be realized in principle by the monolayer modification.[3,34,35] To cope with the unavoidable decrease in the mass transfer rate by adopting the polymer coated electrode, functional electrode systems have been developed which involve the electrolyte solution.[36–39] However, this system requires the addition of catalysts or functional compounds and requires their separation from the products.

In the following sections, we will classify past investigations of modified electrodes according to the regions in which they function (regions I to V on Fig. 2). A complete survey of the literature on modified electrodes will not be attempted here because of the tremendous body of knowledge which has already been accumulated in this field and, also since several excellent review articles have recently been published. The reader who is interested in a comprehensive review of the field should consult these.[16,40–47] A full description[2] of the methods of modification, the characterization of the modified surfaces, and the application of modified electrodes to the various fields has appeared. This chapter will focus on application of modified electrodes and will examine its prospects for the future.

2. MODIFICATION OF THE ELECTRODE PHASE (REGION I)

It is evident from Fig. 1 that materials which comprise region I must be electronically conductive. Whether solid or liquid, most of the metals are suitable materials for the electrode phase (region I). Semimetals and semiconductors also can be used although they are less conductive. The more conductive material is economically preferable due to the lower ohmic loss and, hence, reduced power consumption. For some special purposes, such as photoconversion[48] and optical transparency,[49,50] semiconductive properties of the electrodes are often preferable.

Among the electrode materials, the semiconductor holds a very unique position. The fundamental characteristics of semiconductors, such as the bandgap, the flat-band potential, the carrier density, distribution of the impurity levels, and the surface states, etc., are very sensitive to their chemical composition.[5] (See Fox, Chapter 4.) Consequently, such modifications are likely to play an important role in dark and light semiconductor electrochemistry as these materials become better understood.

Conventional methods used for solid-state devices,[51-53] such as doping by thermal diffusion, ion implantation, etc., can be applied. In addition to the modification of the known materials by such techniques, one can seek new materials from the wide range of sources which include organic semiconductors[54,55] and conductive polymers.[56-58] Electronically conductive polymer materials deposited electrochemically on the substrate electrode are most attractive. The modification of semiconductor electrodes and the improvement of these electrode materials for industrial use has been recently reviewed.[43]

The bandgap, E_g, and the flat-band potential, E_{fb}, of the semiconductor photoelectrodes are the most important factors which determine the characteristics of the photoelectrochemical cells.[59-62] The photovoltage and the ability to evolve hydrogen from water are greatly dependent upon the values of E_g and E_{fb}. However, for most of the oxide semiconductors, these parameters are not independent and are related according to the following equation:[63,64]

$$E_{fb} = 2.94 - E_g \tag{2}$$

This relation appears to limit these materials; however, several other materials have been found not to be governed by this relation. An understanding of the independent control of these parameters and the development of new materials is a most interesting and important subject in this field.

The quantum yield of the charge separation depends greatly on the interrelation of the slope of the space charge region and the absorptivity of the semiconductor. Among these properties, the band shape of the space charge region can be determined mainly by the carrier density and by the applied potential, i.e., indirectly by the flat-band potential. The carrier density can be controlled by the concentration of impurity, and various doping methods already developed can be used for this.

Another unique subject related to the modification of electrode phase I is associated with corrosion of the electrode materials. The electrode is often used under very severe conditions as in the case of electrolysis of brine. The chemical stability of the electrode is, of course, required. Corrosion science offers many answers, but several interesting ideas have come out recently in connection with the modified electrodes. In several studies, ion implantation was used to create a resistive alloy film on an electrode surface.[65-69] Hydrophobic organic redox films, such as polyferrocenes, have been used to inhibit corrosion of photoanodes, which, otherwise, are easily corroded by the reaction of photo-generated holes and solvent water molecules.[70-72] This technique permits use of semiconductor photoanodes, such as silicon, which is sensitive to the visible light, but which heretofore could not be used because of its corrosive properties. In this example, the modified layer acts as a mediator in region II to transfer active holes immediately to the redox species in solution competing with the corrosion reaction. The result of the modification is to suppress the corrosive properties of the electrode phase I.

Other examples of region I materials include amorphous metals,[73] one-dimensional conductive polymers,[74] and intercalation[75] compounds. The reversible electrochemical doping and undoping of conducting polymers, such as polyacetylene and polypyrrole, and their application as secondary cells are now under extensive investigation in industry.

3. MODIFICATION OF REGIONS II AND III

The rate of the electron transfer reaction at the electrode-electrolyte interface can be described by the driving force of the reaction which is varied continuously by the change in the applied potential. In some processes, the electron transfer can proceed directly between the electrode and the redox species in solution, but in other processes the electrons cannot be transferred directly but can be transferred via a so-called mediator.[46] When the electron transfer is accompanied by the rearrangement of the constituent atoms, the process proceeds generally via single or multiple intermediates, and the activation overpotential of the electrochemical reaction can be greatly diminished by catalysts which lower the activation energy of the reaction by stabilizing the intermediates.[76] The surface of the electrode itself may furnish the catalytic site. In some cases, adlayers, such as porphyrins or phthalocyanines, are known to be effective catalysts, for example, in the reduction of molecular oxygen. When we try to carry out selective electrosynthesis, the charge transfer layer II of the modified surface must provide the molecular properties to control the selectivity of the reaction.[4] Investigations of the modification of region II can be classified into two groups. One is the modification by immobilizing molecules which themselves have the inherent properties needed. The other is the modification of the substrate metal with different metal atoms which are catalytically active. Both the molecules and the metal atoms are bound in monolayer or in submonolayer coverage. In molecular modification, the functions displayed by the surface are determined mainly by those of the bound molecules. We can, therefore, expect as many kinds of functions of the modified electrode as possessed by the molecule. Two-dimensional order on the surface is very important and useful for catalyses with bound metals, but this has not been utilized with molecular modifications. The instability and poor surface coverage of the bound molecules is a problem which needs to be solved, as the catalytic efficiency decreases with a decrease in the surface coverage.

As suggested above, the transfer and the diffuse double layer are the primary centers for the electrochemical reactions, and the kinds of modifications affected and the results of these are quite extensive. Representative examples of such modified electrodes follow. These are classified according to their function.

3.1. Charge-Transfer Catalysis

The charge-transfer catalysis shown in Scheme 1 is one of the most important subjects of modified electrodes[30] and has been studied most

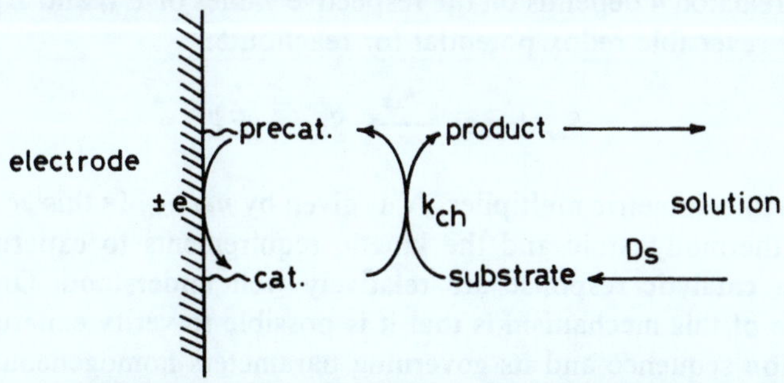

Scheme 1. Charge transfer catalysis on a monolayer modified electrode. Reproduced with the permission of The Royal Society from Reference 30.

extensively. The essential idea of this modified electrode is that a bound mediator molecule shuttles electrons between the electrode surface and the redox species in solution, as shown in Scheme 1. The mediator is recycled between the states of active catalyst (cat.) and precatalyst (precat.).

According to the Marcus theory,[77] when the electron transfer proceeds by an outer sphere mechanism for both the bound mediator and the substrate and between the unmodified electrode and the substrate, the electron transfer reaction cannot be accelerated merely by the monolayer modification.[78] An outer-sphere mechanism may not guarantee a reaction. It is known, for example, that cytochrome *c* can exchange electrons by an outer sphere mechanism with a redox couple in solution,[79,80] while the molecule cannot be reduced or oxidized electrochemically on most of the electrodes with one rare exception.[81] A mediator can be used for the redox reaction of cytochrome *c* on the electrodes.[82,83] The lack of reactivity on the electrode may be attributed to steric hindrance of the macromolecule when it is adsorbed in a specific orientation. In spite of the typical outer-sphere mechanism for cytochrome *c*, its electrochemistry can be accelerated by the presence of mediators. This does not conflict with the idea of the Marcus theory, because the catalysis results from the steric inhibition of the heterogenous electron transfer.

When the electrochemical processes become complex and they do not obey the Marcus theory, there have been many cases reported where we can appreciate the catalytic effects of the modified layer. The mechanism of the catalyses in most cases can be interpreted by an *ec* catalytic regeneration mechanism proposed by Kuwana[46]:

$$M_{ox} + n_M e^- \xrightarrow{k_{s,M}} M_R \qquad E_M^{0\prime} \tag{3}$$

$$NM_R + S_{ox} \xrightarrow{k_f} NM_{ox} + S_R \tag{4}$$

where M represents the mediator and S, the solution species. The driving force for reaction 4 depends on the respective values of $E_M^{0\prime}$ and $E_S^{0\prime}$, where $E_S^{0\prime}$ is the reversible redox potential for reaction 5:

$$S_{ox} + n_s e^- \xrightarrow{k_{s,S}} S_R \qquad E_S^{0\prime} \tag{5}$$

The stoichiometric multiplier, N, is given by n_S/n_M. In this *ec* mechanism, the thermodynamic and the kinetic requirements to experimentally observe a catalytic response are relatively well understood. One major advantage of this mechanism is that it is possible to verify experimentally the reaction sequence and its governing parameters homogeneously prior to the immobilization of the mediator onto the electrode surface. Thermodynamically, S_{ox} should be reduced prior to M_{ox} at $E_M^{0\prime}$, which is more negative in potential than $E_S^{0\prime}$. However, in the presence of a large overpotential due to the heterogeneous rate constant, $k_{s,S}$, being much less in magnitude than $k_{s,M}$, the reduction of M_{ox} proceeds prior to S_{ox}, and hence the turnover of S_{ox} to S_R via reaction 4.

The catalytic reduction of molecular oxygen with water soluble iron porphyrin on a glassy carbon electrode[36] is shown in Fig. 4, trace C. Similar behavior has been observed when the iron porphyrin is bonded to the glassy carbon electrode.[84] In the absence of iron porphyrin, molecular oxygen is reduced with a high overpotential, as shown in trace *B*. The voltage savings here is about 0.46 V. The potential for this *ec* catalysis is governed by the porphyrin redox couple (trace *A*). The *i–E* curve observed for 10^{-3} *M* porphyrin (trace *D*) is characteristic of a reversible, diffusion controlled electrode reaction. At or above *ca.* 2×10^{-4} *M* porphyrin, the peak current for the catalyzed two-electron reduction of oxygen becomes diffusion controlled. The cyclic voltammetric behavior can be described by the following Equations:

$$\mathrm{Fe(III)Por} + e \rightarrow \mathrm{Fe(II)Por} \tag{6}$$

$$\mathrm{Fe(II)Por} + \tfrac{1}{2}\mathrm{O_2} + \mathrm{H^+} \xrightarrow{\text{fast}} \mathrm{Fe(III)Por} + \tfrac{1}{2}\mathrm{H_2O_2} \tag{7}$$

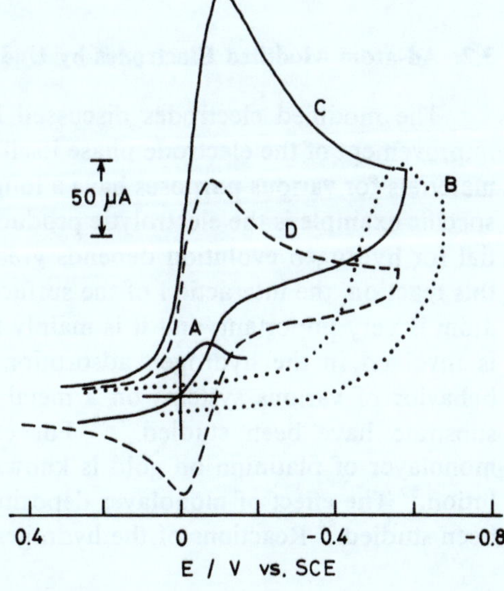

Figure 4. The catalytic reduction of molecular oxygen with water soluble iron porphyrin on a glassy carbon electrode. A, 2×10^{-4} *M* Fe(III)TMP^{5+}; B, only O_2 (air-saturated); C, mixture of 2×10^{-4} *M* Fe(III)TMP^{5+} and O_2 (air-saturated), and D, 10^{-3} *M* Fe(III)TMP^{5+}, no O_2. All solutions contain 0.05 *M* H_2SO_4. Scan rate = 0.11 V s^{-1}. Reproduced with the permission of Elsevier Scientific Publishing Company from Reference 36.

More dramatic catalytic effects for oxygen reduction are observed with face-to-face cobalt porphyrins,[85,86] where the overall process consumes four electrons and produces water.

Other catalytic reductions via modified electrodes[2,46] have been reported. Recently, Murray *et al.*[87] used a rotating disk electrode to study the catalytic reduction of 1,2-dibromophenylethane to styrene on an electrode with a monolayer of cobalt (*p*-aminophenyl) porphyrin. The dependence of the limiting current, i_1, on the angular velocity, ω, follows the Koutecky-Levich plot ($1/i_1$ vs. $1/\omega^{1/2}$), as we would expect by the theoretical Equation[32]:

$$\frac{n\mathrm{FA}}{i_1} = \frac{1}{k_{ch}\Gamma C_s} + \frac{1}{0.62 D_s^{2/3} \nu^{-1/6} \omega^{1/2} C_s} \tag{8}$$

where Γ, C_s, D_s, are the surface concentration of mediator, the concentration, and the diffusion coefficient of the substrate, respectively. The rate constant k_{ch} is assumed to follow the rate equation:

$$-\frac{d\Gamma}{dt} = k_{ch}\Gamma C_s \tag{9}$$

The rate constant, k_{ch}, can be obtained from the intercept of a plot of $1/i_1$ vs. $1/\omega^{1/2}$ with the value of the surface concentration, Γ, determined from cyclic voltammetry. The rate constants are about $10^5\ M^{-1}\ s^{-1}$ and are independent of Γ over a limited range. This is in contrast with the results of the multilayer modifications described later.

3.2. Ad-atom Modified Electrodes by Under-Potential Deposition

The modified electrodes discussed in this section are based on the improvement of the electrode phase itself.[76] Efforts to seek better electrode materials for various purposes have a long history in electrochemistry. One specific example is the electrolytic production of hydrogen. The overpotential for hydrogen evolution depends greatly on the electrode material. In this reaction, the interaction of the surface with the intermediate hydrogen atom is very important, and it is mainly the surface metal monolayer that is involved in the hydrogen adsorption. Studies of the electrochemical behavior of various systems on a metal monolayer over a foreign metal substrate have been studied.[76,88] For example, the monolayer or submonolayer of platinum on gold is known to be active for hydrogen evolution.[89] The effect of monolayer deposition of gold on platinum has also been studied.[89] Reactions of the hydrogen type:

$$\mathrm{A} + \mathrm{A} \rightarrow \mathrm{A}_2 \tag{10}$$

as well as those between different atoms (Equation 11) have been examined.

$$A + B \rightarrow AB \tag{11}$$

For example, the oxidation of CO by a surface oxygen atom[90] to give CO_2 is from the latter category. A method for preparing this monolayer or submonolayer metal coating on a substrate is by underpotential deposition (upd). Simply stated, it is energetically easier to deposit the metal atoms on the substrate than it is for them to deposit on one another, just as a monolayer of hydrogen is formed on platinum before hydrogen evolution commences.[91] Thus by upd, the layer will not grow more than one monolayer thick, because of the stronger interaction between the ad− and the substrate than between the deposited atoms themselves. They become homogeneously dispersed on the surface, rather than condensing into islands. This feature of the upd is important for the geometrical regulation of the two dimensional arrangement of the deposited metal atoms.[95] For more detail of the upd, see the excellent review by Kolb[92] and references therein.

The deposited atom in the upd is called an ad-atom. Electrodes modified with ad-atoms are useful for oxidation of organic fuels,[93,94] such as methanol, formaldehyde, and formic acid, and for reduction of oxygen. For oxidation of organics, Pb, Bi, and Tl have been investigated as ad-atoms on Pt and are found to be active in this order.

3.3. Asymmetric and Other Selective Electrosyntheses

The fundamental idea of tailoring the molecular environment for selective electrosyntheses is briefly discussed here (Fig. 2). Details of selective electrosynthesis will be given later, but let us consider first the kinds of selectivities which can be expected in a chemical reaction. Breslow[96] has classified enzyme catalyzed reactions into the following four categories:

a. stereoselectivity;
b. selectivity in the choice of substrates;
c. selectivity in the type of chemical reaction performed; and
d. selectivity in the region of the molecule attacked when there are several possibilities (regioselectivity).

To accomplish selective electrosyntheses on modified electrodes, we must design the surface to facilitate a specific mechanism. Monolayer modifications are often not very successful due to thinness of the modified region.

As we just saw, substrate selectivity may be accomplished by control of the activation overpotential via monolayer modification of region II or by use of a polymer film in region IV. Selectivity may also be realized in the reaction layer (region IV) through a chemical reaction preceding or following the electron transfer.

3.4. Modification of the Electrical Double Layer (Regions II and III)

The correlation between electrode kinetics and double-layer structure was analyzed by Frumkin[97] as early as 1933. A probable structure of a metal-electrolyte interface proposed by Bockris *et al.*[98] is shown in Fig. 5. A metal electrode has adsorbed upon it a monolayer of orientable solvent dipoles. Beyond this layer is an electrolyte. The plane of closest approach for our unadsorbed ion is known as the outer Helmholtz plane (OHP). The plane passing through the centers of specifically adsorbed ions is the inner Helmholtz plane (IHP). The region beyond OHP is referred to as the diffuse double layer. In contrast with the metal-electrolyte interface, the double layer at a semiconductor-electrolyte interface expands into both semiconductor and electrolyte phases. In this situation, four different types of charges contribute to the double layer[99]:

a. space charge of electrons and holes and of fixed, immobile donor or acceptor states in the lattice, q_{sc};
b. trapped charge in surface states, of both possible signs, q_{ss};

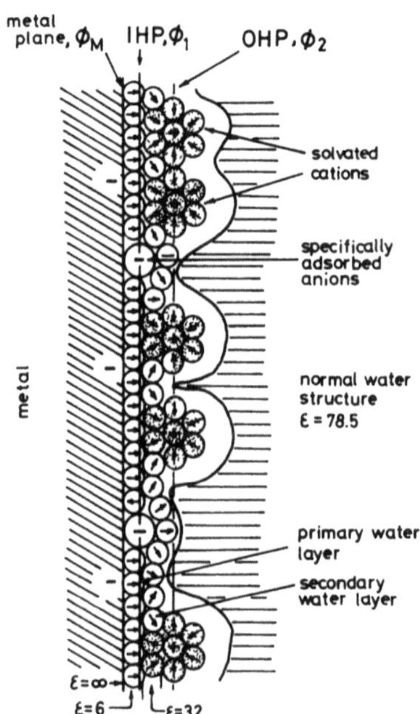

Figure 5. Probable structure of a metal electrolyte interface. Reproduced with the permission of the Royal Society from Reference 98.

c. charges of adsorbed ions or ionized surface groups on the crystal, again of both possible signs of amount q_{ad}; and
d. the ionic counter charge in the electrolyte, q_{el}.

Electroneutrality demands

$$q_{sc} + q_{ss} + q_{ad} + q_{el} = 0 \tag{12}$$

With the help of Poisson's Equation

$$\frac{\partial}{\partial x}\left(\varepsilon(x)\frac{\partial \phi(x)}{\partial x}\right) = -\frac{\rho(x)}{\varepsilon_0} \tag{13}$$

and the statistics of the distribution of charges, the correlations between charge density $\rho(x)$ and electrostatic potential $\phi(x)$ can be derived.

By chemical modification of the electrode phase (region I) and the electrical double layer (regions II and III), the spacial distribution of charge density $\rho(x)$ and the dielectric constant $\varepsilon(x)$ may be to some extent controlled in order to regulate the electrostatic potential profile $\phi(x)$.[100] Surface states, dissociation constants of ionized surface groups, density of specifically adsorbed ions (or covalently bound ions), thickness and dielectric property of the monolayer of orientable dipoles, etc., can be molecularly tailored on the electrode surface (regions II and III).

Alkenes and alkynes form irreversible chemisorbed species on platinum.[101] Through this procedure, Lane and Hubbard[5] have tethered ionic species within the double-layer region in order to prove the mechanisms of electrode reactions. Some of the reagents which have been examined are illustrated in Fig. 6.

Chelates which are chemisorbed allow metals to be selectively bound at the electrode surface. The stability of the chemisorbed metal complex is shown to vary with electrode potential, due to the influence of the electrical double layer, which allows the chelating ability to be effectively switched on or off. For example, chemisorbed 3-allylsalicylate is inactive toward scavenging iron from solutions at potentials more positive than *ca.* +0.2 V vs. SCE, but fully binds one Fe per molecule of adsorbed salicylate at potentials more negative than 0.0 V. This result implies that the electrode, when positively charged, attracts the salicylate anion into the compact layer (while repelling Fe^{2+} and Fe^{3+}), and in so doing decreases the tendency for Fe complexes to coordinate with the adsorbed salicylate.

Studies of electrode reactions of Fe and Pt complexes, and of organic depolarizers, such as substituted polyhydric phenols, indicate that only in certain instances are the adsorbed reactants able to achieve the reactive orientation which is readily acquired by dissolved species. For example,

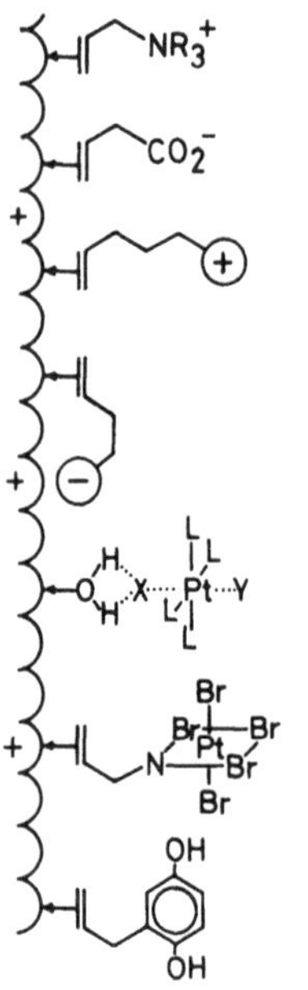

Figure 6. Adsorption of substituted olefins at platinum electrodes. The cases illustrated are, from top to bottom, allyltrialkylammonium ion; vinylacetic acid anion; alkenyl cation and anion oriented in the field of a positively charged electrode; halide bridged reactant complex, Pt(II)L_4; $PtBr_5$(allylamine)$^-$; allylhydroquinone. Reproduced with the permission of the American Chemical Society from Reference 5.

2-allylhydroquinone, which is electroactive in solution, becomes unreactive in the chemisorbed state, while hydroquinones with longer side chain alkenes were reactive in the adsorbed state. In the former case, the quinone moiety is held outside of the double layer, while more degrees of freedom are available to those with longer side chains.

Lane and Hubbard[6] describe the influence of chemisorbed olefins having ionic substituents on the electrode reaction rates of several platinum complexes. Introduction of charged, chemisorbed species onto the electrode surface provides a means of varying the electric field in the interfacial region independent of the electrode potential. This alternation of the potential at the reaction plane causes ionic complexes with opposite charge to react more rapidly at coated surfaces than at clean surfaces.

In addition to these chemisorbed olefins, Hubbard *et al.* studied chemisorbed F^-, Cl^-, Br^-, and I^- [18] and the influence of chirality on the orientation and electrochemical oxidation of *l*- and *dl*-DOPA chemisorbed on Pt.[102]

Murray *et al.*[3] and Osa and Fujihira[13,103] have reported the effects of silanization on the oxide electrode double-layer capacitance. The double-layer capacitance of glassy carbon electrodes[15] modified with amide bonds with *n*-butyl and *n*-octylamines has also been examined. Cyclic voltammograms of propylamine silanized SnO_2 electrodes show lower charging currents than that at the clean electrode. However, the flat-band potentials, which were determined from the inflection point of C^{-2} vs. E plots, are not changed appreciably by modification. This result contradicts the expectation that the change in pK_a of the surface ionized group may change the flat-band potential of unmodified SnO_2 which is pH dependent. No shift in flat-band potential of chemically modified TiO_2 surfaces was observed by Tomkiewicz,[104] but his impedance and photocurrent measurements indicate that surface states are being introduced.

Incomplete silanization of surface hydroxyl groups on SnO_2 and TiO_2 may account for the lack of shift in E_{fb}, but the difference in the Helmholtz capacitance of the ionized acid-base sites is probably what is responsible for the results.[100] As illustrated in Fig. 7, on unmodified SnO_2 oppositely charged counterions can approach the planes δ and δ' from the fixed charge plane defined by the surface ionized sites (A and B) depending upon low and high pH of the solution, respectively. In this region, the Helmholtz capacitance is several tens of $\mu F\ cm^{-2}$. Because of the moderate values of capacitance, dissociation of a part of the surface ionizable sites (*ca.* $10^{15}\ cm^{-2}$) will lead to a change in potential in the Helmholtz layer:

$$\Delta\phi_H = \frac{q_{ad}}{C_H} \tag{14}$$

and the flat-band potentials of these naked oxide semiconductors[99] show a pH dependence of 60 mV pH^{-1}. The difference in the Helmholtz capacitance at low and high pH observed by Tomkiewicz[104] may be attributable to the difference between δ and δ'. In contrast, the aminopropyl modified SnO_2,

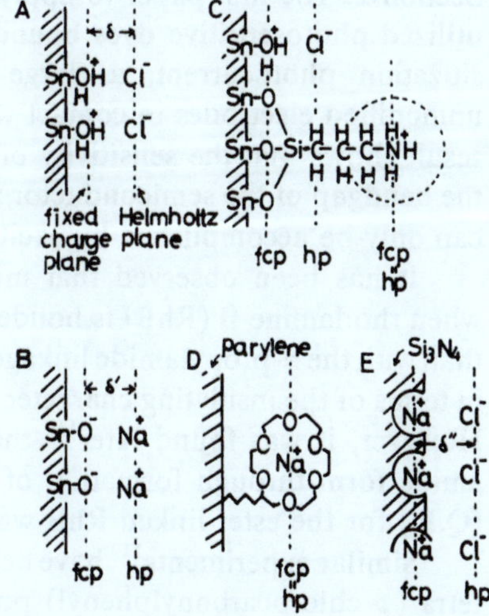

Figure 7. Structural representations of Helmholtz layers at various semiconductor (or insulator) and electrolyte interfaces. A, unmodified SnO_2-electrolyte of low pH; B, unmodified SnO_2-electrolyte of high pH; C, γ-aminopropyl modified SnO_2-electrolyte of low pH; D, parylene film modified by crown ether-electrolyte; E, silicon nitride with templates for Na^+ ion-electrolyte. Reproduction with permission of the Electrochemical Society of Japan from Reference 100.

shown in Fig. 7C, has the amino group out in solution surrounded by an almost spherical ionic cloud with the plane of the counterions almost overlapping with that of ionized ammonium ions.[100] Here, the Helmholtz capacitance is so large that ionization of the ammonium ions causes little change in potential at the interface, which is still determined by the unreacted surface oxides.

These concepts[105] may be applied to the preparation of ion sensitive devices, so-called ion sensitive field effect transistors (ISFET). Fujihira *et al.*[106] have modified the pH-insensitive gate of such a device with a crown ether derivative in order to form a Na^+- or K^+-sensitive ISFET. Low sensitivity of the ISFET is due to a small change in E_{fb} (Fig. 7D). Matsuo *et al.*[107] measured the Na^+ response of an ISFET with Si_3N_4 gate covered with NAS glass. They found that even after removal of the NAS glass layer by prolonged use, the ISFET still showed Na^+ ion response. The result suggests that template cavities for Na^+ ions were formed at the surface of Si_3N_4 during its preparation,[100] and these are sensitive to Na^+ ions in solution by the mechanism shown in Fig. 7E.

In addition to these examples, other chemically sensitive FETs have been fabricated for various analytical purposes.[108-114]

3.5. Dye Sensitization, Photoelectrocatalysis, and Electrogenerated Chemiluminescence by Surface-Bound Molecules

The conversion of light energy to electrical energy and vice versa is currently one of the most exciting fields of electrochemistry. Protection of the semiconductor photoanode by surface modification was introduced in Section 2. The first paper to apply modified electrodes to photochemistry utilized photosensitive dyes bound to the electrode surface.[14] A dye sensitization photocurrent, as large as those previously observed at the unmodified electrodes in contact with the dye solution, was obtained. The result shows that the sensitivity of the cell to photon energies lower than the bandgap of the semiconductor is possible. It is known that sensitization can only be accomplished by excited adsorbed dye molecules.[115,116]

It has been observed that much higher photocurrents are obtained when rhodamine B (RhB) is bonded to SnO_2 through a direct ester linkage than with the *n*-propylamide linkage. This difference was initially interpreted in terms of the insulating character of the propyl chain in the amide system. However, it was found later[117] that RhB loses color when bound in the amide form through formation of a lactam ring. The quantum efficiency (Q.E.) for the ester-linked RhB was determined[118] to be about 9 percent.

Similar experiments[119] have been repeated using ester and amide linked tetra (*p*-chlorocarbonylphenyl) porphyrin on SnO_2. Good action spectra due to the porphyrin were obtained, and again the ester linked form was more efficient (Q.E. = 25%) than the amide form (Q.E. = 10%). A discussion of the barrier height and thickness of the linking chain[120] for the amide proved inconsistent with a fully extended alkylamine chain. A folded chain conformation, as illustrated in Fig. 8, proved consistent with the data. A more quantitative discussion may be possible in the future through preparation of a more defined surface using the Langmuir-Blodgett method.[121] Chlorophylls[122,123] and surface active Ru complexes[124] have been attached in this form, and their photocurrent action spectra have been measured. Tokuda and Matsuda find that the excited state of Ru(III) bipyridine complex monolayer is more active than the excited state of Ru(II) form toward oxidation of water. Their result has been confirmed recently for the reaction in solution.[125]

Other groups[126-134] have also linked various sensitizers on SnO_2 and TiO_2 and have observed the anodic photocurrent action spectra of the bound molecules. Because of low absorptivity of the dye monolayer and low efficiency for the dye sensitization of the excited state beyond the first 1-2 monolayers in the multilayer, the conversion efficiencies based upon incident photons are generally low, although the Q.E.s calculated, based upon the absorbed photons, are respectable. To get around these problems, electrodes have been prepared with several dyes to increase the spectral range,[125,135] or they have been irradiated in a multiple internal reflection mode.[136]

Photoenergy may also be converted to chemical energy in an electrochemical system, for example, an excited surface bound anthraquinone on carbon can act as a catalyst for the oxidation of

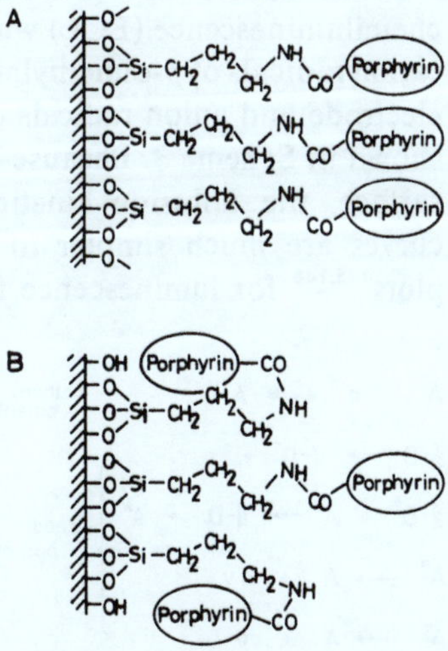

Figure 8. Structural representations of amide-linked porphyrin. A, rigid straight chain model; B, floppy chain folding model. Reproduced with permission of Elsevier Scientific Publishing Company from Reference 119.

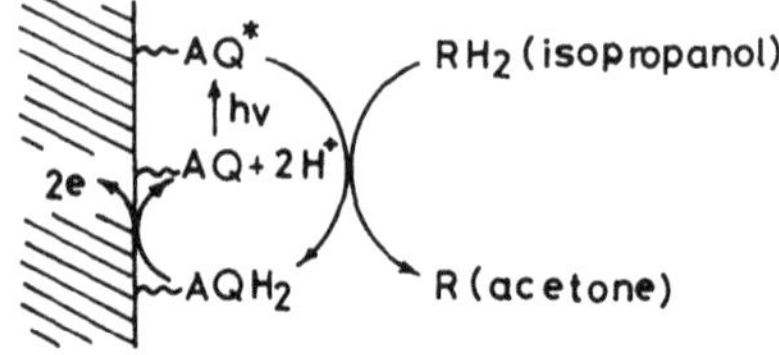

Scheme 2. Photoelectrocatalysis of oxidation of electroinactive compounds by excited quinones bound on a glassy carbon electrode. Reproduced with the permission of Elsevier Scientific Publishing Company from Reference 143.

alcohols.[137] Scheme 2 shows that alcohol is photoxidized by quinone to afford hydroquinone, which in turn is oxidized by the electrode. This kind of "photoelectrocatalysis" has also been studied with quinones in solution.[138-142] The quantum efficiency of the reaction depends on electronic structure of anthraquinone derivatives. The reactivity of the excited state of quinone towards hydrogen abstraction is attributed to the n, π^* character of the lowest triplet state. However, for certain quinones which possess strong electron releasing groups, for example, aminoanthraquinone, the lowest excited state is a charge transfer state and is not reactive toward hydrogen abstraction. Recently,[143] the quantum efficiency for isopropanol oxidation by bound quinones has been compared to that of quinones in solution with the result that the solution species was more efficient (Q.E. = *ca.* 100% vs. *ca.* 1%). It was concluded that most of the excited molecules which are chemically bound are quenched at the carbon electrode before reacting with isopropanol. In this case, the monolayer modification is not efficient. Use of a semiconductor electrode or a multilayer modification will improve the efficiency. The photoelectrochemistry of bound molecules does, however, provide a new method to prove the heterogeneous electron and energy transfer processes at the electrode–electrolyte interface.

The author and co-workers[144] have also observed electrogenerated chemiluminescence (ECL) which resulted from the reaction between stable cation radicals of *p*-dimethylaminobenzoic acid covalently bound on a SnO_2 electrode and anion radicals of fluoranthen in solution. The mechanism is shown in Scheme 3. Because one reactant is immobilized on the electrode surface, the diffusion-kinetic problem and experimental intensity-time curves are much simpler to determine as compared with the Feldberg plots[145,146] for luminescence from the conventional potential step experi-

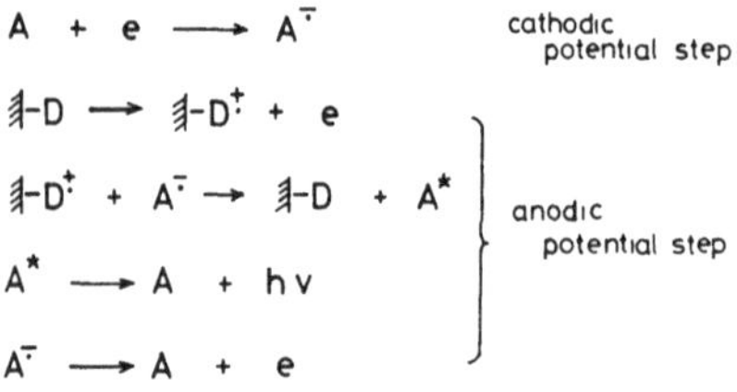

$$A + e \longrightarrow A^{\overline{\cdot}} \quad \text{cathodic potential step}$$

$$⫽\text{-D} \longrightarrow ⫽\text{-D}^{+\cdot} + e$$
$$⫽\text{-D}^{+\cdot} + A^{\overline{\cdot}} \longrightarrow ⫽\text{-D} + A^*$$
$$A^* \longrightarrow A + h\nu$$
$$A^{\overline{\cdot}} \longrightarrow A + e$$
anodic potential step

Scheme 3. Electrogenerated chemiluminescence (ECL) on a chemically modified electrode.

ments, where both cation and anion radicals can diffuse into the solution. The ECL intensity-time curve, recorded using a triple potential step mode, resembled closely the electrochemical *i-t* curve. The ECL spectrum agreed well with the fluorescence spectrum of fluoranthen. ECL on chemically modified electrodes may provide a useful probe for electrocatalysis and heterogeneous electron transfer mechanisms, and could find applications in display devices.

4. MODIFICATION OF THE REACTION LAYER AND THE DIFFUSION LAYER (REGION IV)

Since the thickness of region IV is far beyond the monolayer dimension, organic or inorganic polymer materials are used to design these regions.

4.1. Charge Transfer Catalysis on Polymer-Modified Electrodes

Charge transfer catalysts immobilized in a polymer matrix as shown in Scheme 4 have been studied by many investigators. Two potential differences between electrocatalysis by monomolecular and by multimolecular layers[30] are evident: (i) the quantity of mediator, or catalyst sites, and thus the electrocatalytic rate, can be much greater for multilayers, and (ii) in multilayers, the electrocatalytic rate is potentially moderated by the rate at which catalyst sites migrate through the polymer film (D_{ct}) and the rate at which substrate diffuses through the polymer film ($D_{s,pol}$). Theoretical discussions of this problem have been published by Saveant *et al.*,[147-149] Anson

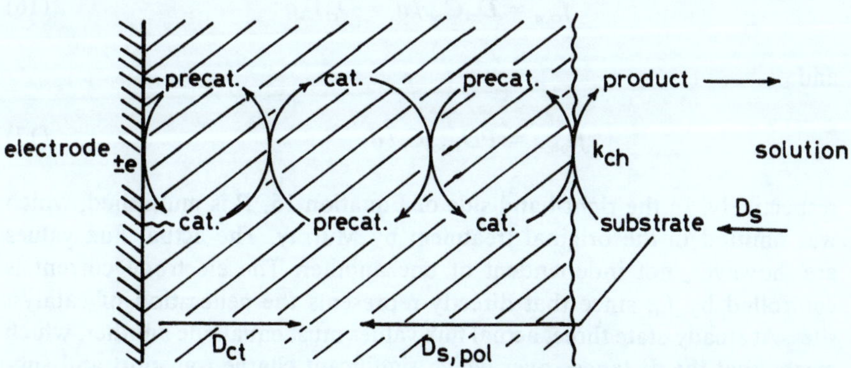

Scheme 4. Charge transfer catalysis on a polymer modified electrode. Reproduced with the permission of The Royal Society from Reference 30.

et al.,[150-152] Murray *et al.*,[30,87,153] Laviron,[154] and Matsuda *et al.*[155] A semi-quantitative discussion of electrocatalysis at the rotating disk electrode[30] will illustrate the significant features of the problem. This is a modified version of Anson's model[150] which takes account of mass transfer of the substrate in the polymer layer.

For a submonolayer or monolayer of catalysis,[87] the limiting electrocatalytic current at the rotating disk was given in Equation 8. This equation contains two electrocatalytic rate elements, the rate of the chemical reaction between catalyst and substrate, and the rate of hydrodynamic mass transport of substrate from the solution to the catalyst surface (the right-hand, Levich term). When the electrode is covered by multimolecular layers of catalyst sites in a polymer film, as in Scheme 4, two additional rate elements can appear. These are (i) the rate at which catalyst sites migrate (charge transport) by electron self-exchange between catalysts within the film, described by the diffusion constant D_{ct}, and (ii) the product of the rate of the diffusion of substrate ($D_{s,\mathrm{pol}}$) and the partition coefficient P from solution to film, i.e., $PD_{s,\mathrm{pol}}$, the permeability. Depending on the relative values of k_c, D_{ct}, and $PD_{s,\mathrm{pol}}$, these two new factors can slow the overall catalytic rate. The charge transport and substrate permeability factors appear as a modification to the intercept term in Equation 8. We can assume that mass transport of substrate from the bulk solution to the film–solution interface is very fast.

The rates of the three kinetic elements, the chemical reaction, charge transport, and substrate diffusion, are conveniently expressed as nominal flux values (i.e., without regard to effects of one flux on another):

$$f_{\mathrm{chem},n} = k_{ch}\Gamma_t PC_s \tag{15}$$

$$f_{ct,n} = D_{ct}C_{\mathrm{cat}}/d = D_{ct}\Gamma_t d^2 \tag{16}$$

and

$$f_{\mathrm{sub},n} = PD_{s,\mathrm{pol}}C_s/d \tag{17}$$

respectively. In the right-hand side of Equation 15, P is multiplied, which was omitted in the original treatment by Murray. The actual flux values are, however, not independent of one another. The electrode current is controlled by f_{ct}, since that directly represents the generation of catalyst sites. At steady-state three, actual flux values must equal one another, which means that the distances over which significant charge transport and substrate diffusion gradients exist, may be less than the total film thickness, d. Flux limitation by substrate diffusion has a special quality in that it is

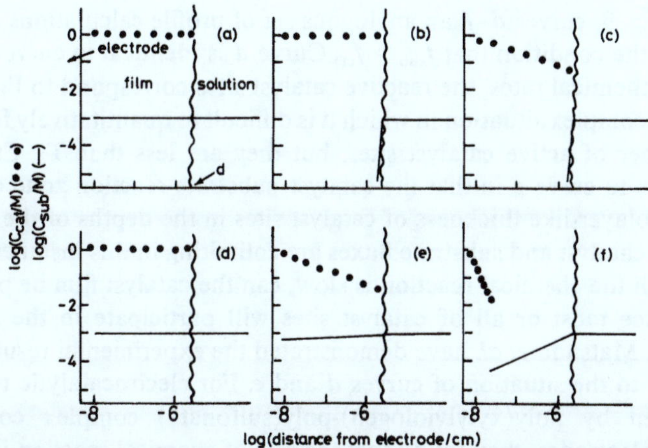

Figure 9. Estimated catalyst and diffusing substrate concentration-distance profiles within polymer film on electrode, assuming a fast supply of substrate from solution to film surface, for specified orderings of nominal fluxes, and $\Gamma_t = 2 \times 10^{-9}$ mol/cm^2, $d = 2 \times 10^{-6}$ cm and $C_{cat} = 1$ *M*. (a) $f_{ct} > f_{sub} > f^*_{chem}$; (b) $f_{ct} > f^*_{chem} > f_{sub}$; (c) $f_{chem} > f^*_{ct} > f_{sub}$; (d) $f_{sub} > f_{ct} > f^*_{chem}$; (e) $f_{sub} > f_{chem} > f^*_{ct}$; (f) $f_{chem} > f^*_{sub} > f^*_{ct}$. Reproduced with permission of The Royal Society from Reference 30.

physically reasonable to assume that the substrate always has access to the outermost catalyst sites in the polymer film, and f_{sub} does not become limiting even if $PD_{s,pol}$ is close to zero. These situations are shown in Fig. 9 by using a representative and a realistic range of choices for k_{ch}, D_{ct}, and $PD_{s,pol}$.

Figure 9 shows that choosing $f_{ct,n}$ and $f_{sub,n}$ and allowing $f_{chem,n}$ to have small, intermediate, and large values (curve *a–c*) has three important consequences: (i) rate control shifts from control by the chemical reaction (in curves *a* and *b*) to control by charge transport (curve *c*), but (ii) owing to our assumption about access to the outermost layer by the substrate, rate control does not pass to f_{sub}, and (iii) the gradient for the substrate extends in curves *b* and *c* over a monolayer thick interval, so that the quantity of catalyst sites active in the chemical reaction is only that on the outermost boundary of the film. In fact, curve *b* offers an explanation for experimental results in which electrocatalytic rate is proportional to C_s, but not proportional to or independent of Γ_t. On the polymer films of poly-Ru$(VB)_3^{+2}$ (where VB is 4-vinyl-4′-methyl-2,2′-bipyridine), the rates of the chemical reaction were independent of the poly-Ru$(VB)_3^{2+}$ thickness from $\Gamma_t = 10^{-9}$ to 10^{-8} mol cm^{-2}, and the electron exchange reactions involve only the outermost monolayer of poly-Ru$(VB)_3^{+2}$ sites in the film.[156] It was also demonstrated that the electron-transfer reaction at the polymer–solution interface was via an outer sphere mechanism.

In Fig. 9, curves *d*–*f*, an analogous set of profile calculations has been done for the condition that $f_{sub} > f_{ct}$. Curve *d* is identical to curve *a*; again for slow chemical rates, the reactive catalyst sites correspond to Γ_t. Curve *e* is a more complex situation in which it is difficult to quantitatively formulate the number of active catalyst sites, but they are less than Γ_t. Curve *f* is analogous to curve *c* in that the catalyst–substrate reaction zone collapses to a monolayer-like thickness of catalyst sites in the depths of the polymer film. The catalyst and substrate fluxes are colimiting in this instance. Again, only when the chemical reaction is slow, can the catalyst film be profitably large, since most or all of catalyst sites will participate in the reaction. Recently, Matsuda *et al.* have demonstrated the experimental results corresponding to the situation of curves *d* and *e*. For electrocatalytic reduction of oxygen by poly(xylylviologen)-poly(sulfonate) complex coated on graphite electrodes, they found that the rate of chemical reaction increased with Γ_t, and that the rate of transport of O_2 through the film was significantly greater than that through the Levich layer.[157,158] Methylviologen incorporated into Nafion also showed similar results. The current-potential curves observed at the rotating disk electrode were also analyzed to determine the chemical rate constants. For other examples of charge transfer catalyses on monolayer or multilayer modified electrodes, consult the recent review by Kuwana.[46]

4.2. Charge Transfer Rate of Redox Catalyst and Mass-Transfer Rate of Substrate in the Film and Application of the Mass-Transfer-Controlled Modified Electrode

When redox catalysts with more than several tens of monomolecular layers undergo the electrochemical reaction, it is not realistic to expect that all of the redox sites exchange electrons directly with electrode. Consequently, charge transfer by successive exchanges of electrons between neighboring redox sites, as shown in Scheme 4, was proposed by Kaufman and Engler.[159] A flow of counterions is necessary to compensate the redox charge change, and there will be polymer lattice motion to accommodate the counterion and solvent flow, and the motions of neighboring sites toward one another. The overall process is actually very complex. Although the detailed mechanism is not clear yet,[160–163] the rate of charge transfer follows Fick's law, and its effective rate is measurable as a diffusion coefficient, D_{ct}.[154,164–167] Experimental values of D_{ct} are scarce, but those available fall into the range 10^{-13} to 10^{-8} $cm^2\,s^{-1}$. The values depend upon the mode of immobilization, such as covalent binding, complex formation, and electrostatic binding, and the degree of cross-linking.

Mass transfer rates also cover a wide range. In ion exchange membranes, polyvalent ions with the same sign as the fixed charge on the polymer are excluded, and do not permeate, but, of course, the ions with opposite sign are generally permeable.[29,168] The degree of cross-linking affects the permeation rate, especially for bulky substrates.[33] Porous or zeolite-like structures in the film facilitate rapid mass transfer of the substrate. By designing the film appropriately, high substrate selectivity for electrosynthesis may be achieved.

Another interesting class of polymer electrodes are the bilayers.[169,170] Diode-like *i*–*E* curves are observed on certain redox polymer bilayers. Figure 10 shows an example of a bilayer of redox polymers, in which the outer layer of polyvinylferrocene (PV Fer) is isolated from the electrode by

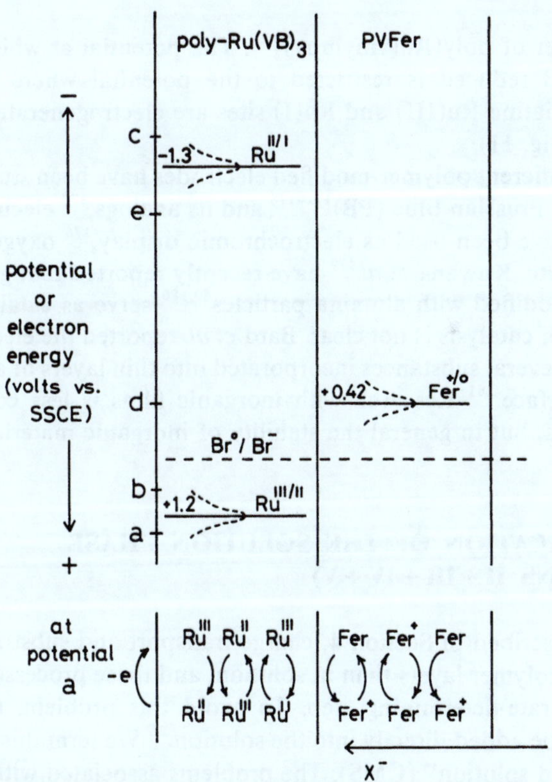

Figure 10. Schematic representation of electron energy levels for a Pt/poly-$Ru(VB)_3^{3+}$/PV Fer bilayer electrode. Reproduced with the permission of the American Chemical Society from Reference 170.

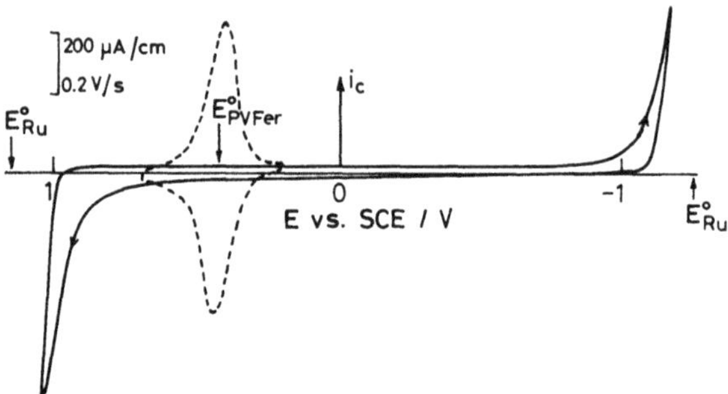

Figure 11. Diode-like cyclic voltammogram of redox polymer bilayer Pt/poly(Ru(vinylbpy)$_3^{2+}$)/PV Fer in 0.1 *M* Et_4NClO_4 acetonitrile. Dashed line, cyclic voltammogram of PV Fer on Pt. Reproduced with the permission of The Royal Society from Reference 30.

an inner layer of poly(Ru(vinylbpy)$_3^{+2}$). The potential at which PV Fer is oxidized and reduced is restricted to the potential where the electron transfer-mediating Ru(III) and Ru(I) sites are electrogenerated within the inner film (Fig. 11).

Many different polymer-modified electrodes have been studied to date. Thin films of Prussian Blue (PB)[171-178] and its analogs,[177] electrochemically deposited, have been used as electrochromic display,[175] oxygen reduction catalysis,[178] etc. Kuwana *et al.*[179] have recently reported that glassy carbon electrodes modified with alumina particles[40,180] serve as catalysts, but the mechanism of catalysis is not clear. Bard *et al.* reported the electrochemical behavior of several substances incorporated into thin layers of a clay-coated electrode surface.[181] Research with inorganic films is less common than with organics, but in general the stability of inorganic materials is greater than organics.

5. MODIFICATION OF THE SOLUTION PHASE (REGIONS II + III + IV + V)

As is described in Section 4, charge transport and substrate diffusion is slower in polymer layers than in solution, and these processes very often become the rate-determining step. To avoid this problem, the catalytic species may be added directly into the solution.[36] We term this the "chemically modified solution" (CMS). The problems associated with CMS have been pointed out in Section 1, but as Andrieux and Saveant[78] have found, homogenous (CMS) outer-sphere electron transfer reactions are generally more efficient than the derivatized electrode analog. Although not an outer-sphere mechanism, the reduction of molecular oxygen with a water soluble porphyrin (Fig. 4) is an example of a CMS catalytic system. Electrochemical reactions in solutions containing an optically active solvent, surfactant micelles, cyclodextrins, crown ethers, etc., have been reported. Gross *et al.*[182] find that $Fe(CN)_6^{-4}$ and $Ru(CN)_6^{-4}$ form a 1-1 complex with polyaza-macrocycles, which changes the redox potential. Gross *et al.*[183] have also studied the electrochemical reduction of mononuclear copper cryptates with diaza-polyoxa-polythia-ether ligands and have shown that the standard redox potential of the Cu(II)/Cu(I) system varies from −0.10 to +0.49 V vs. SCE in aqueous medium, depending on the ligand. Electrochemical reactions on organic compounds can be influenced by the addition of surfactants which affects the microenvironment of the substrate,[184–191] as occurs in photochemical systems.[192,193] A review of this area prior to 1978 has appeared.[188] Advances in water electrolysis technology with an emphasis on the use of the solid polymer electrolytes (SPE) have been reviewed.[194] The principle of the SPE water electrolysis system is shown in Fig. 12a. Recently, the SPE electrolysis method has been applied to organic electrochemistry[195,196] as shown in Fig. 12b. Using this method avoids the use of the supporting electrolyte, which can lead to difficulties during product separation and purification or through unwanted side reactions. SPE cells offer a wide variety of spacial arrangements for the electrode-electrolyte interfaces which can facilitate construction of the electrochemical cells.

So far we have discussed the molecular design of modified electrodes with special emphasis on the relation of elemental processes of the electrode reaction to the boundary layer as seen in Fig. 2. As functional devices which consist of the electrode in the wide sense including further liquid junction, there are many kinds of reference electrodes, pH electrodes, ion electrodes, immuno-electrodes, enzyme electrodes, etc. To design these devices the consideration for whole regions I–V is required. Miniaturization of electrodes such as chemically sensitive FET[100,110–114] and colloidal electrodes[192,197] of semiconductor photocatalysts is one of the current topics in the related area. In these studies, careful design of the systems will certainly

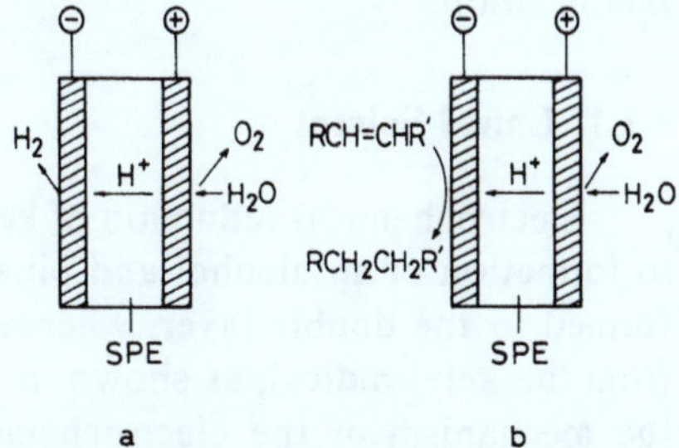

Figure 12. Principles of SPE electrolysis. a, water electrolysis; b, reductive hydrogenation of olefinic double bond. Reproduced with the permission of the Electrochemical Society of Japan from Reference 195.

TABLE 1
Various Types of Morphologic Construction of Electrodes Outside of the Definition of Electrode Shown in Fig. 1

1. Metal/Ionic conductor/Ionic conductor,
   e.g., Coated wire electrode.
2. Metal/Ionic conductor/Ionic conductor/Ionic conductor,
   e.g., Various types of ion selective electrodes.
3. Metal/Insulator/Semiconductor,
   e.g., MOSFET(*M*etal-oxide-semiconductor *F*ield *E*ffect *T*ransistor)
   Ionic conductor/Insulator/Semiconductor,
   e.g., ISFET(*I*on *S*ensitive *F*ET), CHEMFET.
4. Semiconductor/Metal/Ionic conductor,
   e.g., Modified electrochemical photocell, photocatalyst.
5. Semiconductor/Gas,
   e.g., Various types of gas sensor.

increase their specificity and efficiency. Finally, the various types of electrodes which draw the author's attention outside of the definition given of an electrode in Fig. 1 are listed in Table 1.

6. MODIFIED ELECTRODES FOR ORGANIC ELECTROCHEMISTRY

6.1. Modified Electrodes for Stereoselectivity

Eberson and Horner[198] have reviewed the literature through 1970 on the stereoelectrochemistry of organic compounds. It was not until 1975 that Miller *et al.*[4] used chemically modified electrodes to effect an asymmetric synthesis. Recently, van Tiborg and Smit[199] classified the ways of achieving asymmetric induction in electrochemical reactions. The four methods are: (i) through use of a chiral solvent, (ii) the use of chiral supporting electrolytes, (iii) the use of trace amounts of surface active chiral compounds, and (iv) through creation of a chiral electrode surface through chemical modification.

6.1.1. Chiral Solvent

Electrochemical reduction of ketones, such as acetophenone, can lead to formation of an alcohol and pinacols. It is assumed that the alcohol is formed in the double layer, whereas pinacols are formed in the solution from the ketyl radical, as shown in Scheme 5. Seebach and Oei[200] studied the mechanism of the electrochemical pinacolization by comparing the

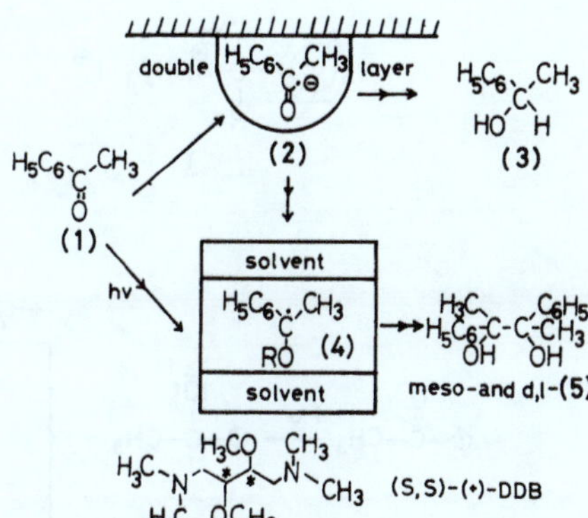

Scheme 5. Electrochemical reduction of acetophenone. Reproduced with the permission of Verlag Chemie GmbH from Reference 200.

electrolysis and photolysis in the chiral amino ether, DDB ((+)-1,4-*bis*-(dimethylamino)-2,3-dimethoxybutane). The electrolysis and photolysis performed under identical conditions should afford the pinacols with the same *meso*/*dl* ratio, with the same isomer and optical yield in the *dl* fraction. Their results showed that this was, indeed, the case, and it is concluded that pinacolization in the electrochemical reduction of arylalkylketones proceeds in the solution.

6.1.2. Chiral-Supporting Electrolyte

Before the appearance of the chemically modified chiral electrode,[4] chiral supporting electrolytes had been used for the molecular building of a chiral double layer. Horner *et al.*[201–204] were the first to demonstrate that supporting electrolytes influence the stereochemical course of the reaction. When (−)- or (+)-ephedrine hydrochloride (Eph.HCl) is used as an electrolyte, methylphenylcarbinol, formed from the reduction of acetophenone, is optically active. (−)-Eph.HCl gives 44% of R-(+)-methylphenylcarbinol in 4.2% optical purity, and (+)-Eph.HCl leads to 38% of S-(−)-methylphenylcarbinol in 4.6% optical purity. The pinacol is produced in 41% yield and is optically inactive and consists of a mixture of *dl* and *meso* forms.

Van Tilborg and Smit[199] have studied the effect of chiral electrolytes in electropinacolization of arylalkylketones in acetonitrile. Asymmetric induction in the pinacol of up to 20 percent can be achieved. The extent of asymmetric induction depends on the following factors: (i) the nature of the supporting electrolyte, (ii) the nature (protic vs. aprotic) of the medium, and (iii) the applied potential. Quaternized ephedrine (**2** in Scheme 6) used as a supporting electrolyte induces more asymmetry in the products than trimethyl[S]-1-phenylethylammonium iodide (**1** in Scheme 6), probably because it has two asymmetric centers instead of one. In the presence of

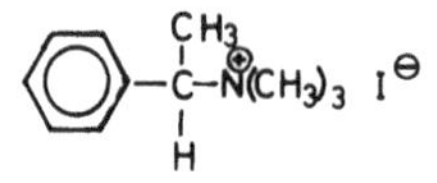

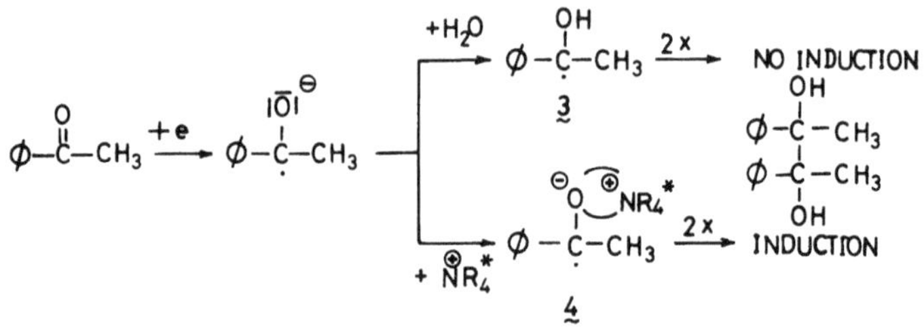

Scheme 6. Electrochemical asymmetric induction in pinacols obtained from acetophenone in the presence of chiral salts (1 and 2) in acetonitrile. Reproduced with the permission of the Royal Netherlands Chemical Society from Reference 199.

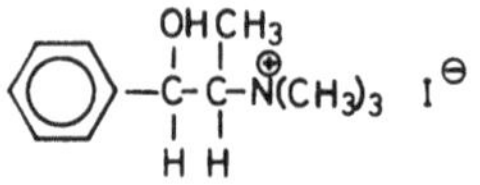

water, even under basic conditions where the yield of pinacol is quantitative, the asymmetric induction almost disappears. They concluded that asymmetric induction originates from recombination of two ketyl radical anions, both in an intimate ion pair state with the chiral ammonium ion. In the protic solvent, the cation is replaced by hydrogen and an optically inactive material is obtained (Scheme 6). The potential dependence was explained in terms of a different mechanism for pinacolization. As shown in Scheme 7, at more negative potentials, the concentration of the ketyl dianions is increased and affords the pinacol via nucleophilic attack on free ketone rather than via radical recombination of two ketyl radical anions. This alternative route is less affected by asymmetry of the electrolyte, since just one of the reacting species is involved with the counterion.

6.1.3. Surface Active Chiral Compounds

Grimshaw *et al.*[205,206] reported the first asymmetric electrochemical synthesis in 1967. They found that the reduction of 4-methylcoumarin in

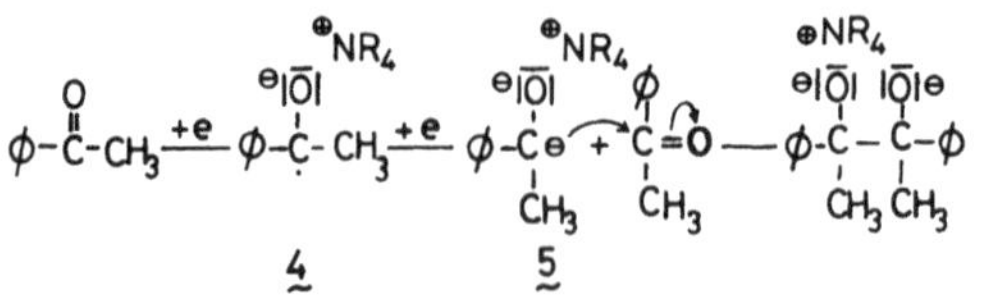

Scheme 7. Pinacol formation via nucleophilic attack of ketyl dianion 5 on free acetophenone. Reproduced with the permission of the Royal Netherlands Chemical Society from Reference 199.

the presence of an optically active tertiary amine affords 3,4-dihydro-4-methylcoumarin in 19% optical yield. They believed the asymmetry arose due to the intervention of a chiral nitrogen radical generated electrolytically:

$$R_3N^+H + e \rightarrow R_3\dot{N}H \rightarrow R_3N + \tfrac{1}{2}H_2 \quad (18)$$

The nitrogen radical was thought to act as a hydrogen atom donor, and alter the course of the electrochemical reduction of coumarin and substituted coumarins. Hydrogen transfer between R_3NH and the initial product of the reduction of coumarin is the critical step of the reaction.

Grimshaw's work has been followed by Gileadi,[207] Kariv,[208] Miller,[209] and their co-workers. This work has focused mainly on the reduction of isomers of acetylpyridine in the presence of different alkaloids on a mercury cathode. In a recent report by Kariv *et al.*,[209] they obtained a 47.5% optical yield of 2-pyridyl-1-ethanol in the reduction of 2-acetylpyridine in the presence of strychnine. Brucine also gives a high-optical yield. The reduction of 3-acetylpyridine gives optically inactive alcohols in all cases studied, whereas 4-acetylpyridine gives the optically active alcohols. The pinacols also formed were optically inactive in all cases. The optical yield of alcohols from the 2- and 4-acetylpyridine was examined as a function of temperature, solvent, pH, potential, and strychnine concentration. They concluded that the mechanism for asymmetric induction involves protonated and adsorbed strychnine, which acts as a chiral acid.

6.1.4. Monolayer Modified Chiral Electrode

Miller and his co-workers[4] covalently bound (S)-(−)-phenylalanine methyl ester ((S)-(−)-PheM) on a carbon electrode and carried out an asymmetric reduction on 4-acetylpyridine. The optical rotation of the enantiomeric product changed sign when the carbon electrode was modified with the *R* configuration of the amino acid ester, and no enantiometric excess was obtained when S-PheM was not bonded to the surface. They concluded that the origin of the optical activity was from the reduction on the chiral electrode surface and not from a solution phase reaction. The reduction of ethylphenylglyoxylate gave the corresponding alcohol in 94% chemical yield and 9.7% optical yield. The oxidation of *p*-tolylmethyl sulfide on this chiral electrode produced an excess of one enantiomer over the other. Metal oxide electrodes, such as SnO_2 and dimensionally stable anodes (DSA), were also modified with (−)-camphoric anhydride through a γ-amino propylsilyl linkage. These electrodes[9] were used to oxidize sulfides to afford optically active sulfoxides in low-optical yield. The mechanism of the asymmetric induction remains to be elucidated. As pointed out in Section

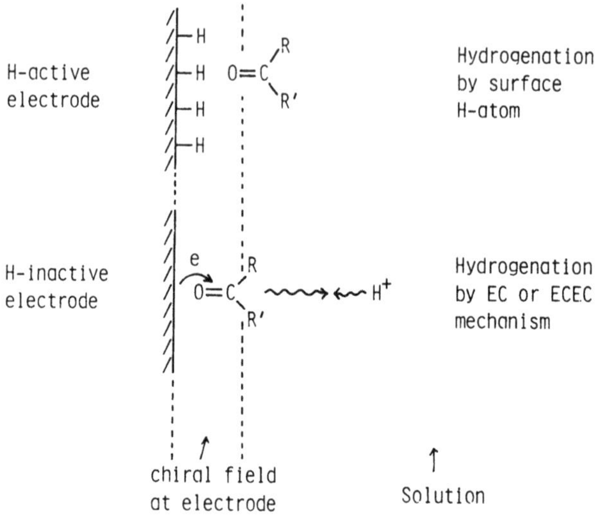

Figure 13. Electrochemical chiral hydrogenation at catalytically active and inactive electrodes for hydrogen.

3.3, the thinness of the reaction layer is a limitation in some reactions and could be responsible for low-optical yields in these experiments. Here, multistep processes may be involved which make the reaction layer thickness more critical (see Fig. 13). This is not the case for the adsorbed alkaloid radical as Grimshaw *et al.*[205] pointed out. We can expect that control of the stereochemical course of the reaction by chemical modification of the electrode surfaces is more effective when the surface plays an active (catalytic) role, than on the passive electrodes as illustrated in Fig. 13. From a consideration of this problem, Osa and the author[210] developed a new enantioselective electrochemical hydrogenation of achiral ketones by Raney nickel powder electrodes modified with optically active tartaric acids. 2-Hexanone was reduced to 2-hexanol at ca. −1.0 V vs. SCE with a current efficiency of more than 67% and an optical purity of 2–6%. Similar results were obtained for the reduction of 2-heptanone and 2-octanone. The Raney nickel modified electrode was applied to asymmetric reduction of methyl acetoacetate.[211] The optical yield of the product was studied as a function of the pK_a of the proton donor (carboxylic acids), temperature, current density, etc., and a maximum optical yield of ca. 20% was obtained. The optical yields are somewhat less than obtained by using modified Raney nickel without electrolysis.[212]

6.1.5. Polymer Modified Chiral Electrode

Recently Nonaka *et al.*[213–216] developed an optically active polymer coated electrode and succeeded in several asymmetric syntheses. In the first paper,[213] they described the asymmetric reduction of citraconic acid and 4-methylcoumarin at a poly-*L*-valine-coated graphite cathode. Maximum optical yields of methylsuccinic acid and 3,4-dihydro-4-methylcoumarin were 25 and 43%, respectively. The controlled potential reduction of citraconic acid was studied in a wide pH range, and the highest optical yield (25%) was obtained in a weakly acidic buffered solution (pH 6). At more negative potential, a greater charge was passed which resulted in decreased optical yields. When a large amount of charge was passed at a constant current density for a shorter time, the optical yield decreased slightly. It was observed that the dip coated electrodes lost their activity after several electrolyses.

Nonaka *et al.*[214] also examined the influence of electrolytic conditions on the optical yields of methylsuccinic acid on graphite coated with various polyamino acids. Asymmetric reduction of mesaconic acid, which is the geometrical isomer of citraconic acid, was also studied.

Three types of poly-*L*-valine-coated platinum electrodes[215] were also studied. The first was prepared by dipping a Pt plate into 0.5 w/v% of poly-*L*-valine. The second electrode was first coated with polypyrrole by anodic polymerization and then with poly-*L*-valine by dip-coating. The third electrode was also prepared by double coating with polypyrrole and poly-*L*-valine. The polypyrrole film in this case was covalently bound by Pt-OSi$(CH_2)_3$N(pyrrole nucleus). The optical yields on the three types of electrodes were 28.0, 40.0, and 54.0%, respectively. The higher optical yields may be due to a tighter adhesion between the polymer and the electrode. Recently,[216] they extended the substrates to prochiral carbonyl compounds, oximes, and gem-dihalides. Phenylglyoxylic acid and its ethyl ester were reduced to give the corresponding alcoholic compounds, but asymmetric yields were not high (0–6.7 percent). The absolute configurations of the excess enantiomer of the alcohols were opposite each other for the acid and ester reduction. 1,1-Dibromo-2,2-diphenylcyclopropane was reduced to 1-bromo-2,2-diphenylcyclopropane at pH 4.7 to give 16.6 percent of asymmetric yield. At present, the mechanism of the asymmetric inductions on these polymer-coated electrodes is not known.

6.2. Modified Electrode for Selectivity in the Choice of Substrates

In Section 5, we described the shift of the standard redox potential of metal ions by complexation with receptor molecules, such as cryptands and crown ethers. We will now examine this same phenomena for organic compounds. Osa and Fujihira[38] examined the effects of added α- and β-CDs ($n = 6$ and 7, respectively) on the half-wave potentials and the diffusion coefficients of *o*-, *m*-, and *p*-nitrophenols and described a polarographic method for determining the dissociation constants of CD-substrate complexes. This method is applicable to many complexes whose dissociation

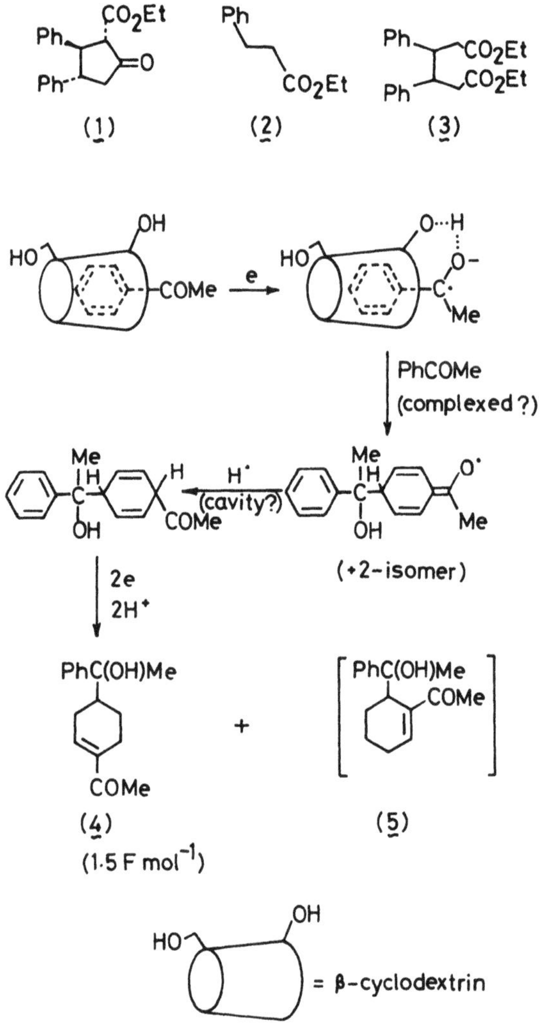

Scheme 8. Modification of pathways for cathodic reduction via complexation with β-cyclodextrin. Reproduced with the permission of the Chemical Society from reference 218.

constants cannot be determined by spectrophotometric methods. Osa *et al.* demonstrated the cyclic voltammetric determination[39] and selective reduction[217] of *o*-nitrophenol in the presence of *p*-nitrophenol using selective complexation of the *para* isomer with α-CD. α-CD adsorbed or immobilized on graphite electrodes has a similar effect; however, selectivity is lower. The decrease on the diffusion coefficient by complexation plays an important role in the separation as does the shift in the reduction potential.

6.3. Modified Electrode for Selectivity in the Type of Chemical Reaction

Utley *et al.*[218] have studied the cathodic reduction of cyclodextrin (CD) complexes of ethyl cinnamate, benzaldehyde, and benzophenone. Protonation of the radical anions is highly efficient for the complexes, presumably because of the availability of protons from peripheral hydroxyl groups of the CD. Reductive coupling is also influenced by complexation, as is mass transport. The hydrophobic cavity of the CD forms 1:1 complexes with ethyl cinnamate and acetophenone, whereas benzophenone and benzaldehyde form 2:1 complexes (CD:guest). The cyclic voltammetry in *N,N*-dimethylformamide (DMF) of the guest compounds gives small shifts in the reduction potential, with greatly reduced anodic peaks on the return scan due to the rapid protonation of the anion radicals.

Scheme 8 illustrates the dramatic changes which occur in the course of controlled potential preparative electrolysis due to complexation. Ethyl cinnamate normally affords the all-trans cyclic hydrodimer ((**1**) in Scheme 8), while complexation with β-CD gives the dihydro-product (**2**) in 71% yield together with 19% of the linear hydrodimer (**3**), and 4% of (**1**). These results show that in the complex, protonation is fast compared with the normally rapid dimerization of the ethyl cinnamate radical anion. Complexation would also expect to slow the dimerization by slowing the diffusion of the intermediates.

For acetophenone, its behavior on cathodic reduction is strikingly different. As illustrated in Scheme 8, a previously unreported coupling takes place to give a 1:1 mixture of the isomers (**4**) and (**5**) in 90% yield. In another report, Utley *et al.*[219] described the catalytic cathodic cleavage of benzyl esters via complexation with a α-CD, which were substituted at the 6-position with *o*-benzoylbenzoate.

6.4. Modified Electrodes for Regioselectivity

Regioselective anodic chlorination of some benzene derivatives using a CD chemically modified electrode has been described by Osa and coworkers.[220–222] α-CD was chemically bound to graphite electrodes via ester

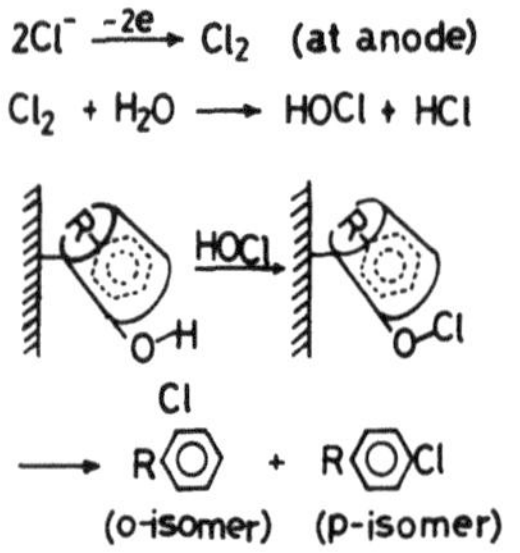

$$2Cl^- \xrightarrow{-2e} Cl_2 \quad \text{(at anode)}$$

$$Cl_2 + H_2O \longrightarrow HOCl + HCl$$

HOCl

(o-isomer) (p-isomer)

Scheme 9. A mechanism of the regioselective anodic chlorination with α-CD-CME. Reproduced with the permission of the Electrochemical Society from Reference 222.

linkage (α-CD-CME) and through adsorption (α-CD-AE). The immobilized α-CD had a pronounced effect on the regioselectivity of chlorination of anisole and toluene, but not in the chlorination of chlorobenzene. Breslow[96] had discussed the regiochemistry of the solution phase reaction, Scheme 9. The homogeneous chlorination by NaOCl in the presence of α-CD showed a relatively high p/o substitution ratio. The electrochemical method using α-CD-CME gives slightly higher ratios, and produced high regioselectivity even after a month of immersion of the electrode in water. However, the α-CD-AE shows a decrease in the p/o chlorination ratio with increasing immersion time. The electrochemical method has the advantage over the solution phase method in requiring only a small amount of α-CD and eliminating the need to separate this catalyst from the reaction medium.

The influence of α-CD addition to the electrolyte solution for the hydroxylation of toluene by an electrogenerated hydroxyl radical[223] shows only small regioselectivity. This suggests that the ionic relay chlorination via CD hypochlorite, as described by Breslow,[96] plays a critical role.

REFERENCES

1. R. Defay, N. Ibl, E. Levart, G. Milazzo, G. Valensi, and P. van Rysselberghe, *J. Electroanal. Chem.* 7, 417 (1964).
2. R. W. Murray, in *Electroanalytical Chemistry*, edited by A. J. Bard (Marcel Dekker, New York, 1984), Vol. 13.
3. P. R. Moses, L. Wier, and R. W. Murray, *Anal. Chem.* **47**, 1882 (1975).
4. B. F. Watkins, J. R. Behling, E. Kariv, and L. L. Miller, *J. Am. Chem. Soc.* **97**, 3549 (1975).
5. R. F. Lane and A. T. Hubbard, *J. Phys. Chem.* **77**, 1401 (1973).
6. R. F. Lane and A. T. Hubbard, *J. Phys. Chem.* **77**, 1411 (1973).
7. National Science Foundation Workshop Report, Chapel Hill, North Carolina, 1974.
8. B. E. Firth, L. L. Miller, M. Mitani, T. Rogers, J. Lennox, and R. W. Murray, *J. Am. Chem. Soc.* **98**, 8271 (1976).
9. B. E. Firth and L. L. Miller, *J. Am. Chem. Soc.* **98**, 8272 (1976).
10. C. M. Elliott and R. W. Murray, *Anal. Chem.* **48**, 1247 (1976).

11. R. R. Moses and R. W. Murray, *J. Am. Chem. Soc.* **98**, 7435 (1976).
12. N. R. Armstrong, A. W. C. Lin, M. Fujihira, and T. Kuwana, *Anal. Chem.* **48**, 741 (1976).
13. M. Fujihira, T. Matsue, and T. Osa, *Chem. Lett.* **1976**, 875.
14. T. Osa and M. Fujihira, *Nature* **264**, 349 (1976).
15. M. Fujihira, A. Tamura, and T. Osa, *Rev. Polarogr.* **22**, 87 (1976); *Chem. Lett.* **1977**, 361.
16. M. Fujihira and T. Osa, *Denki Kagaku* **45**, 270 (1977).
17. R. J. Burt, G. J. Leigh, and C. J. Pickett, *J. Chem. Soc. Chem. Commun.* **1976**, 940.
18. R. F. Lane and A. T. Hubbard, *J. Phys. Chem.* **79**, 808 (1975).
19. A. P. Brown, C. Koval, and F. C. Anson, *J. Electroanal. Chem.* **72**, 379 (1976).
20. A. Merz and A. J. Bard, *J. Am. Chem. Soc.* **100**, 3222 (1978).
21. L. L. Miller and M. R. Van De Mark, *J. Am. Chem. Soc.* **100**, 3223 (1978).
22. M. S. Wrighton, R. G. Austin, A. B. Bocarsly, J. M. Bolts, O. Haas, K. D. Legg, L. Nadjo, and M. C. Palazzotto, *J. Electroanal. Chem.* **87**, 429 (1978).
23. R. Nowak, F. A. Schultz, M. Umana, H. Abruna, and R. W. Murray, *J. Electroanal. Chem.* **94**, 219 (1978).
24. P. J. Peerce and A. J. Bard, *J. Electroanal. Chem.* **112**, 97 (1980).
25. A. F. Diaz, W. Y. Lee, J. A. Logan, and D. C. Green, *J. Electroanal. Chem.* **108**, 377 (1980).
26. K. K. Kanazawa, A. F. Diaz, R. H. Geiss, W. D. Gill, J. F. Kwak, J. A. Logan, J. F. Rabolt, and G. B. Street, *J. Chem. Soc., Chem. Commun.* **1979**, 854.
27. A. F. Diaz, K. K. Kanazawa, and G. P. Gardini, *J. Chem. Soc., Chem. Commun.* **1979**, 365.
28. A. F. Diaz, J. I. Castillo, J. A. Logan, and W. Lee, *J. Electroanal. Chem.* **129**, 115 (1981).
29. N. Oyama and H. Matsuda, *Denki Kagaku* **49**, 396 (1981).
30. R. W. Murray, *Phil. Trans. R. Soc. London* **A302**, 253 (1981).
31. E. Gileadi, E. Kirowa-Eisner, and J. Penciner, *Interfacial Electrochemistry* (Addison Wesley, Reading, 1975).
32. A. J. Bard and L. R. Faulkner, *Electrochemical Methods* (John Wiley, New York, 1980).
33. C. D. Ellis, W. R. Murphy, Jr., and T. J. Meyer, *J. Am. Chem. Soc.* **103**, 7480 (1981).
34. A. F. Diaz, F. A. O. Rosales, J. P. Rosales, and K. K. Kanazawa, *J. Electroanal. Chem.* **103**, 233 (1979).
35. A. F. Diaz and K. K. Kanazawa, *IBM J. Res. Develop.* **23**, 316 (1979).
36. T. Kuwana, M. Fujihira, K. Sunakawa, and T. Osa, *J. Electroanal. Chem.* **88**, 299 (1978).
37. N. Kobayashi, M. Fujihira, K. Sunakawa, and T. Osa, *J. Electroanal. Chem.* **101**, 269 (1979).
38. T. Osa, T. Matsue, and M. Fujihira, *Heterocycles* **6**, 1833 (1977).
39. T. Matsue, M. Fujihira, and T. Osa, *Anal. Chem.* **53**, 722 (1981).
40. K. D. Snell and A. G. Keenan, *Chem. Soc. Rev.* **8**, 259 (1979).
41. W. R. Heineman and P. T. Kissinger, *Anal. Chem.* **50**, 166R (1978); **52**, 138R (1980).
42. R. W. Murray, *Acc. Chem. Res.* **13**, 135 (1980).
43. Special Issue, *Novel Electrodes, Denki Kagaku* **49**, No. 7 (1981).
44. J. Weber and L. Kavan, *Chem. Listy* **74**, 803 (1980).
45. W. E. Van der Linden, and J. W. Dieker, *Anal. Chim. Acta.* **119**, 1 (1980).
46. J. Zak and T. Kuwana, *J. Electroanal. Chem.* **150**, 645 (1983).
47. J. S. Miller, ed., *Chemically Modified Surfaces in Catalysis and Electrocatalysis*, ACS Symposium Series 192, Washington, 1982.
48. S. R. Morrison, *Electrochemistry at Semiconductor and Oxidized Metal Electrodes* (Plenum Press, New York, 1980).
49. T. Kuwana and N. Winograd, in *Electroanalytical Chemistry*, edited by A. J. Bard (Marcel Dekker, New York, 1974), Vol. 7.
50. W. R. Heineman, F. M. Fawkridge, and H. N. Blount, in *Electroanalytical Chemistry*, edited by A. J. Bard (Marcel Dekker, New York, 1984), Vol. 13.

51. A. S. Grove, *Physics and Technology of Semiconductor Devices* (John Wiley, New York, 1967).
52. G. Dearnaley, J. H. Freeman, R. S. Nelson, and J. Stephen, *Ion Implantation* (North-Holland, Amsterdam, 1973).
53. J. W. Mayer, L. Erikson, and J. A. Davies, *Ion Implantation in Semiconductors* (Academic Press, New York, 1970).
54. H. Tachikawa and L. R. Faulkner, *J. Am. Chem. Soc.* **100**, 8025 (1978).
55. F. R. Fan and L. R. Faulkner, *J. Am. Chem. Soc.* **101**, 4779 (1979).
56. A. F. Diaz, J. M. V. Vallejo, and A. M. Duran, *IBM J. Res. Develop.* **25**, 42 (1981).
57. G. B. Street and T. C. Clarke, *IBM J. Res. Develop.* **25**, 51 (1981).
58. G. Wegner, *Agnew. Chem. Int. Ed.* **20**, 361 (1981).
59. H. Gerischer, in *Solar Energy Conversion*, edited by B. O. Seraphin (Springer-Verlag, New York, 1979).
60. A. Heller and B. Miller, *Electrochim. Acta* **25**, 29 (1980).
61. R. Memming, in *Electroanalytical Chemistry*, edited by A. J. Bard (Marcel Dekker, New York, 1979), Vol. 11.
62. A. J. Nozik, ed., *Photoeffects at Semiconductor-Electrolyte Interface*, ACS Symposium Series 146, Washington (1981).
63. M. A. Butler and D. S. Ginley, *J. Electrochem. Soc.* **125**, 228 (1978).
64. D. E. Scaife, *Solar Energy* **25**, 41 (1980).
65. R. E. Benson, E. N. Kaufmann, G. L. Miller, and W. W. Scholtz, ed., *Proceeding of the Second International Conference on Ion Beam Modification of Materials* (*Nucl. Instrum. Methods, 182/183*, 1981), North-Holland, Amsterdam (1981).
66. V. Ashworth, W. A. Grant, R. P. M. Procter, and T. C. Wellington, *Corros. Sci.* **16**, 393 (1976).
67. V. Ashworth and R. P. M. Procter, in *Ion Implantation in Treatise on Material Science and Technology (18)* (Academic Press, New York, 1980).
68. B. S. Cavino, Jr., P. B. Needham, Jr., and G. R. Conner, *J. Electrochem. Soc.* **125**, 370 (1978).
69. Y. Okabe, M. Iwaki, T. Takahashi, H. Hayashi, S. Namba, and K. Yoshida, *Surface Sci.*, **86**, 257 (1979).
70. M. S. Wrighton, *Acc. Chem. Res.* **12**, 303 (1979).
71. M. S. Wrighton, in: *Chemically Modified Surfaces in Catalysis and Photolysis and Electrolysis*, edited by J. S. Miller (ACS Symposium Series 192, Washington, 1982).
72. T. Skotheim, L. G. Petersson, O. Inganäs, and I. Lundström, *J. Electrochem. Soc.* **129**, 1737 (1982).
73. K. Hashimoto and T. Masumoto, *Treatise Mater. Sci. Technol.* **20**, 291 (1981).
74. H. B. Mark, Jr., A. Voulgaropoulos, and C. A. Meyer, *J. Chem. Soc. Chem. Commun.* **1981**, 1021.
75. H. Shirakawa, S. Ikeda, M. Aizawa, J. Yoshitake, and S. Suzuki, *Synth. Met.* **4**, 43 (1981).
76. *Electrocatalysis on Non-Metallic Surfaces*, NBS Special Publication 455, U.S. Department of Commerce (1976).
77. R. A. Murcus, *J. Chem. Phys.* **43**, 679 (1965).
78. C. P. Andrieux and J. M. Saveant, *J. Electroanal. Chem.* **93**, 163 (1978).
79. N. Sutin, *Chemistry in Britain* **8**, 148 (1972).
80. C. Creutz and N. Sutin, *Proc. Nat. Acad. Sci.* **70**, 1701 (1973).
81. P. Yeh and T. Kuwana, *Chem. Lett.* **1977**, 1145.
82. R. Szentrimay, P. Yeh, and T. Kuwana, in: *Electrochemical Studies of Biological Systems*, (ACS Symposium Series 38, Washington, 1977).
83. K. M. Kadish, ed., *Electrochemical and Spectrochemical Studies of Biological Redox Compounds*, ACS Advances in Chemistry Series 201 (1982).

84. N. Kobayashi, T. Matsue, M. Fujihira, and T. Osa, *J. Electroanal. Chem.* **103**, 427 (1979).
85. J. P. Collman, M. Marrocco, P. Denisevich, C. Koval, and F. C. Anson, *J. Electroanal. Chem.* **101**, 117 (1979).
86. J. P. Collman, P. Denisevich, Y. Konai, M. Marrocco, C. Koval, and F. C. Anson, *J. Am. Chem. Soc.* **102**, 6027 (1980).
87. R. D. Rocklin and R. W. Murray, *J. Phys. Chem.* **85**, 2104 (1981).
88. R. R. Adzic, *Israel J. Chem.* **18**, 166 (1979).
89. N. Furuya and S. Motoo, *J. Electroanal. Chem.* **88**, 151 (1978).
90. S. Motoo, M. Shibata, and M. Watanabe, *J. Electroanal. Chem.* **110**, 103 (1980).
91. M. Fujihira and T. Kuwana, *Electrochim. Acta* **20**, 565 (1975).
92. D. M. Kolb, in *Advances in Electrochemistry and Electrochemical Engineering*, edited by H. Gerischer and C. W. Tobias (John Wiley, New York, 1978), Vol. 11.
93. R. R. Adzic, W. E. O'Grady, and S. Srinivasan, *J. Electrochem. Soc.* **128**, 1913 (1981).
94. D. Pletcher and V. Solis, *J. Electroanal. Chem.* **131**, 309 (1982).
95. N. Furuya and S. Motoo, *J. Electroanal. Chem.* **100**, 771 (1979).
96. R. Breslow, *Acc. Chem. Res.* **13**, 170 (1980).
97. A. N. Frumkin, *Z. Physik. Chem.* **164A**, 121 (1933).
98. J. O'M. Bockris, M. A. V. Devanathan, and K. Müller, *Proc. Roy. Soc.* **A274**, 55 (1963).
99. H. Gerischer, in: *Electrochemistry in Physical Chemistry, An Advanced Treatise*, edited by H. Eyring (Academic Press, New York, 1970), Vol. IXA.
100. M. Fujihira, *Denki Kagaku* **49**, 390 (1981).
101. E. Gileadi, ed., *Electrosorption* (Plenum Press, New York, 1967).
102. V. K. F. Chia, M. P. Sorlaga, A. T. Hubbard, and S. E. Anderson, *J. Phys. Chem.* **87**, 232 (1983).
103. M. Fujihira, T. Matsue, and T. Osa, *Elektrokhimiya* **13**, 1679 (1977).
104. M. Tomkiewicz, *J. Electrochem. Soc.* **126**, 1505 (1979); **127**, 1518 (1980).
105. W. M. Siu and R. S. C. Cobbold, *IEEE Trans of Electron. Devices* **26**, 1805 (1979).
106. M. Fujihira, M. Fukui, and T. Osa, *J. Electroanal. Chem.* **106**, 413 (1980).
107. T. Matsuo and M. Esashi, *Ohyo Butsuri* **49**, 586 (1980).
108. N. Yamamoto, Y. Nagasawa, S. Shuto, M. Sawai, T. Sudo, and H. Tsubomura, *Chem. Lett.* **1978**, 245.
109. N. Yamamoto, Y. Nagasawa, M. Sawai, T. Sudo, and T. Tsubomura, *J. Immunol. Methods* **22**, 309 (1978).
110. J. Janata and S. D. Moss, *Biomed. Eng.* **11**, 241 (1976).
111. P. W. Cheung, D. G. Fleming, W. H. Ko, and M. R. Neuman, ed., *Theory, Design, and Biomedical Applications of Solid State Chemical Sensors* (CRC Press, Boca Raton, 1978).
112. R. P. Buck, *Anal. Chem.* **50**, 17R (1978).
113. R. G. Kelley, *Electrochim. Acta* **22**, 1 (1977).
114. J. Zemel and P. Bergverd, ed., *Chemically Sensitive Electronic Devices* (Elsevier Sequoia, Lausanne, 1981).
115. H. Gerischer and H. Tributsch, *Ber. Bunsenges. Phys. Chem.* **72**, 437 (1968).
116. A. Fujishima, T. Watanabe, O. Tatsuoki, and K. Honda, *Chem. Lett.* **1975**, 13.
117. M. Fujihira, N. Ohishi, and T. Osa, *Nature* **268**, 226 (1977).
118. M. Fujihira, T. Osa, D. Hursh, and T. Kuwana, *J. Electroanal. Chem.* **88**, 285 (1978).
119. M. Fujihira, T. Kubota, and T. Osa, *J. Electroanal. Chem.* **119**, 379 (1981).
120. M. Sharp, M. Petersson, and K. Edström, *J. Electroanal. Chem.* **109**, 271 (1980).
121. H. Kühn, D. Moebius, and H. Buecher, *Physical Methods of Chemistry*, edited by A. Weissberger and B. W. Rossiter (John Wiley, New York, 1972), Vol. 1, Part 3B.
122. T. Miyasaka, T. Watanabe, A. Fujishima, and K. Honda, *J. Am. Chem. Soc.* **100**, 6657 (1978).
123. T. Watanabe, T. Miyasaka, A. Fujishima, and K. Honda, *Chem. Lett.* **1978**, 443.

124. H. Daifuku, K. Aoki, K. Tokuda, and H. Matsuda, *J. Electroanal. Chem.* **140**, 179 (1982).
125. C. Creutz and N. Sutin, *Seventh DOE Photochemistry Research Conference*, Oakland, California, p. 25 (1983).
126. D. D. Hawn and N. R. Armstrong, *J. Phys. Chem.* **82**, 1288 (1978).
127. V. R. Shepard, Jr. and N. R. Armstrong, *J. Phys. Chem.* **83**, 1268 (1979).
128. N. R. Armstrong and V. R. Shepard, Jr., *J. Electroanal. Chem.* **131**, 113 (1982).
129. R. Schumacher, R. H. Wilson, and L. A. Harris, *J. Electrochem. Soc.* **127**, 96 (1980).
130. H. T. Tien and J. Higgins, *J. Electrochem. Soc.* **127**, 1475 (1980).
131. M. A. Fox, F. J. Nobs, and T. A. Voynick, *J. Am. Chem. Soc.* **102**, 4036 (1980).
132. J. R. Hohman and M. A. Fox, *J. Am. Chem. Soc.* **104**, 401 (1982).
133. P. K. Ghosh and T. G. Spiro, *J. Am. Chem. Soc.* **102**, 5543 (1980).
134. S. Anderson, E. C. Constable, M. P. Dare-Edwards, J. B. Goodenough, A. Hamnett, K. R. Seddon, and R. D. Wright, *Nature* **280**, 571 (1979).
135. M. Fujihira, T. Kubota, and T. Osa, *Rev. Polarogr.* **23**, 87 (1977).
136. T. Osa and M. Fujihira, *Ger. Offen.*, 2,943,672 (1980).
137. M. Fujihira, S. Tasaki, T. Osa, and T. Kuwana, *J. Electroanal. Chem.* **137**, 163 (1982).
138. M. Fujihira, *J. Electroanal. Chem.* **130**, 351 (1981).
139. J. M. Bobbitt and J. P. Willis, *J. Org. Chem.* **42**, 2347 (1977).
140. T. K. Zolotova, I. V. Shelepin, and Yu. B. Vasil'ev, *Elektrokhimiya* **11**, 1442 (1975).
141. T. K. Zolotova, I. V. Shelepin, and Yu. B. Vasil'ev, *Elektrokhimiya, Engl. Ed.* **11**, 1713 (1975).
142. O. A. Ushakov, Yu. B. Vasil'ev, and I. V. Shelepin, *Elektrokhimiya* **12**, 976 (1976).
143. M. Fujihira, S. Tasaki, T. Osa, and T. Kuwana, *J. Electroanal. Chem.* **150**, 665 (1983).
144. M. Fujihira, H. Sagae, T. Osa, K. Itaya, and S. Toshima, *Rev. Polarogr.* **23**, 86 (1977); T. Osa and M. Fujihira, *Jpn Kokai Tokkyo Koho* **79**, 46, 184.
145. S. W. Feldberg, *J. Am. Chem. Soc.*, **88**, 390 (1966); *J. Phys. Chem.* **70**, 3929 (1966).
146. R. Bezman and L. R. Faulkner, *J. Am. Chem. Soc.* **94**, 3699 (1972).
147. C. P. Andrieux, J. M. Dumas-Bouchiat, and J. M. Saveant, *J. Electroanal. Chem.* **114**, 159 (1980).
148. C. P. Andrieux, J. M. Dumas-Bouchiat, and J. M. Saveant, *J. Electroanal. Chem.* **131**, 1 (1982).
149. C. P. Andrieux and J. M. Saveant, *J. Electroanal. Chem.* **134**, 163 (1982).
150. F. C. Anson, *J. Phys. Chem.* **84**, 3336 (1980).
151. N. Oyama and F. C. Anson, *Anal. Chem.* **52**, 1192 (1980).
152. K. Shigehara, N. Oyama, and F. C. Anson, *Inorg. Chem.* **20**, 518 (1981).
153. P. Daum and R. W. Murray, *J. Phys. Chem.* **85**, 389 (1981).
154. E. Laviron, *J. Electroanal. Chem.* **131**, 61 (1982).
155. N. Oyama, Y. Ohnuki, T. Ohsaka, and H. Matsuda, *J. Chem. Soc. Japan* **1983**, 949.
156. T. Ikeda, C. R. Leidner, and R. W. Murray, *J. Am. Chem. Soc.* **103**, 7422 (1981).
157. N. Oyama, N. Ohta, Y. Ohnukio, K. Sato, and H. Matsuda, *J. Chem. Soc. Japan* **1983**, 940.
158. N. Oyama, N. Oki, H. Ohno, Y. Ohnuki, H. Matsuda, and E. Tsuchida, *J. Phys. Chem.* **87**, 3642 (1983).
159. F. B. Kaufman and E. M. Engler, *J. Am. Chem. Soc.* **101**, 547 (1979).
160. D. A. Buttry and F. C. Anson, *J. Electroanal. Chem.* **130**, 333 (1981).
161. J. Facci and R. W. Murray, *J. Phys. Chem.* **85**, 2870 (1981).
162. K. Shigehara, N. Oyama, and F. C. Anson, *J. Am. Chem. Soc.* **103**, 2552 (1981).
163. J. B. Kerr, L. L. Miller, and M. R. Van De Mark, *J. Am. Chem. Soc.* **102**, 3383 (1980).
164. J. Q. Chambers, *J. Electroanal. Chem.* **130**, 381 (1981).
165. N. Oyama, S. Yamaguchi, Y. Nishiki, K. Tokuda, H. Matsuda, and F. C. Anson, *J. Electroanal. Chem.* **139**, 371 (1982).

166. N. Oyama, T. Ohsaka, M. Kaneko, K. Sato, and H. Matsuda, *J. Am. Chem. Soc.* **105**, 6003 (1983).
167. J. S. Facci, R. H. Schmehl, and R. W. Murray, *J. Am. Chem. Soc.* **104**, 4959 (1982).
168. P. Burgmayer and R. W. Murray, *J. Am. Chem. Soc.* **104**, 6139 (1982).
169. A. D. Abruna, P. Denisevich, M. Umana, T. J. Meyer, and R. W. Murray, *J. Am. Chem. Soc.* **103**, 1 (1981).
170. P. Denisevich, K. W. William, and R. W. Murray, *J. Am. Chem. Soc.* **103**, 4727 (1981).
171. V. D. Neff, *J. Electrochem. Soc.* **125**, 886 (1978).
172. D. Ellis, M. Eckhoff, and V. D. Neff, *J. Phys. Chem.* **85**, 1225 (1981).
173. K. Itaya, T. Ataka, and S. Toshima, *J. Am. Chem. Soc.* **104**, 4767 (1982).
174. K. Itaya, H. Akahoshi, and S. Toshima, *J. Electrochem. Soc.* **129**, 1498 (1982).
175. K. Itaya, K. Shibayama, H. Akahoshi, and S. Toshima, *J. Appl. Phys.* **53**, 804 (1982).
176. K. Itaya, T. Ataka, S. Toshima, and T. Shinohara, *J. Phys. Chem.* **86**, 2415 (1982).
177. K. Itaya, T. Ataka, and S. Toshima, *J. Am. Chem. Soc.* **104**, 3751 (1982).
178. N. Shouji, K. Itaya, I. Uchida, T. Iwasaki, and S. Toshima, *Abstract of Annual Meeting Electrochem. Soc. Japan*, p. 163, Tokyo (1983).
179. L. M. Siperko and T. Kuwana, *J. Electrochem. Soc.* **130**, 396 (1983).
180. J. Zak and T. Kuwana, *J. Am. Chem. Soc.* **104**, 5514 (1982).
181. P. K. Ghosh and A. J. Bard, *J. Am. Chem. Soc.* **105**, 5691 (1983).
182. F. Peter, M. Gross, M. W. Hoseini, J. M. Lehn, and R. S. Sessions, *J. Chem. Soc., Chem. Commun.* **1981**, 1067.
183. J. P. Gisselbrecht and M. Gross, *J. Electroanal. Chem.* **127**, 127 (1981).
184. G. Meyer, L. Nadjo, and J. M. Saveant, *J. Electroanal. Chem.* **119**, 417 (1981).
185. G. E. O. Proske, *Anal. Chem.* **24**, 1834 (1952).
186. S. Hayano and N. Shinozuka, *Bull. Chem. Soc. Japan* **42**, 1469 (1969).
187. S. Hayano and M. Fujihira, in: *Proceedings of International Conference on Colloid and Surface Science* (E. Wolfrom, ed.), Vol. 1, p. 609, Akademiai Kiado, Budapest (1975); *Rev. Polarogr.* **21**, 148 (1975).
188. N. Shinozuka and S. Hayano, in *Solution Chemistry of Surfactants*, edited by K. L. Mittal (Plenum Publishing Co., New York, 1979), Vol. 2.
189. Y. Ohsawa, Y. Shimazaki, and S. Aoyagui, *J. Electroanal. Chem.* **108**, 385 (1980); **114**, 235 (1980).
190. Y. Ohsawa, Y. Shimazaki, K. Suga, and S. Aoyagui, *J. Electroanal. Chem.* **123**, 409 (1981).
191. Y. Ohsawa and S. Aoyagui, *J. Electroanal. Chem.* **145**, 109 (1983); **136**, 353 (1982).
192. N. J. Turro, M. Grätzel, and A. M. Braun, *Angew. Chem. Int. Ed. Engl.* **19**, 675 (1980).
193. M. Grätzel, *Acc. Chem. Res.* **14**, 376 (1981).
194. P. W. T. Lu and S. Srinivasan, *J. Appl. Electrochem.* **9**, 269 (1979).
195. Z. Ogumi, K. Nishio, and S. Yoshizawa, *Denki Kagaku* **49**, 212 (1981).
196. Z. Ogumi, K. Nishio, and S. Yoshizawa, *Electrochim. Acta* **26**, 1779 (1981).
197. A. J. Bard, *Science*, **207**, 139 (1980).
198. L. Eberson and L. Horner, in *Organic Electrochemistry*, edited by M. M. Baizer (Marcel Dekker, New York, 1973).
199. W. J. M. van Tilborg and C. J. Smit, *Recl. Trav. Chim. Pays-Bas* **97**, 89 (1978).
200. D. Seebach and H. A. Oei, *Angew. Chem. Intern. Ed. Engl.* **14**, 634 (1975).
201. L. Horner and D. Degner, *Tetrahedron Lett.* **1968**, 5889.
202. L. Horner and D. Degner, *Tetrahedron Lett.* **1971**, 1241.
203. L. Horner and R. Schneider, *Tetrahedron Lett.* **1973**, 3133.
204. D. Brown and L. Horner, *Liebigs Ann. Chem.* **1977**, 77.
205. R. N. Gourley, J. Grimshaw, and P. G. Millar, *J. Chem. Soc., Chem. Commun.* **1967**, 1278.
206. R. N. Gourley, J. Grimshaw, and P. G. Millar, *J. Chem. Soc.* **C1970**, 2318.

207. J. Hermolin, J. Kopilov, and E. Gileadi, *J. Electroanal. Chem.* **71**, 245 (1976).
208. E. Kariv, H. A. Terni, and E. Gileadi, *J. Electrochem. Soc.* **120**, 639 (1973); *Electrochim. Acta* **18**, 433 (1973).
209. J. Kopilov, E. Kariv, and L. L. Miller, *J. Am. Chem. Soc* **99**, 3450 (1977).
210. M. Fujihira, A. Yokozawa, H. Kinoshita, and T. Osa, *Chem. Lett.* **1982**, 1089.
211. T. Osa, Y. Matsue, A. Yokozawa, T. Yamada, and M. Fujihara, Denki.
212. Y. Izumi, *Angew. Chem. Intern. Ed. Engl.* **10**, 871 (1971).
213. S. Abe, T. Nonaka, and T. Fuchigami, *J. Am. Chem. Soc.* **105**, 3630 (1983).
214. T. Nonaka, S. Abe, and T. Fuchigami, *Bull. Chem. Soc. Japan* **55**, 1327 (1983).
215. T. Komori and T. Nonaka, *J. Am. Chem. Soc.* **105**, 5690 (1983).
216. S. Abe, T. Fuchigami, and T. Nonaka, *Chem. Lett.* **1983**, 1541.
217. T. Matsue, M. Fujihira, and T. Osa, *J. Electrochem. Soc.* **129**, 1681 (1982).
218. C. Z. Smith and J. H. P. Utley, *J. Chem. Soc., Chem. Commun.* **1981**, 492.
219. C. Z. Smith and J. H. P. Utley, *J. Chem. Soc., Chem. Commun.* **1981**, 792.
220. T. Matsue, M. Fujihira, and T. Osa, *J. Electrochem. Soc.* **126**, 500 (1979).
221. T. Matsue, M. Fujihira, and T. Osa, *Bull. Chem. Soc. Japan* **52**, 3692 (1979).
222. T. Matsue, M. Fujihira, and T. Osa, *J. Electrochem. Soc.* **128**, 1473 (1981).
223. T. Matsue, M. Fujihira, and T. Osa, *J. Electrochem. Soc.* **128**, 2565 (1981).

Index

www.ingramcontent.com/pod-product-compliance
Ingram Content Group UK Ltd.
Pitfield, Milton Keynes, MK11 3LW, UK
UKHW012159240726
13966UKWH00002B/439

* 9 7 8 1 4 8 9 9 2 0 3 5 5 *